Wild Man from Borneo

Wild Man from Borneo

 A Cultural History of the Orangutan

Robert Cribb
Helen Gilbert
Helen Tiffin

HAWAI

University of Hawai'i Press
Honolulu

© 2014 University of Hawai'i Press
All rights reserved
Printed in the United States of America

19   18   17   16   15   14   6   5   4   3   2   1

Library of Congress Cataloging-in-Publication Data

Cribb, R. B., author.
    Wild man from Borneo : a cultural history of the orangutan /
        Robert Cribb, Helen Gilbert, Helen Tiffin.
    pages   cm
    Includes bibliographical references and index.
    ISBN 978-0-8248-3714-3 (cloth : alk. paper)
    1. Orangutan.   2. Orangutan—Symbolic aspects.   3. Human-
        animal relationships.   I. Gilbert, Helen, author.
        II. Tiffin, Helen, author.   III. Title.
    QL737.P96C75 2014
    599.88'3—dc23   2013031367

University of Hawai'i Press books are printed on acid-free
paper and meet the guidelines for permanence and
durability of the Council on Library Resources.

Designed by Integrated Composition Systems, Spokane, Washington
Printed by Sheridan Books, Inc.

ℛ Contents

✣ Illustrations

꒤ Acknowledgments

Three authors writing a book probably produce three times the need for acknowledgments of assistance, especially when the task of writing stretched over a much longer period than we originally intended. Our greatest debt is to Amanda Lynch, who worked as research assistant for the project from its inception until its conclusion and whose diligence and initiative have been invaluable, not only in routine research and editing tasks but in going a step further to find fascinating and fugitive data that might have otherwise escaped our attention. We also wish to thank Charlotte Hammond, Melissa Poll, and especially Nesreen Hussein for assistance in finding resources in London, and Sally O'Gorman for helping to secure permissions for many of the illustrations included here. Matthew Cohen, Elizabeth Schafer, Gilli Bush-Bailey, Robert Jordan, and Kirsten Holst Petersen gave insightful feedback on draft material, and Dani Phillipson facilitated travel arrangements for some of our research meetings. Robert Ong of Sandakan assisted us in seeing an orangutan in the wild. Special thanks to Cameron Browne for his skillful repair of several fragile images and for general technical assistance. Thanks also to Annelise Friend for typing assistance. We would also like to thank two anonymous readers for the Press for their helpful and incisive comments.

We are particularly grateful to staff at the Singapore Zoo for facilitating a visit to their orangutans and sharing stories about them, and to the staff of the Camp Leakey research station in the Tanjung Puting National Park, Central Kalimantan, Indonesia. John Blower, Kiras Depari, Anton Fernhout, Regina Frey, Colin Groves, Hoegeng, John MacKinnon, Kathy MacKinnon, Jeffrey McNeely, Barito O. Manullang, Herman Rijksen, Rudolf Schenkel, I Made Taman, Tony Whitmore, and Tony Whitten contributed in formal and informal discussions to our understanding of orangutan conservation, though none of them bear responsibility for the conclusions we have presented here.

We would also like to thank the many libraries, archives, and institutions that have allowed access to their collections and assisted us with obtaining photographs and illustrations: National Library of Australia, Australian National University Library, State Library of Queensland, Queensland Museum Library, University of Queensland Library, British Library, Library of Royal Holloway University of London, National Archives (formerly the Public Record Office), University of Bristol Theatre Collection, Herbarium Bogoriense (Indonesia), Library of Leiden University, National Herbarium of the Netherlands, Naturalis Biodiversity Center (Leiden), Nationaal Archief (Netherlands), Gemeentearchief Amsterdam, Bibliothèque Nationale de France, Institute for Cryptozoology at the Musée Cantonal de Lausanne, Worldwide Fund for Nature (Gland, Switzerland), University of California (Berkeley), Cornell University Library, Library of Congress, Smithsonian Institution Library, and Bronx Zoo Archive.

For supplying photographs and illustrations we wish to acknowledge the Highland Archive Service, Natural History Museum (London), Zoological Society London, Houghton Library (Harvard), Biodiversity Heritage Library, the Wellcome Institute, Smithsonian Institution Libraries, Österreichisches Theatermuseum (Vienna), Profiles in History, Bronek Kaminski, Fong Sam, Recorded Picture Company, Singapore Philatelic Museum, British Library, NBC Productions, Koninklijk Instituut voor Taal-, Land- en Volkenkunde (Leiden), the Walburg Press, Muséum d'histoire naturelle de La Rochelle France and Teylers Museum (Haarlem).

The project was generously supported by a Discovery Grant from the Australian Research Council and a British Academy Small Grant.

Wild Man from Borneo

🜚 Introduction

Between the fifteenth century and the eighteenth, a great extinction took place. Unlike the later wave of extinction that would sweep away species after native species in Africa, the Americas, Australia, and above all the islands of the Atlantic, Indian, and Pacific Oceans, this earlier extinction was one of the mind. It did not wipe out living creatures, but rather relegated to the realms of pure fantasy a rich bestiary that had charmed, inspired, and frightened people in ancient and medieval times. Some of these creatures were spectacular but distant perils like dragons, with their glittering scales and flashing wings; some were close by and unobtrusive like the gnomes, sober-suited, taciturn, and inhabiting local woods or caves. Some were monstrous and alarming, like the man-eating cyclops with a single eye in the middle of its head, while others were gentle, noble, and elegant like the unicorn. Some of them were humans in odd forms with tails or with huge pendulous ears drooping to their feet, big enough to be used as umbrellas in time of rain; some seemed to be no more than tribes with strange social customs like sharing one wife among several husbands.

Belief in their existence had been sustained over centuries by the faithful repetition of the statements of ancient authorities such as the Roman naturalist Pliny. More recent travel tales often reworked the pronouncements of these sages, reinforcing their plausibility and occasionally adding new details. According to these stories, fantastical creatures fed on many different things—sometimes humans, or their own young, or honey, or just the air—but the meat and drink of their sustained existence in the human imagination was mystery and symbolism. They were nourished on the one hand by the energizing tension that comes from setting credulity against incredulity, and on the other hand by the ideas and ideals they symbolized for human societies. Western society has persistently drawn strong analogies between humans and animals, though the character of these analogies has changed repeatedly. In ancient times and during the Middle Ages, animals were most commonly seen as providing moral examples to humankind. In Aesop's Fables, as in the *Panchatantra* of India, the actions of animals were used to provide models of good or bad behavior: ants modeled diligence, the tortoise determination, the hare complacency, and so on. A myth that the female pelican would tear its own flesh to provide blood to feed its young affirmed the role that human mothers were expected to play in sacrificing themselves for their offspring. The nobility of the unicorn and the uncompromising savagery of the cyclops served as opposing models, inspiring and warning, for human society.

Modern scientific method cut its teeth driving these creatures to extinction. This new way of thinking that began to emerge in the Renaissance gradually set up new standards of belief. Fascination with fantastical beings did not disappear, but it lost the authority that Pliny and others had once given it, and it retreated to the intellectual borderlands of society as superstition and pseudo-science. The desire to look at nature for instructive analogies to the human condition remained powerful, but over a couple of centuries the creatures that had carried those analogies disappeared. From being real but remote figures in a diverse and wondrous global fauna, they became the stuff of refined allusion, wispy figments of the collective imagination. This process of extinction was driven partly by the European overseas expansion that began in the late fifteenth century. Those who sailed to distant lands found many marvels, but they often failed to find confirmation of the wonderful creatures they had once taken for granted. Some travelers simply wrote reports that pushed this bestiary to still more remote parts of the world, but others informed their readers, sometimes gently, sometimes sternly, that the received wisdom was flawed. And yet, here and there, traveling Europeans encountered something that conjured up ancient imaginations and presented the emerging inquisitive scientific frame of mind with a challenge. Early in the seventeenth century a Dutch physician, Jacobus Bontius, neatly captured the spirit of the age in his short chapter on the "Ourang Outang" when he invoked the memory of creatures that Pliny had called "satyrs," insisted on the importance of a general skepticism, and provided the rudiments of an eyewitness account of something remarkable and perplexing that he had seen for the first time.

In 1699, at the end of Bontius' century, an English investigative surgeon, Edward Tyson, largely completed the disentanglement of orangutans from the mythologies of the ancient and medieval world when he appended to a report on the dissection of a chimpanzee a long essay whose title reveals its conclusion: *A*

Philological Essay Concerning the Pygmies, the Cynocephali, the Satyrs, and Sphinges of the Ancients. Wherein it will appear that they are all either apes or monkeys, and not men, as formerly pretended. "In this opinion," he commented, "I was the more confirmed, because the most diligent Enquiries of late into all Parts of the inhabited World, could never discover any such Puny diminutive Race of Humans."[1] Even after Tyson's intervention, a few writers continued to invoke Pliny and others to entertain a broader public hungry for exaggerated accounts of the exotic. In the early eighteenth century, for instance, the Italian priest Giacinto Gimma published a series of dissertations with titles such as *De hominibus fabulosis* and *De fabulosis animalibus,* which repeated a potpourri of ancient and medieval legends, enlivened with citations from more recent authorities such as Bontius.[2] Travel tales, too, continued to rework old legends, but there was a new and increasingly irresistible insistence on the direct evidence of eyewitnesses and the inspection of specimens, rather than on the pronouncements of venerable scholars. The French traveler Louis Le Comte, writing in the late seventeenth century, repeated the usual collection of remarkable legends about a creature in Borneo that we half-recognize as the orangutan, but he went on to admonish his readers:

> I discovered all these particulars from one of our principal French Merchants, who has lived for some time on the Island. Nevertheless, I do not believe that one should give much credence to such accounts; nor should they be rejected altogether, neither must we altogether reject them as fabulous; rather, we should await the unanimous testimony of other travelers that may confirm its truth.[3]

This modern skepticism replaced the indiscriminate menagerie of the Middle Ages with a new bestiary of real animals, less grotesque than their predecessors but more vivid because they were more accurately described and depicted, and soon to be placed on display, turning into eyewitnesses a public that had once been only a passive audience for countless retellings.

The real orangutan, *Pongo,* is one of only four genera of great ape. Two of the others—*Pan* (chimpanzees and bonobos) and *Gorilla* (gorillas)—are indigenous to Africa, while the fourth, *Homo,* is considered indigenous to nearly every part of the world. Orangutans are found naturally only on the great equatorial islands of Borneo and Sumatra. They are jungle dwellers, tree dwellers with long arms and legs, supple joints, and the strong prehensile fingers and toes needed to support a weight of up to 130 kilograms (285 lbs.). Almost as big as human beings but far stronger, they move with calm assurance in the jungle canopy, from branch to branch, tree to tree, building a nest each night with leaves and branches. They are mainly solitary, ranging over a wide area of jungle in search of fruits, insects, mushrooms, lizards, birds' eggs, and other sustenance. Observers often comment on the hair of the orangutan, which can range from deep brown to a striking red, otherwise a rare color among mammals. Humans who encounter orangutans at close quarters are frequently struck by their specific looks and personalities. Males are often bearded and may have impressive and distinctive facial flanges. Individual orangutans are easy to recognize, and their faces seem to show a range of complex human emotions, including curiosity, wisdom, mischievousness, and melancholy. Although genetic research indicates that orangutans are more distantly related to humans in evolutionary terms than are chimpanzees, bonobos, or gorillas, they appear to be the most *sapiens,* the most thoughtful, of all the great apes apart from humans.

Orangutans today face the threat of total extinction. Only a few tens of thousands are left in the wild, a few hundred more in captivity. They will not disappear from the world this decade, and extinction is not a foregone conclusion. Energetic individuals and institutions are working to save the orangutan, but a powerful array of factors is steadily pushing the species toward oblivion and the eventual outcome is, at best, in doubt. In 2004, orangutan specialists systematically assessed recent reports to estimate the size of the population surviving in the wild. The obstacles to conducting such a census are formidable, dwarfing the logistical challenges of most human censuses. Orangutans roam widely and live in places remote from human lines of communication. For the small number of researchers and conservation officials who take serious responsibility for orangutans, better understanding of their behavior and ecology and promoting measures to protect them more effectively have always had priority over the costly distraction of an accurate census. The review, however, concluded that there were around 7,300 orangutans in Sumatra and 57,000 in Borneo, including some 13,000 in Sabah and 1,000 in Sarawak. More than half the Borneo population (32,000) was concentrated in peat forests of southern Central Kalimantan.[4] There has been no comparable expert attempt to tally orangutan numbers since 2004. Some recent estimates, taking account of newly discovered communities, report similar numbers; others suggest the real figure may be only half this number.[5]

Orangutans have never been of practical importance to humans in the way of domestic animals, prey, or predators—sheep or dogs, bison or cod, wolves or fleas—but along with the chimpanzee, the orangutan has had a special place in the imagination because of its striking similarity to humans. For centuries before European science encountered orangutans, the smaller apes of North Africa and the monkeys of parts of Asia had been kept as pets or put on display for both popular and court entertainments. Their humanlike behaviors gave rise to a rich store of ape lore, associating them with foolishness, sinfulness, slyness, imitation, and lustfulness.[6] But these apes had been miniature, vulgar parodies of humanity that did not raise metaphysical questions of any great depth. The great apes that began to appear in European travelers' tales, and as occasional curiosities exhibited in Europe itself in the sixteenth century, were different, closer to humans in size, in physical form, and in behavior than any creature that had previously come to the attention of the West. At a time when assumptions about the nature of humanity that had governed medieval thinking were under unprecedented questioning, the red ape and its African relatives presented observers with the exciting challenge of assessing, expressing, and judging just how much of humanity there could be in a creature from the jungle and just what it might mean for humans to know that there was something of the jungle in them.

In their tropical homelands, however, orangutans appear to have attracted much less attention from indigenous people than they did from Europeans. For Indonesians, even those who lived closest to the red apes in the jungles of Borneo, the orangutan seems to have conjured up no special metaphysical challenges; it was normally no threat to humans and could only survive far from human settlement. Orangutans appear in some folkloric tales, but the most striking ones seem to have been brought to Borneo by Europeans and then re-exported, as it were, as ostensibly indigenous beliefs.

Orangutans—however quasi-solitary by habit—have, like the other great apes and indeed many animal species, individual and community interactions and habits of daily life in which we can perceive cultural complexity. But this book is less about their actual behaviors (in the wild or in captivity) or their anatomy, tool use, intelligence, mating, and rearing practices than it is about the orangutan-human encounter over four centuries, about the ways in which their actual or imagined presence has impinged upon us, and how we have envisioned, studied, and treated them. These relationships have been protean and multifaceted, fostering an enduring curiosity about the red ape, whether to learn more about ourselves or (and potentially less anthropocentrically) to learn more about them. As the possibility of their extinction galvanizes us into concern for their preservation, we increasingly deploy scientific, diplomatic, and cultural resources on their behalf, yet still to a degree for our own benefit, hoping to arrest a decline for which we have been largely responsible.

Over the centuries, many studies of orangutans have aimed to determine the precise character of their differences from and similarities to humans. Some of the earliest naturalists and their contemporaries asked whether orangutans could walk upright or speak, while others sought in orangutan life moral lessons for the "human condition." Later investigators wanted to know whether humans and orangutans could produce offspring together. Still later, we have asked how much genetic makeup the orangutan has in common with us, how much it shares high-level human emotions, such as the capacity to empathize, how much it can plan complex actions. Scientists offered answers to these questions as they arose, but since the early nineteenth century, as orangutans were increasingly presented to the general public in the West, first in philosophy, menageries, and exhibitions and later through literature, theater, and eventually film, ordinary people also came to firm conclusions about what orangutans could and could not do. Scientific answers, however, did not move in a simple linear trajectory from unknown to known. The question of whether orangutans could speak appeared to be resolved in the negative in about 1770, yet it returned to the scientific agenda in a slightly different form in the twentieth century. In the century following the publication of Darwin's *Origin of Species* (1859), scientific attention shifted significantly from the orangutan to the chimpanzee, partly because chimps were identified as the closest living relatives of humans and partly because they were more convenient laboratory subjects. In the twentieth century, scientific research shifted away from questions related to the definition of the human and the animal to interest in orangutans themselves. Such research was frequently combined with conservation initiatives conducted in the field and thus much more closely linked to orangutan behavior and habitat.

Knowledge about orangutans came not just from the world of science—and scientists themselves were often at odds over what might be the correct answers to questions of the time—but also from the broader world of public and popular culture. Popular enthusiasms and anxieties encouraged authors, playwrights, artists, and impresarios to toy with the implications of granting orangutans, or recognizing in them, still more human characteristics than scientists might have done. In most cases, however, the purpose of this apparent accentuation of the human in orangutans was to find a line along which it was still possible to differentiate the two species, to keep the orangutan just on the other side of the crucial boundary. To identify human characteristics in an orangutan was not to say that orangutans were human, but rather to suggest that the essence of humanity lay elsewhere.

The reflective spirit of the post-Enlightenment era made it of fundamental importance to determine just what constituted humanness. The Swedish naturalist Carl Linnaeus, who devised what became the modern system of classification of all living things, gave humankind the scientific name *Homo sapiens,* but he shied away from defining humanity. Instead he simply instructed his readers in Latin, *nosce te ipsum,* to know themselves. Of course not everyone turned to science for an answer to this question, but contemplating the creature that seemed closest of all to humankind was a constant part of self-definition. Precisely because the intention was to decide the limits of humanness, this contemplation was loaded against the orangutan. For most observers, the answer was already known: orangutans were not humans and the challenge was to put a finger on just what it was that disqualified them. Always, however, and with increasing volume in the twentieth and twenty-first centuries, there have been a few daring figures who have been willing to regard orangutans as human and to consider the implications of doing so.

The cultural history we offer in the following pages begins with the story of how the red ape received its names, both the common name *orangutan* and the scientific name *Pongo pygmaeus*. This naming established the red ape definitively in the eyes of science as a creature distinct from human beings. Yet both while this naming process was under way and ever since, those working outside the rigorous but limiting framework of scientific reasoning probed and tested the species boundary created by the scientists. In the eighteenth century, European philosophers and writers explored the proposition that orangutans were truly human, albeit not quite the same kind of humans as Europeans. This exploration ended for science with the conclusion that orangutans were wholly animal, but the idea of a human-orangutan nexus never lost its fascination for the public or for many artists and thinkers. In this book, therefore, the history of developing scientific knowledge of the orangutan is intertwined with the history of the representation of orangutans in key areas of human culture: travel writing, literature, exhibitions, stage performance, film, zoos, and circuses. The focus is on Western culture, where images of the orangutan have been widely used and extensively preserved, but we have drawn in elements from Asian cultures where possible.

Just as investigating, writing, and thinking about orangutans (and humans) has tended to be a pursuit of Western cultures rather than those of Southeast Asia, it is primarily in the West that the orangutan has become iconic, a charismatic example of an endangered species. Because of its iconicity, the orangutan image has percolated into recent advertisements for luxury cars, electronic communications, pension funds, and other areas that have nothing to do with species endangerment. The scenarios featured in such advertisements are as fanciful in linking orangutans with human worlds as was the famous nineteenth-century fairground exhibit The Wild Men of Borneo,[7] from which this book derives its title. Devised in 1852 by American showman Lyman Warner, the act featured twin dwarf brothers as "savage" creatures taken by sailors, it was claimed, after a great struggle in the jungle. Warner had actually purchased the brothers, Hiram and Barney Davis, from their parents in Ohio; they were renamed Waino and Plutano and became one of the most popular acts on the exhibition circuit, touring for more than half a century in their home country and to most major European cities:

> The boys learned to play their part well, snarling and hooting, capering frantically about the stage, and even feigning attempts at attacking audience members. They performed before a painted jungle backdrop purported to represent the remote island home of their "lost race."[8]

This popular performance later evolved into feats of strength as it became harder to promote the gentle dwarfs as wild and dangerous. The expression "the wild man from Borneo" has since been used in a range of vernacular contexts to signify brutishness, outlandish behavior, recklessness, hucksterism, lawlessness, or simply unkempt looks. In an intertextual move that has particular significance for this book, the term was also reappropriated to refer to the orangutan itself, notably in late nineteenth-century zoo exhibits. Today, travel brochures invoke this cultural history in advertisements for orangutan sanctuaries in Sabah where tourists can encounter the "original" wild man of Borneo in his natural habitat. Though the phrase's usage has waned as a behavioral descriptor, other terms have sprung up to draw new orangutan-human analogies. In contemporary Australia, for instance, people with auburn hair are commonly called "rangas" in acknowledgment of the vivid red hue of the orangutan.[9]

A cultural history is not a history of science, though it inevitably draws on it. Many different approaches, including those science would regard as lacking its own authority, are crucial components. The "truth" of a cultural history of orangutans lies not in biological or behavioral "facts" alone, but in recording the various ways in which orangutans have been perceived and represented in a number of disciplines, discourses, and domains of practice. To alter Claude Lévi-Strauss' phrase a little, orangutans have proved good to think with,[10] and not just about the nature of humanness, our proximity to other great apes, or our treatment of our fellow planetary beings. We have also "thought with them" in less general terms at particular times: about race during the period of European slave trading, the late nineteenth century and beyond; about the nature, status, and place of women, particularly in the era of the "new woman"; about friendship, desire, and sexuality; about the decline and degeneration of humans at the turn of the nineteenth century; about our future as a species; and about our biological ancestors and our myths of human exceptionalism as *the* special creation of God.

We have begun this orangutan-human narrative with accounts of the discovery of the red ape and the Western anatomical and philosophical investigations

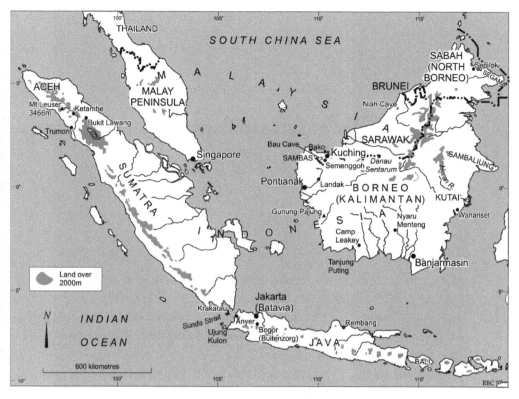

Map of Malaysia-Indonesia region

undertaken to differentiate it from other apes, monkeys, and humans. In the next chapters we move from these early inquiries to consider the public exhibition of orangutans in Europe and the United States as well as their encounters with humans in Borneo and Sumatra. While scientists and philosophers continued their inquiries into the anatomy and status of the orangutan, writers, dramatists, and filmmakers began to include orangutans in their creative works. The following chapters, therefore, give accounts of the ways the red ape has been imaginatively represented in fiction, on stage, and on the screen, from the eighteenth century to the present. Today the targets of scientific interest have become orangutan psychology and behavior in what is left of its natural habitat. Hence, we conclude with a discussion of moves made toward the orangutan's conservation, crucially habitat preservation, rehabilitation, and with the complex emergence of the idea that orangutans possess rights. In so doing, we hope this book may in some way contribute to the orangutan's survival in the wild.

1 ❧ From Satyr to Pongo
Discovering the Red Ape

In 1641, the Dutch anatomist Nicolaes Tulp published a book with the bland title *Observationes Medicæ* (Medical Observations). Tulp is best known now for his depiction in Rembrandt's celebrated painting *The Anatomy Lesson of Dr Tulp,* in which the doctor is shown explaining the features of a human cadaver to an audience of students. Tulp's real-life interests, however, extended beyond human anatomy to the biology of other creatures. Chapter 56 of Tulp's book was titled "An Indian Satyr" and opened with the paragraph:

> Although it lies just outside the field of medicine, I shall weave into my account here a mention of the Indian Satyr which was brought, as I recall, from Angola and presented to Frederik Hendrik, Prince of Orange. This satyr had four feet, but because of the human appearance [*gedaante*] it shows, it is called *Orang Outang* by the Indians, or forest man, and by the Africans *Quoias Morrou.* It is as tall as a child of three years and as stout as one of six.[1]

Tulp's account was the first attempt by a trained scientist to provide a detailed description of a great ape. After the introduction above, he went on to describe the state of the creature's muscles, the texture of its skin, the color of its hair, the shape of its thumb. He had seen the creature when it was still alive, and he described the way in which it drank liquid and how it pulled a blanket over itself to go to bed. The description, though stretching over little more than a page, was a model of direct scientific observation of a kind that had nothing in common with the wild stories of fantastical beasts in distant places that was still the stuff of travelers' tales in the early seventeenth century.

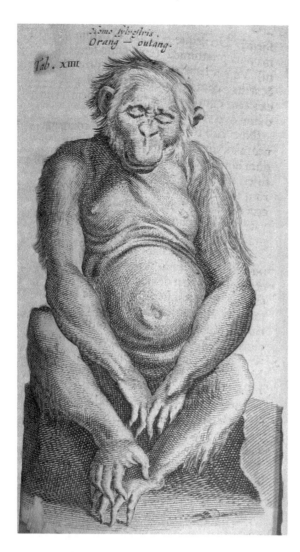

Tulp's "Orang-outang," 1640 (Tulp, *Observationes Medicae*, 271)

Even more impressive, Tulp illustrated his brief account with an engraving of the creature. Like his description, the engraving was clearly done from real life. The folds of skin, the wispy hair, the disconsolate expression on the face, have nothing of the conventional formalism of the previous centuries. This picture was an attempt to bring the reader into the presence of a new creature.

In two respects, however, Tulp's report is puzzling. First, he recalls that the creature came from Angola. In the sixteenth and seventeenth centuries, Angola was a large and well-known state on the western coast of southern Africa. The European sense of geographical boundaries was far less strict at that time than it is now, and Angola in its broadest sense included the habitat of the great ape we now know as the chimpanzee (*Pan troglodytes*). In its general posture, however, and in small but significant details such as the relatively small ears and relatively hairless chest, the engraving more closely suggests the ape that we know as the orangutan (*Pongo pygmaeus*) than it does a chimpanzee. We can probably therefore take seriously Tulp's protestation of uncertainty ("as I recall, from Angola") and accept that the creature in the engraving is indeed an orangutan.[2]

Tulp himself never visited the East Indies, where the Dutch East India Company (VOC) was in the process of building up a commercial empire from its base in Batavia (now Jakarta), but he was aware of the presence of great apes in the Indonesian archipelago, for he states later in the same chapter:

> In fact, the king of Sambaca [probably Sambas, on the northwestern tip of Borneo[3]] once told our brother-in-law, Samuel Bloemaart that these Satyrs, especially the males, on the island of Borneo, have so much confidence of spirit and such strong musculature that they have more than once attacked armed men, not to mention the weaker sex, the women and girls.[4]

The second puzzle, however, lies in name "Orang Outang" that Tulp attributes to the creature in its Indies form and that he accurately translates as "forest man" (in Malay, the lingua franca of the Indonesian archipelago, *orang* means "person" and *utan* or *hutan* means "forest"). There is no literary record, however, of the Malay-speaking peoples of the Indonesian archipelago using the term "orang utan" or one of its variants to refer to the ape before the middle of the nineteenth century. The web-based Malay Concordance Project, located at the Australian National University, contains many references to "orang hutan," but before the mid-nineteenth century they all refer, sometimes in a derogatory tone, to human beings who inhabit the forests.[5] Thomas Bowrey's 1701 *Dictionary English and Malayo, Malayo and English* defines "Ōran ootan" as "a Clown, Boor, a wild Man, a Satyr, a Beast found in the Woods of Borneo,"[6] but it is a record of Malay compiled for the purposes of Europeans and it cannot be taken as a reliable source on local usage. In fact, the first recorded Malay use of a term resembling *orang hutan* to denote the ape identifies the word as a Western term. The *Hikayat Abdullah,* written in the 1840s, recounts that "The Ruler of Sambas sent Mr. Raffles a present of two apes of the kind which the English call orang-utang."[7] Nineteenth-century Western writers on the Malay language and the Malay world were also confident that orangutan was not an indigenous term for the red ape.[8] In contemporary Indonesia, too, "forest dweller" is the most immediate meaning of the term "orang hutan," though the Western meaning is also current. Indonesians sometimes comment that Westerners call the orangutan by that name only because they believe humans are descended from the apes. The most common local name for the ape is *mawas,* or sometimes *mias* (pronounced to rhyme with "bias"), though this name does not appear in the Malay Concordance Project before the nineteenth century either.[9] Since there is no earlier published use of the term "orangutan," most observers have assumed that Tulp derived his nomenclature from Bloemaart, although his account does not specifically say so. It is likely, however, that Tulp had access to another source for the name.

In 1631, a decade before Tulp published his account, a young Dutch physician named Jacobus Bontius had arrived in the town of Batavia, headquarters of the Dutch East India Company. Bontius lived there for four years before succumbing, like so many Europeans, to one of the city's rich variety of tropical diseases. During this period, he was a prolific researcher, both on tropical medicine in the narrow sense and on the natural history of the Indies in general. His writings contained graphic descriptions of the symptoms of tropical diseases and more general accounts of the plants and animals of the Indonesian archipelago. He wrote his reports for the executives of the company, but he had also regularly sent manuscripts to his brother Willem, with the request that he arrange for their publication in Holland. Bontius' father had been the first professor of medicine at the University of Leiden, and Jacobus himself had enrolled there as a medical student at the age of twelve although, perhaps mercifully, his course of study lasted ten years. The family was thus well connected with the flourishing scientific world of the Netherlands, and Bontius probably hoped that a series of pioneering works on the medical risks and potential of the new Dutch interests in the Indies would lay the ground for a return to a chair at his father's university.[10] For some reason, however, Bontius' brother was slow in publishing the material. In the end, the writings of Bontius appeared in print as *De Medicina Indorum* (On the Medicine of the Indies) only in 1642, seven years after Bontius' death and a year after the publication of Tulp's *Observationes Medicæ.*

Although we cannot be sure, it seems probable that Tulp, a powerful broker in the emerging scientific world of the Netherlands, had seen Bontius' manuscript before publication. There is no hint that Tulp plagiarized Bontius' materials, but if we turn to *De Medicina Indorum,* we find that Bontius indeed used the term "ourang outang." Since he died in 1635, six years before Tulp's publication, he is almost certainly the first Westerner to have recorded the name and it is thus entirely possible that Tulp first learned the name from Bontius rather than from Bloemaart. But whereas Tulp is unambiguous in using "orang outang" for an ape, a close reading of Bontius suggests that he may have intended something altogether different.

In the style of the times, *De Medicina Indorum* consisted of numerous short chapters on diseases, medicines, plants, animals, and other curious features of the Indies. Among the accounts of exotic fruits and the behavior of rhinoceros, we find a short chapter that reads in part:

OURANG OUTANG OR HOMO SYLVESTRIS

Pliny, the great naturalist, writes about satyrs in Book 7, Chapter 2: Satyrs live in the tropics and mountains of India and are extremely swift. They sometimes walk on all fours, sometimes upright, they have the appearance and posture of a human. Because of their speed, they cannot be caught unless they are old or ill.

And the surprising thing is, I once saw myself some of both sexes walking upright, and the female satyr (of which I give a drawing here) hid her genitals with great shyness from the men whom she did not know, covering her face with her hands (if I can call them that) and she cried abundantly and sighed and showed other human characteristics, so that one would be inclined say that she was human, but for the fact that she did not speak. According to the Javanese, they are able to speak, but choose not to do so, to avoid being forced to work. This is really a ridiculous suggestion, of course. The name given to them is Ourang Outang, which means man of the woods, and it is said that they are the result of the lust of the women of the Indies, who slake their detestable desires with apes and monkeys.[11]

On reading the chapter title, "Ourang Outang," our instinct is to assume that Bontius was presenting his readers with a description of the ape we now know as the orangutan. Three elements of his description, however, fit poorly with this assumption. First, he comments that he had seen these creatures and that the "ourang outang" walked upright. From medieval times, Europeans placed great emphasis on the ability to walk upright as an important defining feature of humanity. In this passage, and in others in the book, Bontius makes a careful distinction between his own observations and the stories he was told by his Javanese informants, so we can assume that the creatures he saw did indeed walk upright. Yet the orangutan does not normally walk on its hind feet alone.[12] Although it is capable of standing upright, especially if it can grasp hold of something, and although it can even take a few steps on its hind legs, an orangutan on the ground will normally prefer to walk on its hands and feet. It is possible that Bontius was so impressed by the often-upright posture of the "ourang outang" that he recalled them as standing fully upright. Nonetheless, he would have been aware of the convention that apes were imperfectly bipedal, and it is unlikely that he would have made so categorical a statement if there were doubt about the gait of the creature he observed.

Second, Bontius reports that he himself saw "some" of these creatures. He thus implies that he saw them as a group and that the incident took place in Java (since he refers to stories the Javanese told him about them). Orangutans, however, tend to be solitary and were difficult to capture because of their arboreal habitat. Moreover, the animal was not known from Java in historical times. There is no hint of a folk memory of the orangutan in Javanese legends, nor have these apes been recorded as a regular item of trade. In the Chinese records of Sumatra and Borneo, there is not a single reference that can be linked to orangutans among the many exotic products listed as items of trade and tribute.[13] They produced no body parts with pharmaceutical or other uses, and there was evidently no trade in orangutans as curiosities. Bontius may have seen an orangutan in Batavia, but what we know of the trade of the archipelago in those times makes this possibility unlikely.

Third, Bontius comments that the female "hid her genitals with great shyness from the men whom she did not know." In Christian doctrine, sexual modesty was a consequence of the Fall, the biblical episode in which Adam and Eve, having eaten the fruit of the tree of the knowledge of good and evil, discovered a sense of shame and were expelled from the Garden of Eden.[14] Modesty was such a strong sign of human identity that it was hardly ever attributed to apes by medieval Europeans. Far more common was to portray them as wanton and lustful.[15] Any contemporary observer of orangutans will have noted that neither females nor males show any inclination to conceal their genitals from the gaze of either humans or other orangutans. On the contrary, supple joints and a habit of clambering from branch to branch leave the observer in no doubt about their genital physiology. The situation Bontius describes indicates that the female "ourang outang" was naked, as one would expect of an ape, but the behavior that followed was that of a human, rather than an ape. Only Bontius' comment that "she cried abundantly and sighed and showed other human characteristics, so that one would be inclined say that she was human, but for the fact that she did not speak" coincides with our contemporary perception of orangutans as having humanlike emotions.

The problems with Bontius' chapter are compounded when we examine the striking picture that illustrates it. The figure is female and naked, with a pot belly, tufts of hair all over her body, long hair on her head, a substantial beard under her

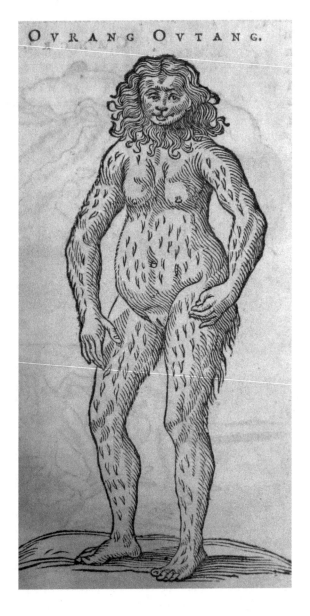

Bontius' "Ourang Ou-
tang," 1642 (Bontius, *On
Tropical Medicine*, 284)

OVRANG OVTANG.

chin, and a slightly leonine face. In all, she looks somewhat like a woman, but
not much like the ape that we now know as the orangutan. The drawing matches
Bontius' written description in having the "ourang outang" standing upright,
but she does not display the modesty that Bontius attributes to her. Nor is there
any mention at all in Bontius' account of the hairiness that is so prominent in the
drawing.[16]

The drawing style of the representation, however, is not conspicuously like that of other illustrations in the book, which tend to show a more mechanical draftsmanship and less sense of character. In fact, Bontius' drawing seems to be a composite of two established traditions in Western representation. One is the so-called hairy woman, a common trope in Renaissance literature and art. She was a cautionary tale, said to have been a consequence of her pregnant mother having looked at a picture of St. John in the desert, dressed in hairy animal skins. It was believed that the form of a baby could be influenced by the mother's experiences at the moment of conception and during pregnancy.[17] In posture and in the representation of hairiness, Bontius' picture seems clearly to draw on this tradition. Medical and quasi-medical works using similar images were in circulation at the time of Bontius' medical training.

The other apparent source is "Von Breydenbach's Baboon," a portrait of a female ape standing upright (albeit with a stick) that appeared in Bernhard von Breydenbach's *Journey,* published in 1486. The lionlike mane of Bontius' "ourang outang" seems to be derived from this picture, drawn by Von Breydenbach's companion on his travels in the Middle East, Erhard Reuwich, and probably based on the Egyptian sacred baboon, *Papio hamadryas.*[18] The picture was reproduced in 1602 in Konrad von Gesner's *Historiae animalium*[19] and is thus likely to have been known to Bontius. In fact we need not suppose that Bontius was familiar with either illustration. The most likely explanation for the illustration is that Bontius' original was lost or was unsuitable for publication and so had to be redrawn by an artist who embellished it in the light of popular models he had at hand.[20]

But if Bontius' "ourang outang" was not an ape, what might it have been? Used to refer to "people of the forest," *orang hutan* was, and is, a disparaging term that implies a lack of social skills. Bontius' insistence that they walked upright and that the female showed sexual modesty is further indication that they were true humans. They may, however, have been human beings who suffered from some illness that cast doubt on their humanity. Among the various debilitating conditions known to have been found in the archipelago at the time, the most likely possibility is endemic cretinism, a complex physical condition caused by iodine deficiency.[21] In the first generation and in mild cases, lack of iodine leads to the development of goiters, massive swellings of the thyroid gland which form pendulous sacks at the neck. The offspring of goitrous women living in iodine-deficient regions may not develop goiters, but may instead suffer from a complex of symptoms including mental deficiency, deaf mutism, short stature, low hairline, bulging eyes, a protruding tongue and an abnormal gait. Endemic cretinism (that is, cretinism produced by the cumulative effect of environmental iodine deficiency over two or more generations) is characteristically a disease of mountain regions where iodine has been leached from the soil by erosion and is absent from drinking water. The condition is rare to nonexistent in coastal regions (saltwater fish are a major dietary source of iodine) while the European Alps, the Himalayas, and the Andes remain the areas of greatest incidence.[22] Although nowhere in Java is far from the sea, salt was a rare and valuable trade commodity in precolonial and colonial times, and it is entirely plausible that remote mountain regions might have been too poor to obtain the salt or fish that would have prevented goiter and cretinism. Goiters were reported among inhabitants of the slopes of Mount Salak, south of Batavia, in the late eighteenth century,[23] and there is no reason to suppose they were not also present in Bontius' time. Because cretinism is caused by external environmental conditions, it tends to be widely present in affected communities, so it is not unlikely that a group of cretins might have ended up for some reason in the Dutch company capital. Bontius, encountering such beings for the first time, may have doubted that they were fully human, while still recognizing their closeness to humanity. A visitor to the European Alps in the late eighteenth century described the cretins he encountered there as "*vix humana*" (barely human) and "*non multo meliores brutis animalibus*" (not much better than brutish animals).[24]

Bontius' account does not give us enough evidence to be sure that what he saw in Batavia were cretinous human beings, but we can be close to certain that the first *orang hutan* he encountered in the streets of Batavia were true humans, people of the forest, whose peculiar appearance was a consequence of some specific affliction, not a bewildered group of apes on their way to an uncertain fate. Tulp thus appears to have conferred the name "ourang-outang" on a red ape from the East Indies as a consequence of misreading the unpublished manuscript of Bontius. But if he lifted the name from Bontius' description of deformed human beings, he did so because its translation—"person of the forest"—seemed so apt for the animal he had newly encountered. The name survived for the same reason. When the English anatomist Edward Tyson published his book on the dissection of a chimpanzee, he reached back more than half a century to choose the name that Tulp had used, "Orang Outang," for the creature he was about to describe minutely. He was able to apply that name to an ape from Africa because Tulp had described his specimen as coming from Angola. But he chose this name, rather than *Quoias Morrou,* which Tulp had after all identified as the correct African name, because the meaning of "orang outang" seemed uniquely suited to the kind of being he was describing.

For the Europeans of Tulp's era, the term "orang outang" was not just an example of nomenclatural exoticism. The name was recognized as the Malay equivalent of a set of terms in European languages—man of the woods, *homo sylvestris, homme sauvage, bosjesman*—all with the same literal meaning and all representing a problem in the understanding of the nature of humanity. Encompassed within this loose translinguistic category were what we now identify as three very different forms of being. One was feral human beings, people who had retreated from or been lost to human society, sometimes as children, sometimes as adults.[25] In the second category were human communities whose culture appeared—in the judgement of Europeans—to be so rudimentary as to call into question whether they were fully human. This category included the indigenous peoples of southern Africa, now referred to as Khoisan, who were known for decades as the Bushmen or Hottentots. The third kind was the "orang outang," identified by Tulp and subsequently credited with an impressive range of almost human characteristics. All three phenomena were unsettling because they raised the possibility that there might be more than one form of human being, a possibility that philosophy, religion, and ethics then, as now, were poorly equipped to consider. Throughout the seventeenth and eighteenth centuries, Western science was moving along a number of paths away from the traditional Christian view that humankind had been created separately from all animals. This movement created the challenge of deciding both how to distinguish humans from animals and just what significance to attach to that distinction. In what ways did the human body differ physiologically and anatomically from those of the nearest animals? In medieval times, the fact of standing upright constituted the human advantage over animals, but feral human beings had sometimes become accustomed to walking on all fours, and so the question needed to be answered in terms of anatomical potential, rather than habit. In what way did the human mind differ from those of animals? Humans spoke, of course, but many feral people had lost the power of speech. Rationality, rather than simple loquacity, seemed to sum up better the human advantage. Yet rationality was unquantifiable and subjective.

Evaluation of the possible human status of the orangutan thus followed two distinct lines of analysis, one of them anatomical, the other behavioral. The anatomical separation of orangutans from humanity (and incidentally from chimpanzees) was endorsed by the emerging science of biology and was largely complete by the 1770s. By a strange twist, the orangutan eventually acquired a scientific name, *Pongo pygmaeus,* combining terms that had originally applied only to African apes. This process is outlined in the remainder of this chapter. The behavioral separation of humans and orangutans, by contrast, was slower and more ambivalent. It is discussed in later chapters.

In 1735, Linnaeus published the first edition of his *Systema Naturae* (System of

Nature). The work was an attempt to identify and name every species of living thing and mineral and to group them into categories based on their physical form or appearance. Thus species were grouped into genera, which in turn belonged to orders within still larger classes. The single most dramatic feature of Linnaeus' system was that he included humans as just another species, *Homo homo*,[26] in the hierarchy of animal kinds. Even in a world that had begun to question the implications of the orthodox reading of Genesis and the story of a separate, divine creation of humankind, this classification was daring and potentially dangerous. Linnaeus himself refrained from providing a physical description of humans comparable to the description he gave of all other animals. Nonetheless, it was implicit in the Linnaean system that a physical difference distinguished humans from apes, even though it was only enough to mark them as separate genera, *Homo* and *Simia,* within the class that was tellingly named Anthropomorpha ("having human form").

The Linnaean injunction "Know yourself" appeared next to the genus *Homo* in each of the twelve increasingly complex editions of the *Systema Naturae* that appeared during Linnaeus' lifetime. Linnaeus, however, was less constant in deciding just what might be regarded as *H. homo* (later *H. sapiens*).[27] In his sixth edition of 1748, Linnaeus identified the creatures of both Tulp and Bontius as apes, *S. satyrus,* describing them briefly as "tailless, hairless underneath." In the 1758 tenth edition, by contrast, *S. satyrus* was identified as the "Indian satyr" of Tulp, while the "ourang outang" of Bontius was located within the genus *Homo* but placed within a new human species called *H. troglodytes,* along with albinos.[28] In this same edition, Linnaeus fixed on what have become the standard terms for higher-level classifications: class "Quadrupedia" became "Mammalia," and order "Anthropomorpha" became "Primata."[29]

As the Linnaean system took hold, it became immensely important that there be a means of distinguishing humans from animals. The vast majority of people—scientists and nonscientists—were reluctant to see humans simply as the first-ranked among animals. They wanted a clear and qualitative distinction between humankind and the animal world. Once this special distinction had been made, however, there was much less at stake in the question of how orangutans should be classified. It mattered little in the grand Linnaean schema whether "orang outangs" of Bontius and Tulp were human or not; whatever their precise nature, they were rare and distant, presenting neither a threat nor an economic opportunity to Western European society. Crucially, recognition of "orang outangs" as human implied nothing for the status of any other animal species: the stories that gave the "orang outang" special status were exclusive to it and did not risk changing the general ethical relationship between humans and animals.

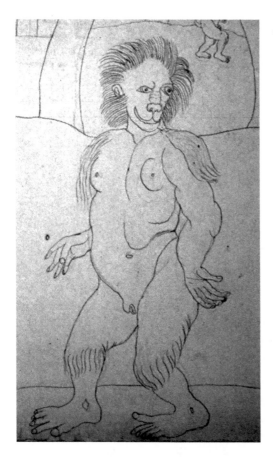

Jan Velten's orangutan c. 1700
(*Wonderen der Natuur*, n.p.)

Linnaean classification, however, was based exclusively on physical characteristics. The term *species* meant "appearance" in Latin, and the Linnaean system
had no place for the consideration of behavior. In this practice, the Linnaean
system was well suited to an age in which scientists for the most part remained
in Europe and examined specimens—mainly skeletons and skins—held in private
and public collections. The uncertainty in Linnaeus' classification arose primarily
because he had no specimen to refer to. No trace remained of the animal Tulp
had sketched. In about 1700 a living orangutan had arrived in Amsterdam, where
it was put on display at Blauw Jan, a menagerie attached to an inn in the center
of Amsterdam. We know of the Blauw Jan orangutan, however, not from the
reports of scientists but rather from an intriguing and recently discovered
collection of paintings and sketches of creatures at the menagerie by an otherwise
unknown visitor, Jan Velten.[30]

Every now and then during the first half of the eighteenth century, orangutan cadavers and body parts arrived in Europe. Francois Valentijn, an encyclopedic author on the people and natural history of the Indies, wrote in 1726 of a "Mr Camper, the seigneur of Ouwekerk aan der IJssel, who . . . [has a] specimen in his possession, now in a jar in spirit at his home in Leiden."[31] The collection of Sir Hans Sloane, which later became the core of the British Museum collection, contained parts of four orangutans at the time of his death in 1753, which were recorded as follows:

> The sceleton of an Orang Outang or Wild Man from Sumatra, in the East Indies, by Capt. Aprice. The hands and feet were thrown overboard in coming from the East Indies when this creature died. It was given to me by Dr. Maidstone.
> The viscera of the Chimpanzee Orang Outang, or Homo Sylvestris.
> The Homo Sylvestris, Orang Outang, or Chiampanze from Borneo. It dy'd in China and was put into rack,[32] and brought over to Mr. Charles Lockyer who gave her to me. She is covered wt longer hair.
> The fore paws of an Orang Outang or Champanzi from the Duke of Richmond's sale by Dr. Stack.[33]

None of these specimens, however, came to Linnaeus' attention. More surprising is that Linnaeus also appears to have been unaware of the 1699 publication of the English anatomist Edward Tyson, titled *Orang-Outang or Homo Sylvestris,* which describes in detail the dissection of a great ape, identifying it as representative of an animal found in both Africa and Southeast Asia.[34] Whereas the identity of Tulp's creature is contested because his description is only brief and his illustration has a touch of whimsy, Tyson's specimen was unambiguously a chimpanzee, which he describes as coming from Angola, although "some Sea-Captains and Merchants who came to my House to see it, assured me, that they had seen a great many of them in Borneo, Sumatra and other Parts."[35]

Tyson's report was a remarkable piece of work. He started from the proposition that science lay in the meticulous collection of apparently tedious data. He compared the observations of Aristotle with those of other Classical authors concerning the anatomical features of apes, here and there expressing puzzlement at some difference between the comments of Aristotle or the others and his own observations, sometimes finding a resolution to the puzzlement, sometimes not. The largest part of the book was a detailed account of the chimpanzee's anatomy, its bone structure and its musculature. Like all taxonomists, Tyson wrestled with the relative significance of different physical characteristics of his ape. He considered, for instance, that the hair of his ape was more like human hair than was the hair of other animals. The ape's body hair was also far more abundant, but Tyson wondered whether humans' lack of hair was a consequence of wearing clothes, and he mentioned cases of people living in the wild without clothes who were reported to have become hairier (8–9). At the conclusion of this detailed examination, Tyson offered the reader a list of forty-eight points on which "the Orang-Outang or Pygmie more resembled a Man, than Apes and Monkeys do" (92–94), followed by a list of thirty-four points on which the creature more closely resembled "the Ape and Monkey-kind" (94–95). The overall verdict of Tyson's investigations was subtle but unambiguous: "I take him to be wholly a Brute, tho' in the formation of the Body, and in the Sensitive or Brutal Soul, it may be, more resembling a Man, than any other Animal; so that in this Chain of the Creation, as an intermediate Link between an Ape and a Man, I would place our Pygmie" (5).

The force of Tyson's comparison came from the abundance of minutely detailed evidence. No single difference that he identified stood out as being diagnostic. Rather, the combined weight of those thirty-four points of difference established that the "orang outang" could not be human, despite the impressive tally of similarities. Tyson reached a similar conclusion in examining his creature's voicebox and brain (51, 55). Everything he found seemed to him to be exactly like that of humans. There was no physical obstacle to "orang outang" speech. Yet, he insisted, the physiological capacity to speak did not mean that the creature ever did so. Speech, and the system of thought that lay behind it, was possible because of what he called "those Nobler Faculties in the Mind of Man," though he did not specify what these faculties might be. Tyson was dismissive of what he called the current "Humour" to "make Men but meer Brutes, and Matter." Physiology did not determine humanity.

The physical specimen that Linnaeus hoped for was eventually provided instead by another Englishman, the naturalist George Edwards, a Fellow of the Royal Society in London, who saw two live orangutans in London in the late 1750s, possibly a pair reported to have been taken to Europe from Batavia by the former harbor master, a Mr. Pallavicini, in about 1759.[36] Both died, however, before Edwards had a chance to commission a picture, and he had to work from one imperfectly preserved specimen: "it was first soaked in spirits of wine, then dried, and set up in the action I have given it, the draught being taken before its parts were too much dried or fallen in."[37] Edwards stated unambiguously, "I believe them all to be natives of Africa, though there are voyagers to India who describe something like them." Nonetheless, the reddish, shaggy hair, small ears, and the shape of the feet of Edwards' portrait all strongly suggest an orangutan. Edwards'

Edwards' great ape, 1758. The caption reads: Synonymus Names: The Satier, Savage, Wild-man, Pigmy, Orang-autang, Chimp-anzee &c. (Edwards, *Gleanings of Natural History,* ch.3, plate 213)

portrait and description circulated widely and was redrawn for other influential natural histories.[38]

In 1760, just two years after the tenth edition of the *Systema Naturae,* a student of Linnaeus, Christian Emmanuel Hoppe of St. Petersburg (often known, like Linnaeus, by his Latinized name, Hoppius) published a thesis titled *Anthropomorpha,* in which he provided descriptions and crude redrawings of earlier pictures. Bontius' woman was shorn of her leonine mane and presented as *Simia troglodyta*. Breydenbach's ape was there as *S. lucifer,* a new category of tailed apes.[39] Tulp's ape, *S. satyrus,* was represented not by the illustration that he had included in *Observationes Medicæ* but rather, bizarrely, by the drawing of a chimpanzee that had been on display in London in 1738.[40] And finally, Edwards' orangutan was there with the name *S. pygmaeus.* Unimpressive in evidence and argument, Hoppius' work later became significant for its timing. As the Linnaean system became the standard for classifying animals and plants, a wide range of names came to be applied to the red ape from Borneo. It was called variously *Simia satyrus, Simia pongo, Pithecus satyrus, Pongo borneo* and, most memorably, *Orangus outangus.*[41] With a similar abundance of names referring to other species, zoologists developed a strict law of priority in zoological nomenclature. This law takes the tenth edition of the *Systema Naturae* as the baseline for scientific naming. In other words, no name used before 1758 has any scientific standing and all names given to species in that edition remain valid, as long as they can be shown to refer to real animals. Thereafter the first valid name applied to a species (validity meaning especially that the name has not been used for another species) has priority. Linnaeus had identified *Simia satyrus* in the tenth edition but his authority was Tulp, whose drawing was ambiguous and who had identified his ape as coming from Angola. Crucially, the systematists Stiles and Orleman, writing in 1927, accepted this interpretation of Tulp's ape, stating categorically that "from the standpoint of Linn., 1758a, the animal known to-day as the Orang-utan was systematically and nomenclatorialy non-existent."[42]

Stiles and Orleman then ruled that Hoppius' redrawing of Edwards' picture was the first valid portrayal of an orangutan (Edwards himself, publishing in 1758, was just within the valid period allowed by the rule of priority, but he had not given his creature a Linnaean name.) Hoppius' name, *Simia pygmaeus,* therefore appeared to have priority, and *pygmaeus* became the specific epithet for the orangutan. The Stiles and Orleman ruling, however, made *Simia satyrus* the correct name for the chimpanzee, even though this name was generally used for the orangutan. Recognizing that this arrangement was likely to cause enormous confusion, they ruled that the name *Simia satyrus* should be suppressed, that is, not be applied to the chimpanzee either. To find the correct generic name for the red ape, they examined early scientific literature meticulously, concluding eventually that the first to provide a valid generic name for the orangutan had been the French naturalist Bernard Germain de Lacépède, who had published a "Table of Mammals" in 1799 during the first decade of the French Revolutionary government, in which he had identified the orangutan by the name *Pongo.*[43] A Scottish sailor,

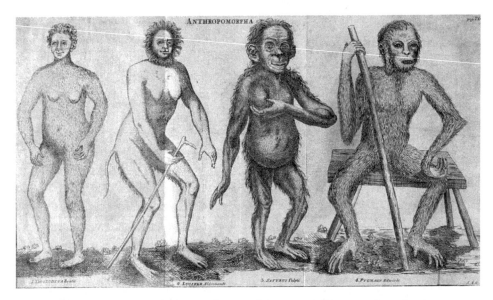

Hoppius' Anthropomorpha, 1760 (Hoppius, *Anthropomorpha*, 76)

Andrew Battell, who lived in Angola in 1600 and 1601, had been the first to record this name. He had written an influential account of the behavior of gorillas[44] (*pongo*) and chimpanzees (*engeco*), which made a major contribution to the wider identification of the great apes as almost human. The brilliant French scientist Georges-Louis Leclerc, the Comte de Buffon, had then used these terms in his massive *Histoire naturelle,* believing that there were two species, the large "Pongo" and the small "Jocko."[45] Published in 1766, Buffon's work came after the 1758 milestone, but he did not use or accept the Linnaean system of classification, and his name became scientifically valid only when his successors finally accepted the Linnaean framework. In this way, a great ape found only in the jungles of Borneo acquired a generic name, *Pongo,* referring to great apes of Africa, and a specific epithet, *pygmaeus,* alluding to Tyson's insistence that the African pygmies known to the ancient Greeks were apes, not humans. Acceptance of the name *Pongo pygmaeus* marked the hegemony of rules within the system of scientific classification, just as the Linnaean system marked the primacy of anatomy and physiology in determining the classification of a species. A scientific name, however, could do nothing to rein in the imagination of the public and of a succession of entrepreneurs who presented the orangutan to the world. Nor could anatomy convince the many who believed that the essence of human identity lay in the mind, and thus in behavior. For them, an animal that behaved like a human might truly be a person.

Whereas Hoppius' clumsy representation of great apes became important only in retrospect, the definitive confirmation that the orangutan was physiologically distinct from humans was the work of the Dutch anatomist Petrus Camper, son of the seigneur who had kept a preserved orangutan in a jar in his home at Leiden. Camper explicitly emulated the investigation by Tyson:

> As the celebrated Dr. TYSON found the organ of voice so similar to that of men in his Pigmy, I endeavoured to get one from the East Indies. For this purpose I offered a good sum of money to my correspondents to have a well-preserved Orang Outang, because none were to be met with in any collection of Natural History in Holland.
>
> I soon got a female one in 1770, by the kindness and generosity of Dr. HOFFMAN, physician at Batavia, formerly my pupil; and in the year 1771 another, by favour of Mr. HOPE, Director of the East India Company of Amsterdam, and Representative of His Most Serene Highness the Prince of Orange in the same Company, who was so good as to order not only a female,

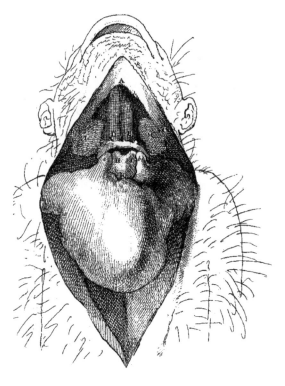

Camper's sketch of the larynx of an orangutan, 1770 (Camper, "Account of the Organs of Speech," facing 158)

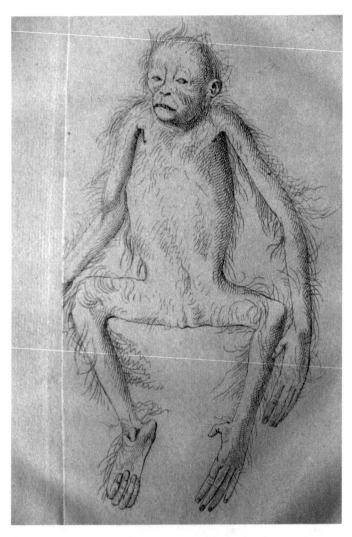

Camper's "orang outang," 1770 ([Camper], *Natuurkundige Verhandelingen,* facing 120)

but a male one for my use; but this last was unluckily lost, with the whole ship, betwixt Java and the Cape of Good Hope.[46]

Camper's conclusion from his examination of the larynx and bone structure of the orangutan was unambiguous: orangutans could not talk and were not suited to walking upright.[47] As soon as he began his dissection, however, Camper was conscious of significant differences between his orangutan and the creature described by Tyson. He had no doubt that the two were different species, and his drawings make clear that his specimen was an orangutan, just as Tyson's drawings are immediately recognizable as those of a chimpanzee. His close anatomical examination of five orangutans, his inspection of another six specimens, and the report he requested from London on the then-still-available specimen recorded by Edwards all confirmed this impression.[48]

Scientific inspection of orangutan cadavers in Europe in the second half of the eighteenth century allowed comparative anatomy to dispel the myth of the almost human "orang outang." Hard on their heels, however, came live orangutans who survived long enough in the unfriendly European climate to generate a new set of observations.

2 🜍 "A More than Animal Intelligence"
Exploring the Species Boundary

Although the Linnaean system proposed that species could be distinguished from one another by their physical characteristics alone, the striking exception that Linnaeus himself made for humans—exhorting them to know their own character—hinted at another way of defining humanity. Linnaeus' instruction could be read in several ways. In Classical and medieval times, the maxim *nosce te ipsum* had been a warning against hubris, advice not to overestimate one's capacities. In the sixteenth century, it was often meant more literally, commonly appearing on so-called anatomical fugitive sheets, which were printed for public sale and purchased by members of the public keen to learn about the arrangement of their own internal organs.[1] But in the eighteenth century it was also an assertion that human intelligence, in the form of rationality or self-awareness, was the essence of human identity. By virtue of its difference from the terms in which all other animals were classified, *nosce te ipsum* asserted a special status for humans.[2]

For those who saw rationality and self-awareness as criteria for humanness, the nonhuman status of the orangutan was much less certain than it was to the anatomists. From the seventeenth century until late in the eighteenth, observations of humanlike behavior on the part of the orangutan, alongside the animal's obvious physical resemblance to humans, seemed to mark it as a being that was something more than an animal, even if it might be something less than a human. During these centuries, scientists used the term "orang outang" for the great apes—both orangutans as we understand them today and the black, African ape we now call the chimpanzee, and later also the newly rediscovered gorilla—but in society more broadly the term referred to a half-mythical creature that might be, or might almost be, another form of human being. The scientific writings of

Bontius and Tulp had lifted the orangutan out of the shadowy world of fantastical beasts and placed it securely in the secular system of nature, yet their own accounts and the brief but suggestive accounts of later travelers wove a new set of scientific myths that seemed to confirm the animal's special similarity to humans. In the generalized image of the great ape conjured up by the reports of travelers to Asia, retold, amended, and interwoven with tales from Africa, the separate identities of orangutan and chimpanzee are only sometimes evident. More important, these stories together created a picture of a being with so many human attributes that it could not simply be regarded as animal.

Bontius had reported that the "ourang outang" walked upright, could speak but chose not to, and felt sexual modesty. If his report concerned human beings, as seems was really the case, then these attributes are unremarkable. Once, however, mistaken identity linked these stories to a creature with the physical appearance of a large ape, they presented a startling challenge to the question of how animals were to be distinguished from humankind. That challenge was all the sharper because of Bontius' suggestion that the "ourang outang" were hybrids between humans and apes and because of a report by Tulp, based on the testimony of Bloemaart, that orangutans habitually raped human women.[3] For a century or so after Bontius and Tulp, these stories circulated relatively unquestioned, sometimes attributed to the two Dutch physicians themselves, sometimes presented as fresh reports. The accounts of Bontius and Tulp, moreover, were soon supplemented by new travelers' tales. These stories were mostly brief, but they sustained a perception on the part of the European public that there were the strange human-like creatures in the tropics and subsequently inspired imaginative writing, as we see in later chapters.

Elias Hesse, for instance, survivor of an unfortunate mining venture launched at Sillida in the interior of the west coast of Sumatra in 1680 by an expedition from the German state of Saxony, repeated the core of Tulp's observation while being the first to locate orangutans in Sumatra: "They walk on their hind legs. . . . In general they are lustful and love women, so that the latter pass through the forests in great danger of being impregnated by them."[4] A French traveler, Louis Le Comte, writing at the end of the seventeenth century, speculated directly on the humanness of "orang outangs," emphasizing both their physical resemblance to humans and their emotional similarities:

> They find in the woods a sort of Beast, called the Savageman; whose Shape, Stature, Countenance, Arms, Legs and other Members of the Body, are so like ours that excepting the Voice only, one should have much ado not to reckon

them equally Men with certain Barbarians in Africa, who do not much differ from Beasts. . . .

As to the rest, it cries exactly like a Child; the whole outward action is so Human, and the Passions so lively and significant, that dumb Men can scarce express better their Conceptions and Appetites. They do especially appear to be of very kind Nature; and to shew their Affections to Persons they know and love, they embrace them, and kiss them with transports that surprise a Man. They have also a certain Motion, that we meet not with in any Beast, very proper to Children, that is, to make a noise with their Feet, for joy or spight, when one gives or refuses them what they passionately long for.[5]

In 1708, another French author, "François Leguat," published an account of an ape that he said he had seen in Batavia, a female who walked on her hind legs and

Conceal'd the Parts that distinguishes [sic] the Sexes, by one of its hands, which was neither hairy without nor within. Its Face had no Hair upon it other than the Eye-brows, and in general it much resembled one of those Grotesque Faces which Female Hottentots have at the Cape.[6]

"Leguat" has since been shown to be a fiction, the construction of another French writer, François Misson, who compiled and elaborated upon elements from earlier travel writers, though his source for the brief orangutan description is unknown.[7]

Other accounts presented by apparently sober observers appear in retrospect to be nothing but fantasy or reported rumor. Alexander Hamilton, a Scottish sea captain who traveled the Asian seas for more than three decades from 1688, reported seeing an orangutan in Java: "He blows his Nose, and throws away the Snot with his Fingers, can kindle a Fire, and blow it with his Mouth. And I saw one broyl a Fish to eat with his boyled Rice."[8] An Englishman, Charles Frederick Noble, who appears to have been in close contact with a young orangutan, concludes his account with a gratuitous swipe at Dutch colonial practices and a slightly different version of Bontius' story of sexual encounters between women and apes:

There is an animal here, which I had the curiosity to view very attentively. It resembled the human form much more than any creature I had ever seen. It was young, had a melancholy look, the face almost quite bare but the head, eye-brows, and lower part of the chin very rough. It made little noise, shewed great fondness, in grasping me around and squeezing me; and sometimes made a low pensive sound, as if whining and crying. It walked upright with great

Leguat's "Ape of the island of Java," 1708 (Leguat,
*De Gevaarlyke en Zeldzame Reysen van den heere
François Leguat*, 140)

ease, and was about three feet and a half high. Some people ... have accounted for its production in the following manner: that the cruelty of the Dutch to their Malayan female slaves, often obliged them to fly into the woods to escape the cruelty of their tyrannical masters; and, being forced to live there solitarily, it was thought that they might, by length of time, turn mad, or insensibly brutish, and might have yielded to an unnatural commerce with some animals in the woods, by which this strange animal I have been speaking of was produced.[9]

These fantastical stories gained plausibility with each telling, but even the credible account of Daniel Beeckman, writing in 1718, conjures up a being that is much more human than any animal his readers would have known:

I bought one for curiosity for six Spanish Dollars; it lived with me seven Months, but then died of a Flux; he was too young to show me many pranks, therefore I shall only tell you that he was a great Thief, and loved strong Liquors; for if our Backs were turned, he would be at the Punch-bowl, and very often would open the Brandy Case, take out a Bottle, drink plentifully, and put it very carefully into its place again. He slept lying along in a humane Posture with one hand under his Head. He could not swim, but I know not whether he might not be capable of being taught. If at any time I was angry with him, he would sigh, sob, and cry, till he found that I was reconled [*sic*, i.e. reconciled] to him; and though he was but about twelve Months old when he died, yet he was stronger than any Man in the Ship.[10]

These stories blended easily with tales from Africa. Battell's account of the "Pongoes" in Angola had them building shelters against the rain, hunting "Negroes"

Beeckman's "oran-ootan," 1718 (Beeckman, *A Voyage to and from the Island of Borneo*, facing 37)

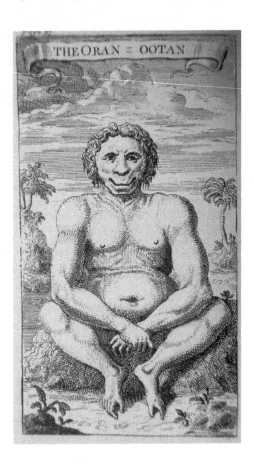

THE ORAN = OOTAN

in the jungle, capturing elephants, and burying their dead under heaps of branches, but unable to make fire.[11] Pyrard wrote of animals in Sierra Leone

> called *baris,* (the orang outang), who are strong and well limbed, and so industrious, that, when properly trained and fed, they work like servants; that they generally walk on the two hind-feet; that they pound any substances in a mortar; that they go to bring water from the river in small pitchers, which they carry full on their heads. But, when they arrive at the door, if the pitchers are not soon taken off, they allow them to fall; and, when they perceive the pitcher overturned and broken, they weep and lament.[12]

The eminent eighteenth-century French naturalist the Comte de Buffon reported information he had received from the earlier botanist Guy de la Brosse that he witnessed two orangutans aboard a ship who ate at the table, using cutlery and goblets, and would communicate their wishes to the cabin staff. On one occasion, one of them was bled for an illness; later, when feeling ill again, he held out his arm to be bled again, believing that the first treatment had done him good.[13]

European philosophers had little or no opportunity to view an "orang outang," dead or alive, but many of them took seriously the cumulative weight of testimony suggesting that these beings might be regarded in some respects as human. In *L'Homme Machine* (1748) Julien Offray de la Mettrie argued that there were no sharp lines of distinction between species and that there was therefore a gradual transition from apes to humans. He suggested that apes failed to speak because of a defect in their organs. Crucially, he tied this speculation to the contemporary philosophical and scientific interest in so-called wild people: "Why would the education of apes be impossible? Why could not an ape through careful effort communicate with sign language just like the deaf? I dare not decide whether an ape would be able to talk after we did something to its speech organs, but I would be greatly surprised if it were absolutely impossible."[14] The possibility was echoed in a 1754 satirical work by the French philosopher and encyclopedist Denis Diderot, who portrayed the Roman Catholic cardinal Melchior de Polignac as encountering an "orang outang" in the royal gardens in Paris. The ape, it was reported, had the air of John the Baptist preaching in the desert, and the cardinal responded by crying out, "Speak, and I shall baptise you."[15]

The most important contribution to the incipient recognition of "orang outangs" as human, however, came from the French philosopher Jean-Jacques Rousseau in his 1755 *Discourse on the Origin of Inequality.* In a digressive note appended to the work, he cited several travelers' accounts of great apes and commented:

We find in the descriptions of these supposed monsters striking similarities to the human species and fewer differences from men than those one could find between one man and another. It is not clear in these passages why the writer withholds the name of savage men from the animals in question, but one may readily guess that the reason is their stupidity and also the fact that they did not speak—poor reasons for those who know that although the organ of speech is natural in man, speech itself is not natural to him, and who recognize how much his perfectibility may have raised the civilized man above his original state.[16]

The widely read collection of voyage accounts to which Rousseau was responding here included not only the usual stories of orangutan/chimpanzee intelligence and social structure, but also a redrawing of Tulp's "Indian satyr" as a wistful-looking woman with full breasts and carefully arranged hair, which may have increased his inclination to stress the humanness of the great apes.[17]

Nonetheless, Rousseau's central argument was not based on physical form at all. He suggested that "pongos and orangutans" might be distinguished from lesser apes by the fact that they shared with humans the capacity to perfect themselves, to learn and to improve their circumstances. He juxtaposed the accounts of orangutans with reports of a "wild boy" captured in 1694 who walked on all fours, had no language, and showed no sign of "reason." After careful instruction, however, the boy learned to speak and could then, presumably, take some part in normal human society. "If, unluckily for him," Rousseau sarcastically commented,

> this child had fallen into the hands of our travellers, we cannot doubt that after noting his silence and stupidity, they would have decided to send him back into the forest or lock him up in a menagerie, after which in their brilliant accounts of their travels, they would have learnedly spoken of him as a most curious beast somewhat resembling a man.[18]

Rousseau made these comments in a note appended to a separate discourse, and he did not see these remarks as a manifesto for extending human rights to great apes. After all, such rights were only just emerging as a distinct concept in Western political thinking. Rather, he was expressing a version of the contested modern view that capacity to behave like a human might be learned and that the capacity to learn should trump mere physical differences.

Rousseau's ideas were developed still further by the Scottish judge James Burnet, Lord Monboddo. In 1773 Monboddo published a book titled *On the Ori-*

gin and Progress of Language that explored in more than a hundred pages the possibility that orangutans were a form of human that had somehow failed to develop the civilization that characterized the rest of humankind. They were trapped in an undesirable primitive state but could, he suggested, be trained to become part of civilization. He reviewed in detail a wide range of travelers' accounts of great apes, accepting the truth of every report that made them appear more human:

> The sum and substance of all these relations is, that the Orang Outang is an animal of the human form, inside as well as outside: That he has the human intelligence, as much as can be expected of an animal living without civility and the arts: That he has a disposition of mind mild, docile and humane: That he has the sentiments and affections common to our species, such as the sense of modesty, of honour, and of justice; and likewise an attachment of love and friendship to one individual so strong in some instances, that the one friend will not survive the other: That they live in society and have some arts of life; for they build huts, and use an artificial weapon for attack and defence, viz., a stick; which no animal merely brute is known to do. . . . They appear likewise to have some civility among them, and to practise certain rites, such as that of burying the dead.[19]

Monboddo's ideas were subject to a good deal of ridicule at the time, less because of his views on the orangutan than because of his parallel belief in the existence of men with tails.[20] Rousseau, however, was one of the most influential and widely read philosophers of his day, and his speculative pronouncement on orangutans drew much attention, scholarly and popular, critical and supportive.[21] Monboddo, although the subject of mockery in a way that Rousseau was not, was also widely read. Neither of them asserted that orangutans were exactly like humans, and they appear to have recognized that there were physical differences between the two. Their argument was rather that the mental ability of orangutans gave them such affinity with humankind that they should be classified as human rather than animal. Given the right opportunities, they could play a valuable and distinctive role in European society, as did the fictional flute-playing, cudgel-wielding Sir Oran Haut-ton, elected to parliament in Thomas Love Peacock's novel *Melincourt* (discussed in chapter 5).

Rousseau and Monboddo wrote at a time when barely half a dozen great apes of any kind, alive or dead, had come to any kind of scientific attention and when no scientific observer since Tulp had seen a living orangutan. They were able, therefore, to recruit their "orang outangs" to make a point about humans, that is, that there was an underlying potential in the most primitive of humans. Rousseau did not celebrate the orangutan as a "noble savage," but rather as a sign of the improvability of humankind. Monboddo (more exuberantly) identified the "orang outang" as an admirable being, contemplation of whom might help to rescue decadent modern society from its plight.

In the middle of the eighteenth century, however, the plausibility of the first generation of travelers' tales that underpinned the writings of Rousseau and Monboddo began to diminish. Anecdotal accounts could not easily be disproved, but they wilted as confirmation failed to appear and when confronted with the evidence of direct observation. Thus, when Buffon turned to the classification of the apes and specifically to the orangutan in volume 14 of *Histoire naturelle,* published in 1766, he presented, one after another, the testimonies of Tulp and Tyson, along with those of travelers he regarded as relatively reliable.[22] He also gave a detailed description of the physiology of a chimpanzee that had been exhibited in Paris in 1740 but that had died the following year while on display in London and whose body, preserved in alcohol, had been presented to the "cabinet," or natural history collection, of the French king. His examination was close and detailed, very much in the tradition of Tyson, and he provided extensive measurements of body parts, commenting time and again on their resemblance to or difference from those of humans. An engraving of this specimen, unambiguously a chimpanzee, appeared at the conclusion of his discussion. Although he agreed with Tyson that there were significant physical differences between humans and orangutans, he made what appeared to be a major concession in accepting that orangutans had the physical potential to stand upright and to talk. Unlike De la Mettrie and Rousseau, however, he was confident that this potential would never be fulfilled because the essence of humanity lay elsewhere:

> The tongue, and all the organs of speech, for example, are the same as in man; and yet the orang-outang enjoys not the faculty of speaking; the brain has the same figure and proportions; and yet he possesses not the power of thinking. Can there be a more evident proof than is exhibited in the orang-outang, that matter alone, though perfectly organized, can produce neither language nor thought unless it be animated by a superior principle?[23]

This conclusion enabled Buffon to state matter-of-factly "of all apes, these ones most resemble humans, and that makes them especially worthy of observation,"[24] and thus to leave unexplored the nature of the "superior principle" that allegedly animated humans.

Within a decade, speculation about the degree to which orangutans might be human was rekindled by the arrival of another living specimen in Europe. Aernout Vosmaer, director of the menagerie of the Dutch Stadholder, Prince Willem V of Orange, had long wanted an orangutan for the collection, and like Camper he wrote repeatedly to contacts in the Dutch East Indies in the hope of obtaining one.[25] Despite his connections, he had to wait a long time. Although he received a couple of preserved specimens in the early 1770s, the first live animal, a female, arrived only in 1776. The animal had been living for a year in captivity in the Dutch colony at the Cape of Good Hope and was thus perhaps better able to survive the Atlantic journey to the Netherlands. Acutely aware of the vulnerability of so exotic a creature to the inclement European climate, Vosmaer first kept the animal in his own home, experimenting with different foods to determine how best to keep her alive. Within a fortnight, however, the situation became impossible: so many visitors, including the Stadholder himself, came to see the ape that Vosmaer had to transfer her to the menagerie.[26] Willem V himself wrote about her "sweet" nature, how she clung to people and hated to be left alone, that she drank mostly water but preferred wine and liquor, especially Malaga, drinking from a glass and eating with knife and fork "just like us." She would make her own bed and seemed, he said, to show signs of "a more than animal intelligence."[27]

Aart Schouman, Orangutan at Het Kleine Loo, 1776 (Teylers Museum, Utrecht, used by permission)

Fascinated by his new charge, Vosmaer made a careful record of his observations of the young orangutan, which he later published. The report shows that he was preoccupied by the question of the ape's relation to humans. He did not follow Rousseau in supposing that the ape might really be human. His account was bracketed by repeated characterizations of the creature as an *aap-soort* (variety of ape), and in his discussion he commented that he was more convinced than ever, after having an orangutan in his house for two weeks, that this was an ape. He commented, as Tyson had done seventy years earlier, that a performing animal showed how clever its trainer was, not that the animal itself was human. And yet, his account drew attention again and again to what he called "traces of humanness" (*sporen van menschelijkheid*) in the creature.[28] He wrote that she loved the company of people and had a special affection for the servant who looked after her, and would throw a tantrum and tear cloth into pieces if he went away.[29] He re-

Houttuyn's modest re-drawing of Bontius'
"ourang-outang," 1761 (Houttuyn,
Natuurlijke historie, facing p. 336)

Vosmaer's orangutan, c. 1777 (Vosmaer, *Beschryving,* following p. 23)

ported that she normally moved about on all fours but was fond of standing on her hind legs, leaning on a wooden stick. Strawberries and parsley were among her favorite foods. She ate mainly vegetables, did not like insects, and rarely ate fresh meat. After eating, she would use a toothpick to clean her mouth. After drinking deeply, he reported, she would wipe her lips with her hands or with a cloth, just like a human. He described how she would sometimes use a piece of cloth to polish the boots of visitors or to clean up her own urine. Sometimes, though, she would urinate into her hand and drink the result. She was expert at undoing the most complex of knots, but seemed unable to tie even the simplest one. She made hardly a sound, at most a kind of whine that Vosmaer compared to that of a wood saw.[30]

Vosmaer's attribution of human characteristics to the orangutan was restrained. Whereas Houttuyn in 1761 had redrawn Bontius' hairy woman as if she were indeed protecting her modesty, Vosmaer's illustrations were sober and animal-like. Nonetheless, he drew a savage attack from Camper, the man who had first dissected an orangutan seven years earlier, in an episode known as the *"orang-oetan oorlog"* (orangutan war) of 1777. Camper was unflinching in criticizing views he considered wrong-headed, no matter how eminent the holders of those mistaken views might be. Perhaps the fact that he had grown up in the company of a pickled orangutan inclined him to the belief that apes could not be human. His eminence as an anatomist naturally put him among those who believed that physical form was all-important. He was thus critical of Buffon for allowing his chimpanzee to be portrayed as standing upright with the help of a stick.[31] He took the German naturalist Georg Forster to task for suggesting that the orangutan had a nose like that of a human: "the nose was precisely as I have drawn it from life in the first figure on plate 4" (34). He was especially critical of Hoppius, whose work he described as being built on a foundation of "very superficial guesses, deficient investigations and false descriptions" (25).

The "orangutan war" began when the young orangutan died in January 1777, in the depths of a Dutch winter, about six months after her arrival in the Netherlands. Vosmaer, wanting the orangutan's body for display in the Stadholder's cabinet, had the corpse gutted, preserved in alcohol, and mounted standing up with the aid of a stick. Unknown to Vosmaer, however, Camper wanted a fresh corpse for dissection so that he could examine the orangutan's brain and compare it with the chimpanzee's brain described by Tyson. Camper was better connected in the Dutch elite than Vosmaer was, and he mobilized powerful allies at court to have the remains handed over to him. A ferocious exchange of correspondence took place, preserved now in an archival file labeled "Pièces relatifs à l'affaire de l'ourang-outang 1777."[32] After days of rage on both sides, the preserved remains were handed to Camper, who noted bitterly that the legs had been sawed off and the neck was hanging loose.[33] The clash was seriously damaging to Vosmaer's professional reputation. Despite his earlier rebuke at Camper's hands over the orangutan's nose, Forster attacked Vosmaer for "gross ignorance and canine malice" and suggested that he deliberately sought to prevent Camper from conducting his dissection because he was aligned with conservative, monarchist Orange supporters against the pro-science Dutch Republicans.[34] Later, however, Camper reported graciously that Vosmaer had merely been mistaken about what had been agreed and that the larynx, which was the main focus of his interest, was still intact. Camper's verdict on the orangutan itself, however, was uncompromising: on the basis of his examination of the larynx and bone structure, he affirmed that orangutans could not talk and were not suited to walking upright. He went still further and pronounced, in a direct contradiction of Vosmaer: "the living orangutan demonstrated no human features, despite the affectations with which it has been portrayed."[35] He seems to have had no other opportunity to dissect a recently deceased orangutan. As far as we know, the next living orangutan to reach the Netherlands arrived in 1792, three years after Camper had died. It survived in Holland for at least two years, and its behavior was described briefly by a Dutch doctor, D. L. Oskamp. Camper would have been delighted with Oskamp's categorical assertion that the orangutan was not a *"midden-dier"* (transitional animal) between humans and animals but simply an ape. Oskamp's posthumously published report, however, failed to include any anatomical details of the creature apart from measurements made of its body when it was still alive.[36]

The vehemence of Camper's response to Vosmaer was partly a consequence of the anatomist's prickly personality, but it also reflected the fact that the status of the orangutan was becoming caught up in a set of serious moral and philosophical issues. Camper was one of many people who felt deeply uneasy over Buffon's proposition that the difference between humans and apes was not physical but rather lay in a "superior principle." What could such a principle be? It might be represented as the soul, but how was one to demonstrate a soul's existence scientifically? Apart from the pique that Camper might have felt in this setting aside of the significance of anatomy, he recognized that acknowledging human elements in the behavior of the orangutan might make it difficult to identify that "superior principle" at all. The consequence would be to come even closer than Linnaeus had done to admitting that humans were simply animals. In 1779 the German naturalist Johann Friedrich Blumenbach, unhappy with Linnaeus' flimsy genus-level distinction between *Homo* and *Simia,* divided the order Anthropomorpha into two distinct classes, Bimanes ("two hands," occupied exclusively by humans) and Quadrumanes ("four hands," comprising the apes).[37] It was another assertion of the primacy of anatomy in classification, but it would count for little if some different, or "superior," principle were admitted. Whereas Rousseau and Monboddo had imagined shifting the human-animal boundary so that orangutans would be located on the human side of the divide, the issue underlying Camper's concern was the extent to which orangutan behavior might push humans themselves into to the animal category.

The risks became apparent again when another living orangutan, again a female, arrived in Europe in March 1808, where she was presented to Napoleon's wife, the Empress Joséphine.[38] Napoleon himself reportedly visited her from time to time and called her Mademoiselle Desbois (literally, the Miss of the Forests).[39]

Mademoiselle Desbois, now preserved in the Muséum d'histoire naturelle de La Rochelle (© Muséum d'histoire naturelle de La Rochelle France, used by permission)

Ill by the time she arrived in Paris, the young orangutan refused the breast of a nursing mother and the teat of a goat, but was gradually nursed back to some health by proper feeding, a warm cage, and regular baths. She died, however, after about five months in Paris.

While she lived, the biologist Frédéric Cuvier visited her, as he reported, almost every day in order to observe her physiological features and behavior. Like Tyson and Camper, Cuvier provided a detailed physical description of the animal. Though without physical measurements, this description left no doubt that physiologically the creature was an animal and not a human. He commented that she was "entirely formed for living among trees" and that she could stand upright only with difficulty.[40] Cuvier was also interested in her mentality, noting that she showed no interest in music, but he dismissed this indifference as irrelevant to the status of the orangutan: "even with mankind it is an artificial want; on savages it has no other effect than a noise would have" (190). On the other hand, she used her hands "in all the essential motions in which men employ theirs" (190). Like Vosmaer, Cuvier commented on the orangutan's affection for those who looked after and paid attention to her, and on her tantrums when denied something she wanted or when she missed her carer. Cuvier found this behavior intriguing:

> Was this ourang outang led to act in this manner from the same motives which actuate us in similar circumstances? I am inclined to answer this question in the affirmative: for in its passion it would occasionally raise its head from the ground and suspend its cries, in order to see if it had produced any effect on the people around, and if they were disposed to yield to its entreaties: when it thought there was nothing favourable in their looks or gestures, it began crying again. (197)

He noted especially the orangutan's capacity to learn by generalizing from past experiences and to plan complex actions. In terms of sense and the capacity to reason, he was unable to find any grounds for distinguishing the orangutan from humans. He described how it had developed the idea of shaking branches at humans it perceived as threatening: "In whatever way we regard the above action, it must be impossible for us to overlook the result of a combination of acute intelligence, or to deny the animal the faculty of generalizing" (195). He argued that the intelligence of orangutans, like that of humans, was a consequence of physical weakness and vulnerability.

> It owed nothing to habit, nothing mechanical entered into its actions, all of them were the simple effects of volition, or at least of nature.... This animal, which is certainly not a man, has received senses equally numerous, and at least equally delicate with ours. (192)

Cuvier's solution to this problem of defining human identity, like Buffon's, was to appeal to an undefined higher purpose. The distinction between orangutans and humans was in the purpose to which they applied their respective intellects, rather than in the underlying quality of those intellects: the orangutan's life was directed at protecting itself from danger and at sustenance (it was too young, he noted, to be driven by what he delicately called "the phaenomena connected with generation, &c.," [198]). Humans, by contrast, busied themselves with a higher set of purposes, though he did not elaborate what these were. Despite his insistence that the orangutan was "certainly not a man," Cuvier did little to identify precisely how humans could be distinguished from apes. His observations left a persistent unsettling suspicion that the moral advantage of humans over the closest animals might be slender indeed.

In parallel with this concern was a preoccupation with the implications of the orangutan's status for Western hierarchies of race and class. Rousseau and Monboddo had identified as crucial the issue of whether such hairy and apparently primitive beings might be educated to take a full, rewarding, and responsible role in human society. By the time Cuvier conducted his observations of Mademoiselle Desbois in Paris, the literal possibility of an orangutan being educated had largely evaporated in scientific circles, but questions about the orangutan's classification still impinged on the issue of whether there was a hierarchy of human races. In response to doctrines of human equality, which reckoned the enslavement of black Africans as an abomination, apologists for slavery countered with the claim that there was a hierarchy of races. Whereas the slave system focused on those non-Europeans who could be made economically useful—the large population groups of Asia, Africa, the Pacific, and the Americas—the idea of hierarchy was made more plausible by identifying a select group of societies as egregiously "primitive." These societies were few in number—typically they included Khoisan in southern Africa, "pygmies" in central Africa and the Pacific, Australian Aborigines, and Fuegans at the southern tip of South America—but they were critically important. Chris Ballard has called them the "sheet anchor for racial hierarchies,"[41] that is to say, they played the role of evidence for the objective existence of a hierarchy of human races. Once a single case of inequality was established, then the logic of consistency permitted a racial hierarchy favoring Europeans. The discourse of orangutan humanness provided another such sheet anchor. Recruiting orangutans as humans helped enable the racial ranking of all other humans.[42]

The reported attitude of orangutans themselves was also presented as evidence of such hierarchy. The popularizing naturalist Edward Donovan, writing in 1824, made much of the aloofness of an orangutan from a group of monkeys with which it was being transported to Europe.

> He could evidently discriminate the supreme importance of the human race, which he testified in every manner by his docility and affection, while the

conscious superiority he seemed to entertain above all others of the brute creation could not be pourtrayed [*sic*] more strongly than in his ever shunning their society.[43]

Ironically, the writings of Camper, the anatomist whose scientific work had vehemently insisted on a wide gulf between the physical appearances of humans and orangutans, played a crucial role in adding a physical dimension to the arguments identifying a racial hierarchy. Deprived by Vosmaer of the chance to investigate the brain of an orangutan, Camper turned characteristically to the skeleton, specifically to the skull, which he believed would show clear evidence of brain capacity and thus of the potential for intelligent behavior in different animals. This investigation led him to a celebrated experiment in which he placed a set of skulls in order of what he called the *gezichtshoek,* translated by most authorities as "facial angle," that is, the degree to which a line drawn from the forehead to the upper incisors of a skull was horizontal or vertical. After comparing the facial angle of a wide variety of animal species, Camper suggested that there was a relationship between facial angle and intelligence: the closer to vertical the facial angle, the greater the intelligence.[44]

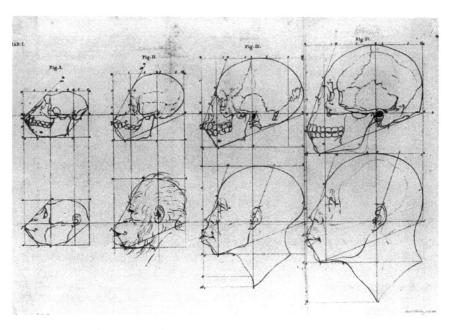

Camper's engraving of the facial angle in a tailed monkey, an orangutan, an Angolan, and a "Kalmuck" (Meijer, *Race and Aesthetics*, 108)

Camper's facial angle has become notorious in the history of so-called scientific racism, because he not only proposed that different races of humankind might have different innate levels of intelligence, but in juxtaposing the orangutan skull with that of an African he appeared to be portraying Africans as apelike. As we have seen, however, the consistent thrust of Camper's work, and indeed that of all the anatomists, was to emphasize the profound difference between apes and humans. His own words are worth quoting:

> The striking resemblance between the race of Monkies and of Blacks, particularly upon a superficial view, has induced some philosophers to conjecture that the race of blacks originated from the commerce of the whites with ourangs and pongos; or that these monsters, by gradual improvements, finally become men.
>
> This is not the place to attempt a full confutation of so extravagant a notion. . . . I shall simply observe at present that the whole generation of apes, from the largest to the smallest, are quadrupeds, not formed to walk erect; and that from the very construction of the larynx, they are incapable of speech. Further: they have a great similarity with the canine species, particularly respecting the organs of generation. The diversities observable in these parts, seem to mark the boundaries which the Creator has placed between the various classes of animals.[45]

Although it is true that Camper's facial angle appeared to give scientific backing to the belief that Africans and Asians were anatomically inferior to Europeans, he had no intention of suggesting any special similarity between non-Europeans and apes.[46] This combination of a racial discrimination among humans with a much sharper distinction between humans and apes was neatly expressed by Oskamp. He insisted that whereas it was perfectly possible for "a Negro or a Hottentot" to produce a Newton or a Hippocrates, it was impossible for an orangutan to produce "even a Negro or Hottentot in their most raw and uncivilized natural condition."[47]

Camper's nemesis in terms of his speculation about racial inequality was the German anatomist Friedrich Tiedemann, who in 1836 published a meticulous study comparing the brains of many humans and a single orangutan, concluding that variation in human brain capacity was not related to race and that the orangutan differed dramatically from humans in its cranial capacity. Tiedemann presented his study as a refutation of Camper and others whom he described as "look[ing] upon the Negroes as a race inferior to the European in organization and intellectual powers, having much resemblance with the Monkey."[48] His study, however, focuses almost exclusively on humans. He reports the weight of fifty-two human brains, compares the cranial capacity of forty-one Africans, seventy-seven Europeans, forty-four Asians, twenty-seven Native Americans, and forty-three Malays, drawn from collections in England, Scotland, Ireland, Germany, and the Netherlands, and he closely examines three preserved African and thirteen preserved European brains. Out of this careful investigation he discovered a marked difference in the size of the brains of men and women, from which he drew no conclusions because his main point was to establish the irrelevance of race to brain size and thus the invalidity of arguments justifying African slavery on a physiological bases.

A major reason for Camper's poor reputation has been the elaboration of his drawing and the reshaping of his ideas by the English physician Charles White in his 1799 *Account of the Regular Gradation in Man &c. &c.* White's argument blurred the species line that Camper had been at pains to draw: "Man, considered in toto, is undoubtedly entitled to preeminence over the animal world; but various tribes of creatures make great advances towards him, in some particular powers and faculties, and even, in some instances, far surpass him. The orang-outang has the person, the manner and the action of man."[49] White also prepared an elaborated version of Camper's chart of facial angles in which there was a difference of only five degrees between an African and a hoolock ape,[50] the same as that between an "Asiatic" and a European and less than that between a European and the ideal of a "Grecian antique." In fact, Tiedemann's careful measurements had confirmed Camper's examination half a century earlier: on the basis of anatomical structure, the orangutan could not possibly be considered human. Tiedemann's specific contribution was to focus on cranial capacity and brain structure as key indicators of underlying intelligence and to confirm that in this respect, too, as well as in gait and laryngeal structure, the orangutan differed from humans in a way that warranted placing it in a separate genus. Anatomists did not stop measuring live orangutans and dissecting them when they had died, and they discovered nothing to cast doubt on Camper's conclusions of the 1770s.[51] In the end, the most comfort that the racialist agenda could get from the anatomical comparison between Africans and orangutans was a hesitant and still tendentious concession that, of all humans, Africans might be closest in form to orangutans. In 1847, for instance, the scientific discoverers of the gorilla, Thomas S. Savage and Jeffreys Wyman, could comment in their conclusion:

> Any anatomist . . . who will take the trouble to compare the skeletons of the Negro and Orang, cannot fail to be struck at sight with the wide gap which

separates them. The difference between the cranium, the pelvis, and the conformation of the upper extremities in the Negro and Caucasian, sinks into comparative insignificance when compared with the vast difference which exists between the conformation of the same parts in the Negro and the Orang. Yet it cannot be denied, however wide the separation, that the Negro and Orang do afford the points where man and the brute, when the totality of their organization is considered, most nearly approach each other.[52]

An underlying difficulty for the anatomists, however, in explaining to the broader public that physiological differences sharply distinguished humans from apes was that for most observers the distinctions were not particularly obvious. There were physical differences between humans that to the untrained eye seemed to matter nearly as much as the differences between humans and apes, especially when seen through a racist lens. It was thus easy for non-anatomists to belittle differences that anatomists took seriously, as the German satirist Adolf Glassbrenner did when he imagined an orangutan demonstrating captive humans to an orangutan audience and commenting, "Man . . . has only two hands; this is the animal that resembles us monkeys the most, though to be sure he climbs but badly, and cannot crack a nut without nut-crackers."[53]

Although Buffon's idea of a nonphysical "superior principle" separating humans from animals had offended the anatomists, he also provided an alternative explanation of species difference that proved more widely acceptable to the general public. Fearing that the Linnaean emphasis on physical appearance would lead to misidentifications based on environmentally determined features, Buffon proposed that each species had an "interior mould" that determined its underlying form. This "interior mould" was transmitted from generation to generation by the reproductive process.[54] In this way, Buffon prefigured what was later to become the orthodox scientific definition of a species as a group of organisms that can reproduce with each other and not with other species. If interbreeding was impossible or produced sterile offspring, then groups of animals could be considered separate species. It was a politically powerful definition, because it established unambiguously that all races of humans belonged to a single (often enthusiastically) interbreeding species, *Homo sapiens*. It reopened, however, the question raised by Bontius of whether humans and orangutans could in fact reproduce. One biblical commentator even speculated that the serpent described in Genesis in the Garden of Eden was actually an orangutan and that it had physically seduced Eve.[55] Given even the little we know about human sexual encounters with animals,[56] it would be surprising if there were not a long and hidden history of orangutan-human sexual relations. Rousseau proposed that the most expeditious way to test whether orangutans were human would be to attempt to breed them with humans. He commented regretfully, however, that the attempt could not be done "innocently" unless one were sure that the orangutans could indeed produce fertile offspring with humans. If the experiment failed, he implied, the great sin of bestiality would have been committed.[57]

Since Rousseau's essay on the human character of orangutans survived especially in discourses supporting slavery, where it could be used to reinforce the idea of a hierarchy of human races, it is not surprising that the idea of interspecies sexuality flourished in this context as well. The best-known exploration of the idea of interbreeding in the eighteenth century was in the rambling and inconsistent writings of Edward Long, especially the key sections of his 1774 *History of Jamaica* that were reprinted in 1788 in *The Columbian Magazine*.[58] Long appears to have viewed humankind as a segmented species in which Europeans, Africans, and orangutans represented separate subspecies, hierarchically arranged so that orangutans were capable of interbreeding with Africans and Africans with Europeans. Long had read and accepted accounts of the orangutan in which it was described as walking upright, living socially in huts, and being readily trained for service, and he followed Rousseau and Monboddo in suggesting that they might well be educated to speak. On this basis he made the notorious comment that an orangutan husband would not disgrace a Hottentot female.[59] He also suggested that many of the Africans brought to the Americas were the offspring of Africans and orangutans. Long's ramblings are difficult to assess in context because they represent an extreme in a range of complex opinion on race and sex. When his suggestion that Africans and orangutans could produce fertile offspring together was republished in *The Columbian Magazine,* another correspondent immediately attacked the view as inaccurate and impious.

Interest in the possibility of human-orangutan hybrids drew the attention of fiction writers and dramatists (see chapters 6 and 7) as well as scientists throughout the nineteenth century. Researchers in midcentury tried to determine, without actually conducting the necessary experiments, whether fertile mating might be physically possible.[60] In Émile Dodillon's 1886 novel *Hémo,* a Dutch amateur scientist considers attempting to impregnate an orangutan, but chooses a female gorilla instead when he realizes that gorillas are closer to humans in evolutionary terms. In the 1920s, the surgeon Serge Voronoff and the Russian zoologist Ilia Ivanov envisaged transplanting glands from great apes into aging human males to restore their sexual vigor. Although Voronoff initially planned to use orangutans, most of his operations were undertaken at the expense of chimpanzees.[61] The underlying assumption in most of this speculation about the possibility of fertile sexual relations between humans and orangutans, however, was that the offspring of such crossings would necessarily be close to inhuman, rather than that orangutans might be considered close to human.

More common than this sexual speculation, however, was an elaboration of Buffon's idea that orangutans must lack a "superior principle." Late eighteenth-century references to the orangutan, which reflect that the name had become a household word even if glimpses of the animal were rare, portray it as a figure of wretchedness. After a poorly planned American revolutionary expedition invaded Canada in 1775, one of the survivors described the condition of the troops as they reached Quebec: "I thought we much resembled the animals... called the Ourang-Outang."[62] In 1780, a strange work appeared with the title *Der Orang-Outang in Europa oder der Pohle, nach seiner wahren Beschaffenheit*[63] (The Orangutan in Europe, or the Poles as They Really Are). The title page declares that the essay received a natural history prize in 1779, and the work is indeed constructed as an anthropological monograph on the Poles. The fact that the fly title of the book carries the warning, "The greatest lies of the 18th century," suggests that the work is satirical rather than serious. The author claims that the little-known writings of a traveler called Hatif show that Poland was settled by a race of apes resembling humans and that they assisted the legendary King Krakus, founder of Krakow, to rid his kingdom of a dragon. Absorbed into the human population, they are the explanation for all "vices and laughable customs" of the Poles.

Most striking, in 1774, the English essayist Oliver Goldsmith launched an extraordinary attack on the orangutan. His book, *The History of Animals,* was one of the many popular works of the age that emulated Buffon in bringing knowledge of natural history to a wider audience. Writing even before Camper had published the results of his dissections, Goldsmith not only contemptuously dismissed the notion that the similarity of orangutans to humans might have any metaphysical significance, but he also attacked the red ape directly for its temerity in appearing to be human:

> From this description of the oran-outang, we perceive at what a distance the first animal of the brute creation is placed from the very lowest of the human species. Even in countries peopled with savages, this creature is considered as a beast; and in those very places where we might suppose the smallest differ-ence between them and mankind, the inhabitants hold it in the greatest contempt and detestation. In Borneo, where this animal has been said to come to its greatest perfection, the natives hunt it in the same manner as they pursue

the elephant or the lion, while its resemblance to the human form procures it neither pity nor protection. The gradations of Nature in the other parts of nature are minute and insensible; in the passage from quadrupeds to fishes we can scarcely tell where the quadruped ends and the fish begins; in the descent from beasts to insects we can hardly distinguish the steps of the progression; but in the ascent from brutes to man, the line is strongly drawn, well marked, and unpassable. It is in vain that the oran-outang resembles man in form, or imitates many of his actions; he still continues a wretched helpless creature, pent up in the most gloomy part of the forest, and, with regard to the provision for his own happiness, inferior even to the elephant or the beaver in sagacity. To us, indeed, this animal seems much wiser than it really is. As we have long been used to measure the sagacity of all actions by their similitude to our own, and not their fitness to the animal's way of living, we are pleased with the imitations of the ape, even though we know they are far from contributing to the convenience of its situation. An ape, or a quadruped, when under the trammels of human education, may be an admirable object for human curiosity, but is very little advanced by all its learning in the road to its own felicity. On the contrary, I have never seen any of these long-instructed animals that did not, by their melancholy air, appear sensible of the wretchedness of their situation. Its marks of seeming sagacity were merely relative to us, and not to the animal; and all its boasted wisdom was merely of our own making.[64]

So strong was this view that in 1824 the popularizing naturalist Edward Donovan commented on "those feelings of dislike . . . with which that tribe [orangutans] in general is regarded by every inconsiderate observer."[65] Much later this same characterization of orangutans emerged in Edgar Allan Poe's disturbing short story "Hop-Frog; Or, the Eight Chained Ourangoutangs," published in 1849, in which the crippled slave Hop-Frog persuades his master and tormenter to dress, along with seven sycophantic courtiers, as chained "ourangoutangs" in a kind of perverse tableau of wretchedness.

This discourse of wretchedness worked as a justification for slavery partly because it suggested that some humans could be so degraded by their very nature that the institution of slavery was no disgrace to them or to their masters. It also (contradictorily) implied that for the most miserable of humans, slavery was a benefit because it introduced them to the improving character of work. The key text in this implication was that of Bontius, which reported a Javanese belief that the "ourang outang" was able to speak but chose not to do so for fear of being put to work. This suggestion is a little perplexing, because horses and dogs are put to work without their being able to talk, and today we might argue that avoiding work by this stratagem shows wisdom rather than wretchedness. In the discourse of the eighteenth and nineteenth centuries, however, human labor was a prerequisite for progress.[66] The orangutan's insistence on indolence was seen as a culpable refusal to take part in the human project of self-improvement. Not only orangutans were accused of this treason against progress: in Western accounts, the alleged unwillingness of Native Americans to work in Western-owned plantations reflected badly on their moral status, despite the appalling conditions that African slaves in those plantations had to experience. Speculation that orangutans could work but chose not to allowed white Europeans and Americans the conceit that their recruitment of African slaves was actually a service to people who would otherwise be mired in primitive indolence in their homelands. Speculation that orangutans might be set to work also offered an intriguing solution to the looming problem of servant and slave assertiveness. Across the cities of western Europe the deferential model of servants was being undermined by new democratic and populist impulses, which members of the elite might both welcome and fear. In the plantation economies of the New World, it was impossible to ignore occasional slave rebellions or to avoid considering the possibility of slave emancipation. In these circumstances, the orangutan offered what appeared to be an intriguing alternative source of labor. Writing in 1795, the English astrologer and polymath Ebenezer Sibly, for instance, drew on Pyrard's account of "orang outangs" as dutiful servants to portray a "domesticated female orang-outang," adding, "She seemed to have a sense of what was required of her, even before the words were spoke; and her chief emulation was that of being more active and adroit than any servant about the house."[67]

But what kind of servants might they be? In his 1829 novel *Count Robert of Paris,* Sir Walter Scott introduced readers to a seven-foot orangutan called Sylvan, who works as a jailer, wearing clothes, serving bread and water, and able to turn a key in a lock. The ape-servant's strength and sense of social place seem to make him an ideal servant, until a misunderstanding leads the orangutan to commit murder. As we shall see in chapters 6 and 7, a variety of novels and plays featured the orangutan as a servant. The twentieth-century wildlife collector Frank Buck tells the story, ostensibly as true reminiscence, of an orangutan, Gladys, who does his laundry as a dutiful, even compulsive, servant.[68] Still later, Paul Carter tells of an orangutan who allegedly ran the bar on a drilling rig: "It was always clean and organized. She made cocktails."[69] At a deeper level, too, orangutans held out the intriguing possibility that they might be an alternative source of labor in which the moral complications of slavery would be diminished. Long commented that

Sibly's domesticated female orangutan, 1795 ([Sibly], *An Universal System of History*, vol. 2, facing p. 25)

A Domesticated Female Orang Outang.

Africans worked on plantations "perhaps not better than an orang-outang might, with a little pains [*sic*], be brought to do."[70]

In the course of the nineteenth century, the quasi-scientific racism of men such as Long was gradually transformed into a racism that was still less scientific and more focused on culture. One of the catalysts for this change was the influx of Irish immigrants into England and the United States as a result of the Great Famine of 1845–1852. Physically European, but miserably poor and widely despised for their perceived lack of education and refinement and for their alleged predilection for gratuitous violence, these Irish were quickly identified as monkeylike. Caricatures drew on a generalized idea of great apes, portraying Irish men with prognathous jaws, large mouths, low eyebrows, and snub noses.[71] The image was enduring. When the Lincoln Park Zoo in Chicago named a new orangutan "Miss Dooley" in 1903, the city's Irish community objected vehemently and threatened an injunction to have the orangutan given a name that was not transparently Irish.[72] T. S. Eliot's character Sweeney—another Irish name—in his poem "Sweeney Erect" (1920) is also given characteristics that critics have described as recalling an orangutan.[73] And the imagery has been sustained by Colin Broderick's 2009 account of an Irishman's recovery from alcohol and drug abuse titled *Orangutan: A Memoir.*

Rousseau, Monboddo, Peacock, and Long, drawing on imaginative travelers' tales, proposed the idea that orangutans might be accepted into the human family, but they did so with very different intentions. Whereas Rousseau did so with a sense of compassion—orangutans might be the most underprivileged of humans and deserved to be rescued from their fate—Monboddo and Peacock came close to prefiguring the late twentieth-century doctrine of multiculturalism, according to which society was enriched by the interaction of different cultural types. Both writers were inclined to suggest that orangutan characteristics were overall an asset to society, not a liability. Long, by contrast, wanted to classify orangutans as humans so as to strengthen the empirical case for the existence of a racial hierarchy among humans. The triumph of the anatomists and the discrediting of the travelers' tales marginalized Monboddo and Long, establishing incontrovertibly that the orangutan was not *Homo sapiens,* probably not even *Homo,* and that orangutans would never be elected to Parliament and would never take the place of servants or slaves. But the underlying difficulty for the anatomists was that the physiological difference between humans and great apes, however easy it might be to catalogue, appeared trivial in comparison with the physiological difference between humans and apes on the one hand and the rest of the animal world. The difficulty was compounded by the humanlike elements of orangutan behavior that had survived the discrediting of the travelers' tales. Pyrard and Hamilton might no longer be taken seriously, but Vosmaer and Cuvier were impossible to ignore.

At stake was the human soul. If a decisive physical difference between humans and great apes could be identified, then it would not be necessary to inquire any further than Buffon had done into the idea of a nonphysical "superior principle" separating humans from animals. If, on the other hand, physical differences were trivial, then the question was whether any true metaphysical difference could be found as a substitute. The issue was stark: just how animal were human beings? In the same way that the implications of including orangutans as humans were interpreted in very different ways by Monboddo and Long, the consequences of there being perhaps no defensible line between humans and animals aroused divergent hopes and fears. On the one hand, the issue had the potential to transform the ethical relationship between humans and the animal world in favor of animals. If humans were not morally distinct from animals, then standards of human morality would have to be applied to animals. The implications of this dramatic claim—

including its consequences for the economy—are still far from being accepted or even comprehended. On the other hand, the issue had the potential to explain all human civilization and culture in the last instance as a product of biological pressures. For some, this prospect was a source of exhilaration, above all because it seemed to promise liberation from the claims of religion to authority over human life. For others, the special status of humans was at risk and the claim seemed to threaten that "animal" (by which they meant coarse and unbridled) behavior would be legitimized.[74] Since early in the nineteenth century, therefore, the orangutan appeared time and again in what Raymond Corbey calls "the misty borderlands between beast and human in Western cultural imagination."[75] In various guises, orangutans have been flashpoints on the shifting front lines of a vast cultural battle over the nature of humanity. Literature and theater played a key role in this battle, because works of fiction, unlike dry summations of physiological characteristics, could explore the social, cultural, and psychological implications of different kinds of resemblance and difference between apes and humans. In 1802 a Dutch popular writer, Martinus Stuart, acclaimed the memory of Camper and told his readers that Camper's work had "freed you of the unbearable humiliation, to which a Monboddo tried to bring you, to call the hideous Orang-utan your brother."[76] Stuart thought the battle was over. In fact, it was just beginning.

3 ❧ Wanted Dead or Alive
Orangutans on Display

In the eighteenth century, European elites began to develop an appetite for natural history collection. Kings and queens, princes and dukes, doctors and merchants, accumulated so-called cabinets of curiosities, some from the natural world (*naturalia*), some made by humans (*artificalia*). In the same spirit, a smaller number of wealthy or powerful individuals maintained menageries of exotic animals, often sponsoring artists to record them in sketches, paintings, and figurines. In the beginning, both kinds of collection were fundamentally miscellaneous. The tooth of a bear might sit incongruously on a shelf alongside a piece of filigreed silverwork or, in the menagerie, a South American capybara might be housed next to an African baboon. Their owners collected what fell into their hands, some employing curators as their holdings grew, supplemented by commissions of especially rare specimens. In this way, several cabinets and menageries began to take on the character of scientific collections, aiming at some degree of comprehensiveness and making themselves available for research. The great collections eventually became the core of national museums and zoological gardens, publicly embedding a spectacular vision of the natural world.

Very little is known about the display of orangutans in Europe before the last decades of the eighteenth century. The Blauw Jan menagerie in Amsterdam ran for more than a hundred years (c. 1675–1784) as a clearinghouse for exotic animals, but the only evidence of an orangutan in its stock is the sketch made by Jan Velten in the early 1700s.[1] The young female who became the focus of the "orangutan war" between Vosmaer and Camper in 1776 was housed in Prince Willem V's menagerie at Het Kleine Loo, where members of the public could come and observe it. Although the court painter Aart Schouman later depicted the orangutan roaming free with other exotic animals in a tranquil Dutch landscape, the reality at Het Kleine Loo seems to have been one of wrought iron cages and wooden shelters.[2] When the orangutan died, the "war" that erupted over whether its corpse should have gone to Vosmaer to be taxidermied or to Camper for dissection was prompted by the rarity of the specimen and the incompatibility of the two intentions. Behind the conflict, however, was Camper's contempt for Vosmaer's preparation of the orangutan for display in a way that emphasized its "traces of humanity." Science might stress the animality of the red ape, but Vosmaer, like many curators who followed him, recognized that the public would be more attracted to a specimen that challenged the human-ape boundary. As Carl Niekerk suggests, there was a tension between "exhibitional" and scientific values: just how much should scientific accuracy trump the public pleasure and interest that was served by an imaginative display?[3]

In the late eighteenth century, live orangutans were still exceptionally uncommon in Europe, and most red apes were held as pickled or stuffed specimens. By 1782, Sir Ashton Lever's Holophusikon, one of the finest London collections of its

Aart Schouman, The Menagerie of Willem V, 1786 (Sliggers and Wertheim, *Een Vorstelijk Dierentuin*, 71, Walburg Press, used by permission)

time, had boasted a male and female orangutan among its "very curious monsters and monkies."[4] Lever's large array of specimens was arranged aesthetically rather than scientifically, the aim being to dazzle visitors with the diversity of nature rather than to educate them in its patterns.[5] Skins, skulls, and skeletons of orangutans could also be found in the British Museum and in a few university and private collections. Well-connected scientists generally had access to these specimens across the European continent, but members of the public saw them mainly in the form of printed, illustrated catalogues. Handbills and newspaper advertisements for menageries promised viewings of rare creatures among their exotic wares, but in the case of anthropoid exhibits, it is often difficult to identify amid the rhetoric of the fairground precisely what was being displayed. A survey of news clippings announcing various human-like "beasts" in long lists of attractions at London shows in the 1790s illustrates the problem. What are probably apes of some kind but not orangutans are cast in these columns as "Wood Monsters," "a little Black Man," "a Man-Tyger," an "Arabian savage," and a "Satyr of the Woods" from Scythia with "Face, Breasts, Hands and Feet, representing human nature."[6] These advertisements gave few clues as to whether the exhibits were alive or dead, which suggests that in the domain of popular entertainment at least, the empirical line between these states began to dissolve in the figure of the specimen. American exhibitors enticed the public with similar cryptic puffs for exotic primates when they first began to appear in traveling menageries and natural history collections. In 1789, for example, the *Massachusetts Centinel* printed an announcement claiming that a "surprising species," an "Ourang Outang" pictured as an erect, human-looking animal, was on show in the "Greatest and Most Curious Natural Collection" ever exhibited in the United States. Although the "ourang outang" initially seemed to be available for live viewing, it was in fact one of a group of three apes—a male, female, and fetus from Africa, not the East Indies—described in a subsequent bulletin as "all in high preservation."[7] Advertisements of a similar timbre continued well into the middle of the nineteenth century in American as well as British papers.

All or most of these displays involved some deception of the public, but there is no evidence that audiences were displeased or disappointed with what they saw. In 1822, however, an American sea captain, Samuel Barrett Eades, tested the limits of public credulity in London with a specimen he had purchased for a large amount of money in Batavia after selling the ship he commanded (without the permission of its owner) to raise sufficient funds. The object appeared to be a dried and shriveled mermaid, about three feet long with a head the size of a two year-old child, while its eyes, nose, chin, and upper body resembled human counterparts. Two fins pro-truded from the abdomen, which extended, apparently seamlessly, into a scaly fish tail. Eades displayed his rare find under a heavy glass dome, called it the Feejee Mermaid although it had nothing to do with Fiji, and charged a shilling for entry. The exhibit caused a sensation, drawing huge crowds until it was exposed as a fake a year later by William Clift, an anatomist at the Hunterian Museum. Clift had examined the creature on its arrival in London but was at first unable to publicize his findings due to a secrecy agreement he signed with Eades in return for viewing access. The release of Clift's report, revealing that the mermaid was made from the cranium and torso of a female orangutan skillfully sutured to the body of a large fish, quickly dampened public enthusiasm for this "maid of the sea," though it continued on rural exhibition circuits for some months afterwards. Decades later it had a second life as the "missing link" between humans and fish when American entertainment magnate P. T. Barnum purchased it for his collection.[8]

As specimen and spectacle, the enigmatic orangutan derived part of its appeal from the power of taxidermy, which can be seen to function with its own particular optics as a subtle drama of theatricalized reanimation. In Karen Wonders' terms, "Unlike embalming, by which the dead body is preserved, taxidermy attempts to restore the form, expression and attitude of the living animal."[9] In the initial phases of Western taxidermy, which had progressed with the growth of natural history during the eighteenth century, rudimentary techniques involved stuffing hides that were often shriveled with preservatives, distorting the animal's features and allowing the taxidermist's imagination to supply missing details of its appearance or posture.[10] Taxidermy's fundamental trick was—and is—to disguise death with the simulation of life. As the techniques of taxidermy were refined, the skillful taxidermist came to be recognized as an artist who could bring inanimate skin, bones, and teeth, along with glass eyes and sundry other props, to such a close approximation of life that only lack of movement and the invariably incongruous location of the specimen made clear that it was a work of art and not a real animal. The art of the taxidermist also concealed the bloody history of the individual animal: slaughter, evisceration, skeletonization, chemical preservation, sewing and stuffing. Its value in the realm of entertainment lay in the frisson of the animal's lifelike presence where no live animal could truly be. Taxidermy also ensured the durability of animal exhibits, greatly expanding potential audiences. It was thus indispensible to display circuits, particularly for species that were not robust outside their natural environments. Specimens could be exhibited, across time, in a greater range of (highly constructed) contexts, each containing traces of the cultural practices that brought specific "sights" into view. The irony inherent in killing animals in order to reconstruct them as living now registers acutely in view of the human activities that threaten the orangutan's survival, but there was little reason to doubt the value of taxidermied exhibits two centuries ago when the treasures of the natural world seemed almost inexhaustible.

An important innovator in the science—and art—of taxidermy was Charles Willson Peale. As well as harvesting new specimens on field trips, he kept a number of birds and wild animals, which he stuffed for display after they died. His pioneering use of arsenic as a preservative produced durable exhibits that could be mounted in lifelike poses against painted backdrops of their natural habitat, creating images designed to both please and educate spectators. An artist of considerable repute and a strong supporter of American independence, Peale started a museum in Philadelphia in 1784 with the mission of providing "rational amusement" as well as "useful knowledge to socialize and improve the population."[11] Whereas many of the early exhibitions of "orang outangs" had no pretensions to anything but entertainment, and Lever's Holophusikon aimed to dazzle rather than to impart scientific information, Peale's museum marked a new confidence in the power of "real" science to inspire and inform the public. Peale arranged his specimens in the museum's galleries in Linnaean order beneath portraits of the heroes of the American Revolution. In this way, his curatorial endeavors subsumed the work of specimen-based natural history into a nationalist celebration of the first New World republic.

The orangutan exhibited at Peale's museum in 1799, probably the first red ape displayed in America, was a young female that had died of "exposure" in Baltimore not long after her arrival. Peale mounted the animal and announced her imminent presentation to the public in a newspaper advertisement titled "Ourang Outang, or Wild Man of the Woods." A few lines of text introduced the orangutan as a "Curious Animal, so nearly approaching to the human species as to occasion some Philosophers to doubt whether it was not allied to mankind." The accompanying woodcut purported to show the orangutan, but was actually based on Buffon's well-known portrait of a chimpanzee in the 1766 *Histoire naturelle,* so it cannot have closely resembled the animal on display.[12] The museum as a whole was suffused with theatrical touches that showed its curator's talents as an artist-naturalist, and it was criticized in its time "for being too much a side show" to count as a serious scientific institution,[13] even though it housed rare collections such as the skeleton of a mastodon that Peale had excavated. Peale's flirtation with populist display is tellingly evident in one experimental exhibit he called a "painter's room of monkeys"; it showed stuffed primates in human dress pursuing various trades and included a monkey painting a figure posed for its portrait.[14] Less obviously staged, Peale's orangutan was intended as an object lesson in "reading" the world from the "open book of nature," a conception of democratic pedagogy that flourished in the philosophical Romanticism that fed the American Revolution. This pedagogy was based on the idea that scientific knowledge ought not be restricted to a scholarly elite but rather should be "accessible to all who would observe and study nature directly."[15] Peale's task was to ensure that the visiting public could read the exhibit in the fullness of its natural associations, as he saw them.

As part of his mission to educate the populace, Peale gave public talks directly from his exhibits, explaining theories of the natural world and the characteristics of particular species. Describing the orangutan in a lecture series delivered in 1799– 1800, he stressed that it possessed "strong mental powers" and "was mild in its manners and gentle in its movements"; the animal also "walked erect, drank from a cup which she put down when finished and covered herself with a blanket when she went to sleep."[16] Peale's theatrical presentation, however, did not just challenge his audience to think about the definition of the distinction between humans and apes. In a museum leaflet, he situated the orangutan next to a racialized order of humans by asking his audiences, "How like an old Negro?" Like the racialists of the late eighteenth century, Peale presented his audiences with an image of orangutan similarity to humans that emphasized the differences among human races.

Peale's exhibitions were more pedagogically driven than those that displayed the red ape in Britain in the early years of the nineteenth century. Whereas the Philadelphia Museum aimed to tell a story of a new nation, most of the London venues exhibiting animals at that time—living or dead—were entrepreneurial ventures with vested interests in peddling spectacle. A centuries-old tradition of showing performing animals had expanded in tandem with traveling menageries to produce "wonders" such as sagacious horses, drumming hares, and the famous "learned pig," who could spell, read, and play cards. Amid these attractions, there was a ready market for unique exhibits, in keeping with the practice of starring noteworthy single acquisitions with special publicity. Periodic accounts of live orangutans being shipped from the East Indies to various (often unspecified) destinations must have generated anticipation among regular visitors to menageries and exhibition halls. An article in *The Observer* in 1803, for instance, briefly tells of an "America China ship" that carried an orangutan of "prodigious strength" and "more than six feet high" who could "hand and reel as expertly as any man on board." He had also apparently mastered "menial offices" such as "cutting wood, carrying water, turning a spit [and] attending a table."[17] Such reports appeared to corroborate the older travelers' tales of orangutans being put to work, and they primed the public to expect human characteristics and behaviors in the animals that were to follow. Satirical sketches added another layer to the textualization of orangutans in the press. In the same year, a handbill circulated claiming that "the most renowned and sagacious MAN TIGER, or OURANG OUTANG, called NAPOLEON BUONAPARTE" was showing at Mr Bull's menagerie in London. According to the spoof, this "wonder" had been "exhibited through the Greatest Part of Europe" before being taken by Admiral Bull. It had the faculty of speech and reason but was inclined to fly at people and, by "a hocus-pocus trick," could change a crown into a guillotine, a clear reference to the French Revolution.[18]

In 1817, a live orangutan was exhibited at London's famed Exeter 'Change Menagerie, which housed exotic species in iron cages in a building on the Strand from 1773 until 1829. The upper rooms of the building were initially let to circus impresarios who used them to winter animals, but it soon became a permanent and popular venue for exhibiting lions, birds, monkeys, and even an elephant. When the young orangutan joined the crowded menagerie, it delighted scientists and curious visitors alike, as *The Observer* reports:

> How often have we heard of the Orang Outang, and how little do we know of that extraordinary link in the creation, that seems to come so near to the human form. A gentleman [Clarke Abel] in the suit of Lord Amherst, has been so fortunate as to succeed in bringing one over at last, and has kindly and liberally sent it to Exeter Change for the gratification of the British Public, where it has already been visited by the greater part of the faculty and naturalists in town. It is a curiosity that impresses every spectator with infinite surprise, and the more it is contemplated the greater the wonder. There is no doubt of its being the first ever seen in this country, being a native of Borneo, and extremely difficult to be taken. It is perfectly harmless and inoffensive, and possesses a degree of natural intellect different from any animal whatever.[19]

The confident claim that the orangutan is the "first" seen in Britain appears in reports of earlier exhibits that seem to have featured chimpanzees or Barbary apes, as well as in accounts up to the 1830s of animals that can be verified as orangutans. These claims point to the intense competition to present new animals on the exhibition circuit, as well as a continued confusion between ape species. The fact that live orangutans seldom survived long in Europe at the time made it easier to reinvent each specimen as an absolute novelty. Within six months of its acquisition, the Exeter 'Change orangutan was ailing. *The Times* reports that an "astonishing" number of inquiries came in from the public during the animal's illness, and after it recuperated it received up to three hundred visitors a day.[20]

Edward Donovan, who visited the orangutan on several occasions and wrote at length about its habits in his *Naturalist's Repository* (1824), remarks that it

Exeter 'Change Menagerie, 1812 (H. Beard Print Collection, Victoria and Albert Museum, London, used by permission)

shunned other animals at the menagerie (excepting a dog) but received human company "with much apparent satisfaction."[21] In the warmer months of its brief life as a curiosity in London, the orangutan went for occasional coach rides about the streets, "stationed with his keeper on the seat" and dressed "in a countryman's smock frock" and hat. He was typically treated to refreshments at an inn during these masquerades and impressed his various hosts greatly with his courteous behavior. On one excursion, a landlord was astonished to realize, after noticing the "rufous hairiness of the orangutan's fingers," that the gracious customer he had served was a "biped of another race." Donovan's account of the orangutan stresses its sagaciousness in dealing with humans, turning frequently to anecdotes of this kind as evidence of its ability to "assimilat[e] to the social habits of man." In terms of appearance, he judges the ape "more extraordinary than displeasing" as he briefly compares its features to "those of the negro, amalgamated with certain peculiarities of the Chinese."[22] When the animal died, the "rufous" ape's skin was stuffed to enhance a small collection of primate remains at the College of Surgeons.

Exeter 'Change was only one of several London venues where exotic fauna "taken" in Europe's distant colonies might be viewed. In 1812, scientist-entrepreneur William Bullock set up a museum known as Egyptian Hall to house artifacts gathered on James Cook's South Seas voyages and an impressive number of stuffed quadrupeds. He presented these specimens in a lofty pavilion to form a tableau featuring an elephant, a rhinoceros, several ostriches, and a kangaroo, grouped in an artificial rainforest, with a twenty-two-foot boa constrictor poised in a tree to devour a petrified wood baboon.[23] Up-to-date taxidermy enlivened the dramatic installation, using the forms of the animals carved from wood in precise poses with their skins fitted over them.[24] Like Peale, Bullock had aspirations to educate as well as entertain the public and saw exhibition practice as a powerful means of doing so. The museum was judged an outstanding success in its first decade, attracting tens of thousands of visitors per month for a shilling entry fee.[25] By the mid-1820s, however, it had changed hands and become a venue for more vulgar entertainments. When a female orangutan and a male chimp were shown live there together in 1831, they joined a succession of unusual humans exhibited as freaks, including a set of Siamese twins and a French man so thin as to be dubbed "the living skeleton."[26]

The young apes must have been important exhibits for Egyptian Hall, as an eight-page leaflet, "Descriptions and anecdotes of the Orang Outangs," was printed to accompany their showing. It came with sketches of the animals by the famous engraver and painter Thomas Landseer, who was known for his parody of human follies in a sequence of etchings published as *Monkeyana* in 1828. The leaflet's unnamed author writes in a quasi-scientific mode about the animals' appearances and habits, using the occasion to ruminate on public entertainment, the state of science, and the character of humanity. The early paragraphs situate the Orang Outang exhibit within the context of "monkeyism," the recent fashionable interest in primates as both pets and subjects of popular theater. What follows weighs the humanity of the apes and their capacity, in that respect, to move spectators and thereby excite their moral sensibilities:

> The reflection which immediately presents itself to the mind upon a first view of them—that we are looking, beyond doubt or dispute, upon the *second link* of animal creation—is alone sufficient to fix our attention, and to prevent our regarding these strangely-endowed creatures with the same feelings with which we survey animals of an inferior grade, or of a form and character farther removed from that of our humanity. We are naturally moved most by whatever most approximates to ourselves; and the resemblance which these animals bear to at least one portion of the human race—ludicrous as the likeness may at first seem to be, and gross as the caricature may appear to our wounded vanity—can hardly fail to excite a deeper and better emotion than that of laughter, or to awaken a sympathy very different from that which is commonly inspired by the brute creation.[27]

Although the leaflet provides no details of how the animals were shown amid the miscellaneous "wonders" of Egyptian Hall, it gives long descriptions of each one's appearance together with precise measurements of their physical parts. At first, the "black orang outang," or chimpanzee, is judged to be nearer to humans in features and form, in part because of his shorter arms, more upright stance, and intelligent expression, which "suggests to every second visitor, a recollection of some face perfectly well known to him." The tawny orangutan, by contrast, is deemed a "libel on humanity" with her pear-shaped head, very long arms, and "grotesque" manner of walking.[28] Yet when it comes to habits, the leaflet tells us, orangutans show themselves in a superior light to chimpanzees because they are "mild and tractable," given to good manners, and consider simian antics unbecoming "in an animal whom nature has not degraded with a tail."[29] Both apes are seen as instructive entertainment for children as they readily "submit themselves to be confined in a little frock which they wear, with a patience quite edifying to the younger generation of their visitors."[30] Interestingly, Landseer's portraits show only the chimpanzee wearing clothes. It is seated in a robe in a contemplative pose holding a glass, while the orangutan lurks beside it empty-handed, unadorned and looking immodest. The final paragraph ends with a direct reference to Comte de Buffon and a modified reprise of his anatomically based conclusion that apes cannot be human: "The tongue is exactly the same, yet this animal does not speak; the brain the same, yet it does not think."[31]

Chimpanzee and orangutan at Egyptian Hall
(Natural History Museum, used by permission)

Manners were a common motif in descriptions of orangutans in Britain in the early nineteenth century and were often used as a readable index of the animal's position in the natural order. Although good manners could be learned in an imitative fashion, what impressed many commentators was the red ape's apparently innate tendency to comport itself in a socially acceptable way. The Exeter 'Change orangutan had been reported as having "very inoffensive manners,"[32] while another male that was sent in 1830 from Calcutta to the fledgling London Zoological Society was praised for his competent use of cutlery and his "conduct at table," which was "decorous and polite." The animal did not live long enough to be exhibited, but it was said to approach death with dignity, evincing "nothing like impatience or ill temper."[33] Good manners, though an admired sign of civilization, could nonetheless be unnerving, as the oft-quoted instance of Queen Victoria's 1842 visit to the London Zoo reminds us: "The Orang Outang is too wonderful preparing and drinking his tea, doing everything by word of command. He is frightful and painfully and disagreeably human."[34]

Whereas in nineteenth-century London orangutans made their first live appearances at establishments primarily engaged in popular entertainment, the Muséum National d'Histoire Naturelle of Paris offered a more scientific context for public encounters with exotic animals, though the dramatic flourishes evident in the commercial menageries and taxidermied installations of the time were never completely absent from its zoological displays. The museum, founded by decree on June 10, 1793, aimed both to develop and to democratize science and was an integral part of the constitution of the new republic born from the French Revolution. As well as encompassing the Jardin des Plantes (formerly the Jardin du Roi), which had long been associated with the work of eminent naturalists such as Buffon, the national museum was tasked with creating a menagerie that would function in an instructive mode as Paris's first public zoo. The menagerie began with a variety of exotic animals confiscated from itinerant circus acts that police cleared from Paris streets in the autumn of 1793 in an effort to curb unruly crowds.[35] The next year saw the addition of a few survivors from the royal menagerie at Versailles—most of its stock had been butchered by revolutionaries who decried the feeding of animals when people were starving—followed by the steady acquisition of new species, mainly in pairs. They were drawn, painted, and studied as well as being displayed to a public that visited in ever-increasing numbers.[36] New facilities built amid the landscaped expanses of the Jardin des Plantes allowed many animals to be presented in picturesque modes as its collection of exotic species grew over several decades to become one of the largest in the world. After they died, the menagerie's rare animals were often dissected in the museum's amphitheater of comparative anatomy or preserved for display in its natural history collections.

Jack, a young male orangutan, arrived at the Jardin des Plantes from Sumatra in 1836, just as the institution's new primate enclosure was being erected. By then, nearly thirty years had passed since Frédéric Cuvier had studied the red ape given to Empress Joséphine in 1808. Records indicate that only one other live specimen had subsequently been seen in France, as a street exhibit in 1809.[37] Public interest in apes had been heightened in the intervening years by the spectacular success of the 1825 ballet-drama *Jocko, ou le Singe de Brésil* and its many derivations. Also circulating at the time, albeit in a different register, were the theories of Jean-Baptiste Lamarck, who was professor of zoology at the national museum from its inception until his death in 1829. Lamarck had specifically mentioned the possibility of apes becoming bipedal like humans in his 1809 treatise *Philosophie Zoologique,* which supposed the adaptive development of species based on the heritability of acquired traits.[38] This theory, termed "transformism," was widely discussed in France after its publication and came to influence the reception of Darwin's theory of evolution by natural selection. Proto-evolutionary discourses also fueled perceived connections between black races and apes, explicitly drawn, for example, in Georges Cuvier's 1817 autopsy report on a young Khoisan woman, Saartje Baartman, who had been infamously exhibited in Paris and London as the "Hottentot Venus." Georges, the better-known brother of Frédéric Cuvier, likened his human "specimen" to an orangutan and, after dissection, had her genitalia, brain, and skeleton sent to the national museum, where they remained on display until 1974.[39]

In the context of such popular and scientific speculations about apes, Jack's installation in the zoological gardens attracted considerable media coverage and prompted several magazine features. The satirical review, *Le Charivari,* published realist sketches of the animal, followed by a set of caricatures by Honoré Daumier titled *Orang-Outaniana* with an editorial article: "Ah, Si j'étais ourang-outang!" ("If only I were an orangutan!"). Daumier's lithographs have been interpreted as lampooning King Louis Philippe's so-called *juste milieu,* begun in 1830 with the promise of a balanced democratic regime but quickly moving away from real social reform.[40] The government had imposed severe censorship on the press, necessitating indirect means of political critique, and the orangutan's appearance supplied a ready vehicle for that purpose. Daumier's sketches are relatively straightforward in their satirical approach, showing orangutans and particular politicians in mirrored poses that emphasize similarities between their physiognomies. The orangutan article, written in two columns, uses a more allusive, allegorical mode.[41] The left column describes a comfortable Jack in his sleek fur coat installed in a huge "apartment" at the Jardin des Plantes, where he is assured of the expert care of his keeper; the right column gives an account of thousands of Parisians in cold, cramped rooms, or homeless on the streets, unable to afford medicine or sufficient food or clothing. The article's subtitle directs this veiled attack at Louis Philippe's regime, suggesting it had failed to meet the needs of its citizens. Translated into English, the phrase reads: "All Frenchmen are equal before the law (subtext: Monkeys are superior to the French) Charter of 1830."

Le Charivari's account of the apparently palatial quarters being readied for the orangutan gives little hint of the mechanisms of display at work in the Jardin des Plantes' new *singerie* (monkey house) or its limited suitability as a shelter for tropical primates. Although the zoo aspired to communicate the scientific value of its exhibits, the imperative to provide visitors with sure sightings of specific animals shaped the architecture of their enclosures around the human field of vision.[42] The monkey house, constructed of glass and iron latticework, exemplified this principle, allowing the animals little reprieve from the visitor's gaze. An 1842 report pointedly describes the circular structure as akin to an amphitheater where boisterous crowds demanded theatrical performances—farce, vaudeville, comedy, or tragedy—from fifty primate ingénues. It also notes that the occupants died quickly in their artificially heated environment and could only be kept in human society for any length of time preserved in the zoological cabinet.[43]

Jack himself lived for just eight months, though he remained in the public eye for more than a century, *empaillé* (stuffed) among the specimens at the Muséum National d'Histoire Naturelle. A taxidermy exhibit still on display in 1979 depicted him tending a stove in the company of Jacqueline, a chimpanzee who came to the gardens shortly after the orangutan's demise. The installation references one of Jack's particular accomplishments: apparently he had learned to light an oven without any cinders escaping and to fetch the baker when the bread was just cooked.[44] Descriptions of the animal during his tenure at the zoo paint him as very affable, with manners at different times "gauche" or "intelligente." He would cross his legs when sitting down and accepted reprimands with resignation, if not good grace. One day, when he found the salad too acidic for his taste, he took its leaves and pressed them, one by one, between the folds of a blanket to remove the vinaigrette dressing.[45]

Jack's presence at the Jardin des Plantes was recorded in a different sculptural form in a compact bronze statue titled *Ape Riding a Gnu*, crafted in 1836 by Antoine-Louis Barye, a celebrated "animalier" (artist specializing in animal subjects) who later became professor of zoological drawing at the museum. The statue shows the orangutan sitting astride his mount in confident control of its muscular energy, with one hand clasping the gnu's mane and the other its tail. Animaliers were given privileged access to the zoo, affording them unique opportunities to render exotic species into art illuminated by "scientific" observation. Barye's successor, Emanuel Frémiet, was able to examine the remains of primates and participate in animal dissections to inform his sculptures.[46] His intimate knowledge of the red ape's physical characteristics is evident in his arresting 1895 commission, *An Orang-utan Strangling a Native of Borneo,* which still graces the foyer of the museum's Galerie de paléontologie et d'anatomie comparée. Possibly staging an epic battle of species survival between man and beast, the sculpture appears to draw a firm distinction between the adult male orangutan, cheeks flaring, and his prostrate human victim. Today, the museum continues to exhibit adult and juvenile orangutan skeletons from its earliest acquisitions, along with a preserved larynx and an anatomical model of a male's head and neck, flayed to reveal the muscular structure.

Emanuel Frémiet's *An Orang-utan Strangling a Native of Borneo,* 1895 (Photo by Robert Cribb)

The London Zoological Gardens, founded in 1826 by Sir Stamford Raffles, was modeled partly on its French counterpart and likewise aspired to be a national institution for scientific advancement. The zoo did not make extensive postmortem use of animals and initially lacked the depth of scientific expertise that lay behind the development of the Muséum complex in Paris. Raffles, an amateur naturalist and botanist, planned an institution that would display the magnificence of the empire he had helped to build during his years as a colonial administrator in the East Indies. Even though he died before the zoo opened in 1828, it was indelibly shaped by his vision. The project of representing the nation's grandeur via its animal acquisitions had distinct class dimensions, as Harriet Ritvo notes. The Zoological Society courted prominent and powerful subscribers, including aristocrats as well as scientific luminaries and entrepreneurs, and was eager to distance the zoo from the vulgarity associated with popular entertainments among the working classes. It declined stock from Exeter 'Change when starting its collection (while accepting animals from the royal menagerie) and at first reserved entry exclusively for members and their guests, upon payment of a shilling.[47] These restrictions, which initially produced a largely elite clientele, were not fully relaxed until 1846, by which time the zoo's most notable individual animals had already made a splash in the popular press.

Among these "stars" was a female orangutan, Jenny, whose arrival at the zoo in late 1837 elicited numerous news and journal articles. Some of these include sketches of Jenny dressed in a long-sleeved tunic and trousers resembling the *shalwar kameez* ensemble worn in northern parts of the Indian subcontinent, and she is frequently described in terms that register her foreignness alongside class-inflected notions of female propriety and good taste. One report states, for instance, that a "keeper has been appointed by the Council to attend exclusively to her accommodation," as if she were an "Asiatic Princess," and that "her ears are much admired for their smallness and neatness."[48] This kind of explicitly gendered description is echoed in the *New Monthly Magazine,* directed to an educated readership:

> The personage who has lately arrived at the gardens of the Zoological Society in the Regent's Park, and is now the 'observed of all observers', is of the softer sex, and very young. She receives company in the Giraffe-house, and appears amiable, though of a gravity and sage deportment far beyond what is usual at her years.[49]

After Jenny's death less than two years later, her replacement, another young female (also named Jenny) was observed to have "spooned and sipped [her tea] in

Jenny the orangutan at the London Zoo, 1837
(Zoological Society London, used by permission)

the most lady-like way" when visited by eminent clergyman and zoologist Richard Owen and his wife.[50]

Dubbed "Lady Jane," Jenny was not the first female ape to function as a site for the projection (and critique) of metropolitan notions of English femininity. Almost a hundred years earlier, in 1738, a chimpanzee shown at two fashionable coffee houses in London drew large crowds even at the then expensive price of a shilling per view. She was said to have "sat for her Picture," acquired a new wardrobe, and been laid in state for viewing before her burial.[51] Captive male apes were likewise subject to being dressed up in the finery of the upper classes. Raffles' 1820 description of an orangutan he kept in Singapore not long before returning to London to set up the zoo provides a telling antecedent to Jenny's recruitment for displays of the British Empire's cultural prowess:

I have one of the most beautiful little 'men of the woods' that can be conceived; he is not much above two feet high, wears a beautiful surtout of fine white woolen, and in his disposition and habits [is] the kindest and most correct creature imaginable; his face is jet black, and his features most expressive; he has not the slightest rudiments of a tail, always walks erect, and would I am sure become a favourite in Park Lane.[52]

Reported in this mode, both orangutans gave apparently natural performances of Britishness in specific gender and class registers.

The first Jenny's arrival at London Zoo also sparked interest among those with a more serious interest in natural history. In February 1838, the *Penny Magazine of the Society for the Diffusion of Useful Knowledge* published an article on the new orangutan, describing its appearance, its habits in captivity, and stories of orangutan behavior in the wild. The article concluded with a firm statement of what was then the scientific consensus about the red ape:

The earlier travellers and voyagers had filled their pages with descriptions teeming with the marvellous; and men of learning had indulged in the wildest speculations respecting its capabilities of progressive refinement, and its affinity to our race. These puerile fancies have all dissipated before true science; and we now know that extraordinary as the Orang may be, compared with its fellows of the brute creation, still in nothing does it trench upon the moral or mental provinces of man.[53]

The society's need to insist that there was nothing human about the orangutan suggests that the conclusions of anatomists, however hegemonic they may have become within the scientific world, had not convinced the broader public. The article nevertheless marked a new phase of scientific interest in human-orangutan resemblance.

In March 1838, just a month after the *Penny Magazine* article, Charles Darwin entered the heated giraffe house to visit Jenny. He had returned to London a year and half early from his five-year journey around the world in the *Beagle,* during which he had visited the Galapagos Islands and had begun to sketch out in his notebooks the ideas that would become his theory of evolution by natural selection. After encountering Jenny, Darwin jotted down a passage that has often been quoted:

Let man visit Ourang-outang in domestication, hear [its] expressive whine, see its intelligence when spoken [to], as if it understood every word said—see its affection to those it knows,—see its passion & rage, sulkiness & very extreme of despair; let him look at the savage, roasting his parent, naked, artless, not improving, yet improvable, and then let him dare to boast of his proud preeminence.[54]

Possibly a private response to the message of the *Penny Magazine* article, this note reads as an echo of the eighteenth-century philosophical speculation about close moral affinity between orangutans and "savage" humans. Writing in a different register about the same zoo visit in a letter to his grandmother, Darwin describes how Jenny had thrown herself on her back in a tantrum, kicking "precisely like a naughty child," when an apple she wanted was withheld. The keeper then talked to her, promising to give her the fruit if she stopped "bawling" and behaved herself. Darwin added that "[s]he certainly understood every word of this, &, though like a child, she had great work to stop whining, she at last succeeded" to eventually eat the apple "with the most contented countenance imaginable."[55] This firsthand observation of behavioral similarities between juvenile apes and children seems to have strengthened Darwin's conviction that humans were biologically linked to animals. After further visits to Jenny he wrote in his notebook, "Man in his arrogance thinks himself a great work, worthy the interposition of a deity. More humble and I believe true to consider him created from animals."[56]

Nearly two decades later, in his 1859 book *On the Origin of Species,* Darwin outlined a persuasive idea about the mechanism for the emergence of new species over very long periods of time. Starting from the observations that individuals in any species vary slightly, Darwin proposed that those variations would influence the capacity of each individual to survive and reproduce. In an immensely slow but ruthless trimming process, external environmental pressures would nudge a species into new forms. A single species might diverge into many, as Darwin's finches did in the Galapagos Islands, or might change its form significantly, or become extinct. Darwin's argument transformed biological science, among other things, because it provided for the first time a powerful reason why distinct species should exist rather than there being an imperceptible gradation between individual organisms of slightly different form: at any one time, environmental pressures would maintain species conformity by weeding out marginal examples. In this respect Darwinism reinforced the primacy of the species distinction that Linnaeus had codified.

On the Origin of Species implied, and Darwin's 1871 book *The Descent of Man* bluntly stated, that humans and apes resembled each other because they shared a common ancestor more recently than they shared an ancestor with other species.

This insight had a dramatic impact well beyond the scientific world because it shattered the assurance with which defenders of the human-animal distinction could call upon the verdict of science—as the *Penny Magazine* had done—to assert that physical similarity was irrelevant. In identifying evolution by natural selection as the means by which species emerged, Darwin displaced divine power from its once-recognized role as the governing force in the cosmos, unleashing a social controversy that continues to burn well over a century after he put the proposition forward. In the twenty-first century the controversy is most commonly interpreted as a confrontation between secular evolutionists and proponents of divine creation ("creationists"). This bitter confrontation, however, has often seemed to overshadow the fact that the implications of Darwinism for secular thinking were just as disquieting. Reference to human descent from "the apes" used those apes as metonyms for the animal world in general; if there was indeed an unbroken line of descent from apelike ancestors to modern humans, then there could be no more than an arbitrary, negotiable basis for drawing a significant species-level boundary between humans and animals.

Debates about evolution were a boon to natural history museums, menageries, and zoos that had orangutan exhibits, especially live ones, and the issue was taken up in imaginative ways by both zoologically oriented displays and those of a more sensationalist bent. For all their apparently human attributes, orangutans were legally chattels, objects owned by humans who could do with them what they wanted. As European imperialism expanded during the nineteenth century, producing a cultural and scientific efflorescence in metropolitan regions, progressive improvements in transport networks made the exotic wildlife of the East Indies more accessible, especially after the advent of the steamship and the opening of the Suez Canal. There was also some systematization in orangutan procurement through a coalition of big-game hunters and animal traders, whose activities we discuss in the next chapter. From the 1850s, newspapers register a significant increase in the number of red apes on display in Europe, though often as stuffed specimens rather than living animals, for adult orangutans remained difficult to capture and manage. Tony Bennett has coined the term "exhibitionary complex" to describe the dynamics of this display culture with its various Western venues operating as "linked sites for the development and articulation of new disciplines (history, biology, art history, anthropology) and their discursive formations (the past, evolution, aesthetics, man) as well as for the development of new technologies of vision."[57]

Many cities in western Europe developed public zoos loosely inspired by the model of the Paris and London institutions. Driven by a logic of comprehensiveness, zoos competed with each other to show the greatest number—and the rarest—of species. Specimens were a source of great civic pride and were regularly promoted in the media for an apparently avid readership that flocked to see the latest new "star." In this environment, orangutans, like other rare animals, were typically shown as solitary representatives of their species rather than as members of social groups, however loose.[58] Their enclosures tended to be small, and innovations such as glass-paned walls were designed less to ensure the animals' comfort than to allow for their visibility, while keeping unpleasant smells at bay. At their best, European zoos juggled the competing demands of science and showmanship to present an accessible "compendium of biological knowledge,"[59] but more often they acceded to the imperatives of public entertainment. The living orangutans that reached Europe in the late eighteenth and early nineteenth centuries were recognized as frail and unlikely to survive more than one or two winters. Although zookeeping techniques gradually improved, it was rare even at the end of the nineteenth century for orangutans in captivity to survive more than a few years. This appalling cost in orangutan life was partly masked by the growing availability of imported orangutans and by the nature of Western display cultures, which avoided drawing attention to animal morbidity. The situation was exacerbated in North America, where orangutans appeared as unashamedly commercial exhibits in traveling menageries and dime museums until the public zoos movement took hold there in the 1890s. In both continents, the apes were also shown alongside ethnological displays of African and Asian peoples that were featured after 1889 in the Midways of World Fairs. These living museums prompted visitors to celebrate humanity's progress from "primitive" societies toward the ideals of modern industrialism expressed elsewhere in the exhibitions.[60]

Live adult orangutan exhibits tended to produce more polarized impressions of the red ape than infants did, particularly as Darwin's theories began to energize popular debates about evolution after the publication of his major works. Full-grown apes were less readily cast as analogous to malleable young humans ripe for acculturation, and they often elicited awe and disgust, as well as occasional admissions that they were "out of place" on display. An 1880 article about a new orangutan at the London Zoo begins, for instance, with a description that deftly introduces the animal as a fit-for-purpose exhibit: "He stands five feet without his stockings, and, being a little bald and well whiskered, is a very respectable specimen of the man ape." Yet unlike previously exhibited juveniles, whom the writer calls "guileless urchins of no decided character," this middle-aged "man of the woods" soon registers unambiguously as an exceptionally powerful animal faced with the indignity of captivity:

The cage in which he travelled was so small that he could not have fair play for his tremendous arms, and the bars so thick that he could not make any impression upon them with his enormous teeth. Impotent, therefore, for mischief, the hairy prisoner sits huddled up and roaring. Any interference with him, however kindly meant, is at once resented by language which might easily be translated into human equivalents, and the vigor with which he shakes his cage proves the sincerity of his ill-feeling. . . . In their native haunts they were never trifled with, for they were monarchs of all they surveyed, and the only neighbors capable of molesting them were the infrequent crocodile and the still rarer python.[61]

A less sympathetic account describes a caged male in Manchester enraged after members of a crowd throw refuse at him. Although the author admits that the ape is regularly tormented and even calls it "poor orang-utan," he nonetheless claims it is "vindictive to an alarming degree," except to its keeper, a "dark-eyed man" with whom it fell in "love at first sight." The article is prefaced with the assertion that the animal is so "repulsively human" as "to make us all feel that we would rather not be related to monkeys, unless the connection is a very distant one indeed."[62] For some spectators, ferocious animality was central to the appeal of orangutan exhibits. On the death of Chief Utan, a male shown in the Javanese village at the Chicago World Fair in 1893 and afterwards at Philadelphia's zoo, one columnist remarked that the orangutan "grew more and more vicious as the time went on," but this only "increased his popular interest." When he died "after indulging in violent gymnastics," he was stuffed for the museum where his "skin [could] grin at his skeleton from neighboring pedestals."[63]

By this time, natural history displays in North America had been irrevocably shaped by P. T. Barnum and entrepreneurs of his ilk, whose overtly theatricalized museums and menageries primed spectators to expect, and desire, sensationalist animal exhibits. Like Peale before him, Barnum played to scientific curiosity but was far more successful in harnessing its vast commercial potential, not least because he flouted propriety by weaving a variety of performance acts into his exhibitions and engaging in unashamed chicanery to present invented "wonders'" alongside bona fide biological specimens. He toured extensively, drew his exhibits from a global catchment, and for many years kept the nation's largest collection of exotic animals, often donating their bodies to science when they died.[64]

In early 1845, Barnum bought a female orangutan, Mademoiselle Fanny, for $3,000 from the Royal Surrey Zoological Garden[65] and returned to the United States to exhibit her as the "Connecting link between Man and Brute!"[66] Named after the German ballerina Fanny Elssler, the orangutan became a crowd favorite, dancing on stage in a pink tulle dress at Barnum's American Museum in New York.[67] She appeared in repertoire with other acts and displays, including two "Mammoth Boys," an "Anatomical Venus," and a "petrologist," who foretold events by looking into a rock.[68] Several months later, a young male orangutan was added to the museum collection and billed as a "perfect personation [sic] of the Wild Man of the Woods." A woodcut sketch shows a hairless, wizened infant with a cloth draped across his groin, faintly evoking the mummified human body touted on the same playbill as another of the venue's wondrous attractions.[69] Among Barnum's subsequent orangutan features was an "immense" male presented in 1852 in tandem with the Ethiopian Serenaders, a popular blackface minstrel troupe.[70] While these exhibits said less about the red ape than about the seductions of spectacle in Barnum's "modernized version of the museum *as* theatre,"[71] they cannily created orangutan performers who, like human actors, could stand in for scientific specimens.

Barnum's pseudoscientific praxis opened up a cultural space for the wayward appropriation of known details about orangutans and their native habitat, particularly in freak shows and fairground entertainments, which then developed performable versions of natural history's enigmas. The Wild Man from Borneo act created for Hiram and Barney Davis in the 1850s was the exemplar in this genre, inspiring many imitations and permutations as the century wore on. Among these, a real red ape, dubbed "Wild Girl of Sumatra," is perhaps the most fascinating. The orangutan's owner, Old Graves, told a *New York Times* reporter in 1882 how years earlier he had exhibited his "little girl" in New England towns, earning up to $500 per day. His success, he intimated in the article, lay in presenting her to the populace as a credible scientific curiosity: "I talked about the 'missing link' long before Darwin ever thought about it and I have an idea that I can explain in my common, homely manner more about what is now called Darwinism than the author himself." Graves added that he had twelve expensive dresses made for the orangutan as well as shoes, stockings, and "pantalettes," and he always kept her "neatly and becomingly attired" for his lecture-demonstrations. On one occasion, when a news editor suspected the Wild Girl was simply an "orang-outang or some other fraud" and refused to run Graves's advertisement, a professor of anatomy was called in to verify that she was indeed a genuine curiosity. Something of a raconteur, Graves insisted on the special status of his "little girl" as "fellow-creature" *and* genuine hybrid, though he also subtly alludes to a fear that his ruse would be discovered.[72]

By the time the first American zoos opened in the 1870s, orangutans had been exhibited as freaks, tramps, children, brutes, marvels, monsters, ancestors, missing links, and clowns, but only occasionally in their zoological context as primates

Playbills featuring Barnum's orangutan exhibits, c. 1845
(Houghton Library, Harvard Theatre Collection, used by permission)

removed from their natural habitat to feed a human appetite for entertainment. Ironically, it was a taxidermy exhibit, *A Fight in the Treetops,* that provided a powerful new optic for viewing the orangutan as an ape in its own environment. The exhibit's creator, William Temple Hornaday, had conceived his pioneering display while on a two-year excursion in Borneo, where he observed and sketched live orangutans as well as hunting them to collect specimens. When the expedition ended, he aimed to work "artistically and exactly" from his field notes to "mount a very striking group" of the largest apes, which he hoped would be "talked about" in prestigious circles.[73] *A Fight in the Treetops* was completed in 1879, measuring eleven feet long, six feet wide and more than ten feet high.[74] It won Hornaday the admiration he sought among scientists and museum aficionados alike with its precise blend of expressive theatricality and zoological realism. The arresting orangutan installation not only elevated the scientific and artistic status of taxidermy but also established the habitat diorama as the dominant modality for displaying wild animals in American natural history museums over much of the following century.

Male orangutans in Hornaday's *Fight in the Treetops,* 1879 (Biodiversity Heritage Library and Smithsonian Institution Libraries, used by permission)

A Fight in the Treetops staged a savage battle between two full-grown males mounted on lifelike branches above vegetative undergrowth, with moss, orchids, and pepper vines professionally made for the exhibit. Teeth bared and one arm extended in loosely mirrored poses, each orangutan fixes his gaze on his adversary, seeming to ignore a partially hidden female (presumably the winner's prize) in a nearby nest with her infant. One male is captured in the act of biting off his adversary's forefinger; the other snarls in protest while higher up in the canopy a juvenile gazes warily down at the fracas. Hornaday admitted the exhibit was "a trifle sensational" but emphasized its accurate detail in terms of the apes' form, behavior, and habitat.[75] Shortly before finishing his tableau, he took the two mounted males to illustrate a lecture on "Bornean Orangs" for the American Association for the Advancement of Science. Like Peale's stuffed orangutan eighty years earlier, Hornaday's unusual props, dramatically posed on the speaker's platform beside him,[76] authenticated his own performance as a self-taught naturalist adept at understanding scientific principles gleaned from observations of the natural world. On this occasion, the bellicose apes dramatized an epic struggle for the "survival of the fittest," in keeping with Hornaday's evolutionism. His lecture concluded with an explicit challenge to those who doubted Darwin's theories, inviting them to go to the forests of Borneo and see for themselves the red ape's "strangely human form in all its various phases of existence."[77]

Hornaday's approach to specimen preservation helps to explain the apparent inconsistencies between his acknowledgment of human-orangutan kinship and his ready participation in the brutalities of orangutan hunting (he personally shot most of the forty-three specimens he sent back to the United States for distribution; see also chapter 4). For Hornaday, taxidermy bridged the gap between art and science to capture the pure essence of an animal: "Perhaps you think a wild animal has no soul. But let me tell you that it has. Its skin is its soul. And when mounted by skillful hands, it becomes comparatively immortal."[78] Ironically, the profound presence he attributed to taxidermied animals rested on aesthetic refinements to the clay manikin models that gave them their form while most of the animal itself was discarded. Yet Hornaday's vision of taxidermy's resonance accrues force in light of the subsequent histories of the orangutans he preserved. *A Fight in the Treetops* was purchased in 1883 by the Smithsonian's National Museum of Natural History in Washington and displayed intact until the 1950s, when the individual animals were integrated into the World of Mammals Hall. New York's American Museum of Natural History likewise secured one of Hornaday's large habitat groups, which he created for $1,500 on commission in 1880. Titled *Orangs at Home,* it posed three adults peacefully eating durian at center stage, with a juvenile sleeping in her nest nearby and an infant hanging from a branch in the background. The museum refurbished its primate collection in 1965, keeping Hornaday's mounts for a new gallery that offered visitors a "quick evolutionary tour" with its "cast of 97 characters."[79]

Hornaday's orangutans were seen by millions of viewers over the decades and quickly gained iconic status as authoritative representations of their species. The habitat group concept also came to influence live animal display in a range of zoos across the United States and beyond. Hornaday turned his own attention to American mammals and eventually spearheaded efforts in the 1890s to create a national zoological park to preserve threatened wildlife, though his concern with conservation never extended to exotic species. By this time, orangutan numbers had already diminished in Borneo after they were hunted for the lucrative trade in exhibits in which Hornaday had participated. His stunning dioramas, at once combining the fiction of immortal life with the specter of violent death, encompassed the savagery of this encounter, if perhaps cryptically. Today, the affective

Taxidermied orangutans in the Sarawak Museum, c. 1912 (Green, *Borneo*, frontispiece)

powers of such exhibits are eerily captured by the new, exceptionally lifelike orang-utan on show at the Smithsonian: "It's like you can see into his soul," one onlooker has remarked.[80] Taxidermied orangutans in museums in Geneva, London, Minneapolis, Melbourne, and Kuching, among other cities, likewise hint at a much larger gallery of specimens that have become protagonists of natural history narratives embedded in the logics of human progress.

Throughout the nineteenth century, the presentation of orangutans as specimens—dead or alive—lent itself to discourses about race, class, and species while suppressing narratives of animal dispensability and death in transit. Science had striven to draw a clear line between human and other primate species, if not always managing to discredit racialized explanations of human-ape likenesses, but display cultures thrived as much on provocation as they did on theatricality. With the material animal as their stock-in-trade, they were less inclined than fiction and drama to probe interspecies encounters or affinities and much more interested in suggestive juxtapositions. A "missing link" in this context could conjure up the specter of cross-species sex, or a "wild man from Borneo" might seem to confirm connections between orangutans and racialized humans positioned at the margins of European modernity, but it was largely up to spectators to judge the import of what they saw. While the more scientific displays adopted theatrical techniques to help communicate first biological and then ecological perspectives, popular entertainment, well attuned to the pleasures of ambiguity, declined to deliver a verdict on the orangutan's essential characteristics or its humanity. In that respect, it was endorsed as simply another fascinating performer on a public stage accustomed to the spectacle of transient oddities.

4 ✺ Darkest Borneo, Savage Sumatra

The orangutans presented to Western audiences throughout the seventeenth and eighteenth centuries were creatures detached from their natural habitat both literally and symbolically. Linnaean classification and comparative anatomy paid as little attention to where an animal came from as they did to its behavior. The earliest carers for captive orangutans in Europe, apart from trying to keep their charges warm, made no effort to replicate the natural environment from which the animals had come. When philosophers and writers elaborated on a half-mythical "orang-outang" that behaved like a human and resembled certain human races, they imagined a creature that was potentially at home anywhere humans lived, despite the already dismal record of orangutan mortality in Europe and the United States. And the scientists and entrepreneurs who displayed orangutans for the education and amusement of European and North American publics necessarily presented them in circumstances that bore little resemblance to their natural habitat until Hornaday prepared his groundbreaking orangutan dioramas.

Ironically, the orangutan loomed larger in the imagination of the West than it did in its native Borneo and Sumatra. By the beginning of the seventeenth century, the red ape had long disappeared from the vicinity of the major settled areas on the two islands. Brunei in the north of Borneo and Banjarmasin in the south were the heartlands of ancient kingdoms that had long fed the China trade, delivering gold, diamonds, camphor, beeswax, bright birds' feathers, fragrant wood, gums, and resins.[1] The northern coast of Sumatra had been dominated for two millennia by small but glittering cities that serviced trade through the Melaka Strait. The orangutan, dependent on an abundance of fruit trees in tropical jungles and vulnerable to hunting, could not survive in the vicinity of such human settlement. Nor were the cosmopolitan cultures of the coast, predominantly Muslim since at least the fifteenth century, intellectually interested in the orangutan for its resemblance to humans. Islamic thinking shared with late medieval Christianity a conception of apes as having a high rank on the Great Chain of Being, and of their behavior as offering powerful allegories of human ways.[2] The intellectual turmoil that beset the West as a consequence of questioning the biblical Creation narrative, however, had not yet touched the Muslim world. Javanese, Malays, and Acehnese tended to shun the jungle as remote, uncomfortable, and dangerous. The red ape was neither hunted nor traded, neither prized nor despised nor feared, in the dominant cultures of the region.

In the interior of both islands were people of the forest: Dayaks in Borneo and Bataks and Gayo in northern Sumatra. They lived in small communities, sometimes loosely under the suzerainty of coastal sultanates but often largely independent. They supplied forest products for the China trade, receiving cloth, implements, and ceremonial goods in exchange. In some regions they formed small permanent agricultural settlements; in others they moved from place to place using swidden ("slash and burn") farming. For all these inland peoples, hunting was also an important source of food. Wild honey and the wild pig that thrived on the jungle margins were a staple,[3] but these *orang hutan* (people of the forest) also opportunistically hunted other animals they came upon, both for food and for trade to the coast. Dayak blowpipes using poisoned darts were effective in sedating even adult orangutans so they would lose their grip and tumble to the ground, where they could be captured. At least some Dayak communities have been reported to prize orangutan meat as especially delicious or as a means of enhancing male sexual potency.[4] Evidence of the cultural and religious significance of orangutans to forest people, however, is unreliable. Animal totems are important in many parts of Indonesia, but the crocodile, the tiger, and the buffalo have had the greatest religious symbolism,[5] and orangutans are mentioned only occasionally in the standard ethnographic literature, though Evans tells a story in which an old, talking orangutan teaches medicine to Dayak women.[6] The most vivid assertions of an indigenous spiritual significance for the orangutan seem to involve ethnographers' conflation of the orangutan with one or other of the malevolent spirits of the forest that are a strong element in Dayak belief, but it remains uncertain that Dayaks, when they encountered an orangutan, interpreted the ape in this way.[7] Many writers have noted that orangutan body parts have been used in potentially significant contexts—bones as sword handles, skins as war jackets, hair wrapped around the staves of shamans, and skulls as substitutes for human heads in head-hunting rituals[8]—but in the absence of stronger ethnographic evidence we have to be cautious about attributing more than incidental meaning to such uses.

This caution is necessary because the two forest peoples involved were deeply enmeshed in a process of exoticization at the hands of Western writers, especially during the nineteenth century. Eighteenth-century Western scholars took for granted the "orang-outang" stories described in chapter 2 and they had no great interest in the cultural or mental world that might have produced such tales. In the nineteenth century, by contrast, Europeans in Borneo collected indigenous lore as an insight into the local people. Europeans regarded both head-hunting in Borneo and cannibalism in Sumatra as especially savage habits, arousing horror and fascination that fed the enduring debate over whether primitive human society had been noble or brutish.[9] Identifying a spiritual relationship between indigenous peoples and their environment was important to both sides of the debate, and the assumption that Malays had named the red ape "*orang hutan*" because of its similarity to humans was grist for this mill. The 1937 colonial *Album van Natuurmonumenten* (Album of Nature Monuments) could thus tell its readers that "The indigenous population believes in a close relationship between humans and orangutans, and apes in general. People can become apes and apes can become people,"[10] even though there appears to be no other ethnographic evidence for this claim.

The Western habit of attributing to Dayaks and Bataks superstitious beliefs about the orangutan tends also to assume that local informants were the naïve subjects of investigation, rather than active participants in a myth-building process. In the twentieth century, Hose and MacKinnon described as "most frequently recurring" among the Dayaks a story in which a human woman, kidnapped and kept by an ape, gave birth to his son. When she seized a moment to escape, the orangutan chased after her with the child. Seeing, however, that she had managed to reach a friendly boat and was lost to him forever, the distraught ape tore the child in two, flinging one half after the mother and the other into the jungle as a symbol of the child's divided origin.[11] The story has been repeated in many forms, including one in which the genders are reversed: a human male is kidnapped and kept as a sex slave by a female orangutan, who bears his child but kills her offspring in despair when he escapes.[12] Although apparently embedded in recent Dayak oral culture, this story appears to have much older, non-Asian origins. In 1573, Antonio de Torquemada published a work titled *Jardín de Floras Curiosas* (Garden of Curious Flowers) in which he recounted the story of a women exiled as punishment to a deserted island apparently near Africa, where she was captured by apes. Under *force majeure*, she "suffered the Ape to have vse [*sic*] of her body, in such sort that she grew great, and at two seuerall times was deliuered of two Sonnes." She was rescued by visiting sailors, but the ape-father, distraught at her departure, drowned the two boys as she watched from the boat.[13] In the stories told to Hose and MacKinnon, thus, we seem to have an insight not into primordial Dayak beliefs about the orangutan but rather into the facility with which Dayaks could assimilate stories from outside and then retell them to curious Westerners. Something similar seems to apply to Bontius' report that his "orang outangs" refused to speak for fear of being enslaved; the story is first recorded of "Babownes" by Richard Jobson, writing about the west coast of Africa in 1623.[14] During the past century, Bataks and Dayaks have been adept at feeding the Western obsession with cannibalism and head-hunting,[15] and it would be surprising if they had not done the same with the orangutan.

More likely to be authentic is a Dayak story reported by Tom Harrisson in which the woman has become pregnant because of some forest magic and is expelled into the jungle by her ashamed father. She is taken in by an orangutan, whose own wife also happens to be pregnant. The outcomes of the pregnancies, both boys, grow up as friends. By chance, they meet a man who, in the end, marries the woman and takes her back to his village.[16] In its gentle character and redemptive ending, the story is reminiscent of a genre of Southeast Asian tales in which misfortune is erased or misconduct is expiated by a period of exile. The orangutans in this story are neither sexual objects nor sources of spiritual power, but simply kind hosts, so human that they could in fact have been real people of the forest.

We know now that hybridization between humans and apes is significantly hampered, though perhaps not made entirely impossible, because humans have only twenty-three pairs of chromosomes, whereas all great apes have twenty-four pairs.[17] We have to seek the origin and meaning of hybridization stories therefore in social and cultural anxieties and prurience rather than in biology, though there is persistent empirical evidence of the sexual attraction of male orangutans to human women. In fact the oldest Western reference to the orangutan is a brief description by the third-century Greek naturalist Aelian, who refers to "reddish" apes that are "too fond of women" and which the local people put to death "as adulterers."[18] Tulp's report of the testimony of his brother-in-law Samuel Bloemaart provides more detail:

> The desire for the latter [women and girls] sometimes burns so strongly that they capture and sometimes even rape them. For they are very fond of sexual intercourse (like the licentious Satyrs of the ancients). In fact they are sometimes so shameless and full of lust that the women of the Indies avoid the woods and forests where these animals hide.[19]

Similar characterizations of the orangutan were made throughout the eighteenth and early nineteenth centuries, never more than as distant reports, though the indefatigable Ebenezer Sibly provided a speculative illustration of a kidnapping.

Bloemaart's account was echoed in 1851 in a report published in German newspapers about a fifteen-year-old Fräulein Schoch, who was allegedly seized by orangutans while she was traveling a public road in West Sumatra and carried to the top of a tree, where they tended her wounds and offered her coconuts until the girl's father summoned a posse of villagers to rescue her.[20] This widely reported

The Orang-Outang carrying off a Negro Girl ([Sibly],
An Universal System of History, vol. 2, frontispiece,
Wellcome Institute, used by permission)

story may have been the inspiration for the kidnapping of a Dutch woman by an orangutan, described in Edgar Rice Burroughs' *Tarzan and "The Foreign Legion"* (1947). The only serious recent record of rape by an orangutan is in the memoirs of the Canadian primatologist Biruté Galdikas, who established and ran an orangutan rehabilitation program in central Borneo:

> One day I went to the [feeding] platform with a visitor from North America and one of the [Indonesian] cooks. When Gundul [a wild-born, subadult male orangutan] arrived, he ate a little but seemed distracted. Suddenly Gundul grabbed the cook by the legs and wrestled her down to the platform, biting at her and pulling at her skirt.... I began to realize that Gundul did not intend to harm the cook, but had something else in mind. The cook stopped struggling. "It's all right," she murmured. She lay back in my arms with Gundul on top of her. Gundul was very calm and deliberate. He raped the cook. As he moved rhythmically back and forth, his eyes rolled upward to the heavens.[21]

Although this incident has been widely discussed in literature on the nature of rape within human societies, the fact that only one such instance has been recorded casts doubt on its representativeness.

Even if the most lurid stories of orangutan-human sexual relations were ultimately foreign in origin, their immediate Southeast Asian provenance gave them plausibility, and they provided material for fictional works that explored the issue further (see chapter 6). Something of the same applied to stories of orangutan ferocity. Several stories from Africa that circulated in the eighteenth century as applicable to all "orang outangs" described them as ferocious in their own defense. Donovan wrote in the *Naturalist's Repository* in 1824:

> The adult Orang-outang, with a stick or branch of a tree in its hands, defends itself until it is killed, or scaling the rocks hurls stones down upon its agressors [sic], never yielding to any force but in death. In close attacks, when disarmed, it must be no less formidable, for then, opposing its antagonist with main strength and casting to the earth, it seizes upon him with its teeth, and while holding him firmly down at arms [sic] length, can with perfect ease and safety tear him to pieces.[22]

By contrast, the living orangutans that reached Europe in the late eighteenth and early nineteenth centuries were above all genteel creatures. As noted in chapter 2, they ate strawberries and cream, dressed and slept demurely, looked sad and sagacious, and died swiftly and gracefully. If they were given to naughtiness or pranks, such as throwing tantrums or drinking Madeira, these missteps were forgivable signs of incomplete training rather than of deep personality flaws. When Hornaday constructed his 1879 tableau *A Fight in the Treetops,* however, he conjured up a fundamentally different image of orangutan savagery. To be sure, the two male orangutans in the tableau were fighting each other and not directly menacing the public, but it was difficult to look at the representation without feeling a frisson of alarm at the apparent viciousness and strength of the two central animals. This change in the reputation of the orangutan came about as Europeans began to encounter adult orangutans in the jungles of Borneo and Sumatra. Full-grown and on their home ground, orangutans were very different creatures from the timid and clingy infants who reached Europe. Although the reports of Bontius and Tulp had alluded to its strength and swiftness, the new eyewitness accounts of Europeans made the ferocity of the orangutan a sudden, vivid reality for Western publics.

The first European eyewitness report of a wild orangutan was published in 1779 by J. C. M. Radermacher, a leading light in the Batavian Society for Arts and Sciences, who reproduced a letter from Willem Adriaan Palm, head of the colonial administration in Rembang on the north coast of East Java. Palm had just returned from Pontianak in western Borneo, where he had been attempting to sort out the affairs of a troublesome company employee,[23] but his news had nothing to do with politics:

> Accompanying this letter, Your Excellency will find (against all expectations, for I have made it known to the Borneans and Banjarese for some time that I would pay more than a hundred ducats for an orangutan measuring 4 to 5 feet) a specimen which I first saw at 8 in the morning. For a considerable time we pondered what means we could use, deep in the wilderness half way to Landak, to capture this fearsome [*verschrikkelijke*] animal alive. In order to keep him occupied, we did not even think of eating, while we took great care not to let him harm us; while he continually broke off heavy pieces of wood and green branches and hurled them at us. This game continued until about 4 in the afternoon, when we decided to give him a bullet. This succeeded, in fact turned out well, because I shot at an angle and the bullet entered him from the side, so that it caused little damage. He was still alive when we brought him to the boat, tied him up to the bollards. He died of his wounds the next morning. The whole of Pontianak came down to the wharf when we arrived in order to see this monster [*gedrogt*]. I then found it necessary to open him up and remove the intestines.[24]

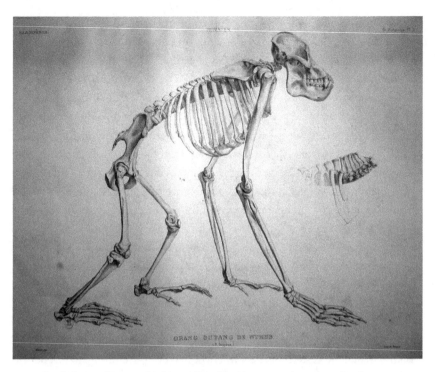

The skeleton of "Wurmb's Pongo," used by De Lacépède to describe the genus
(Ducrotay de Blainville, *Ostéographie*, plate 1)

Palm's specimen, preserved in arak, was subsequently described by the German
botanist Baron Friedrich von Wurmb as "the Borneo orang outang or the East
Indies pongo."[25] Shortly before his death, Von Wurmb sent his specimens to the
Netherlands as a contribution to the cabinet of the Prince of Orange. The ship was
wrecked and its cargo was lost, but in about 1784 another skeleton of a mature
orangutan reached the Netherlands, where it was seen by Camper and widely
understood to be "Wurmb's Pongo." This skeleton was then seized by French forces
when they entered the Netherlands in 1795 to support the Dutch revolutionary
movement and transferred to the Museum of Natural History in Paris. There it
inspired the formal generic name of the orangutan when it was recorded by taxonomist De Lacépède as belonging to the genus *Pongo*.[26]

The most important account of apparent orangutan savagery was that in 1825
of Clarke Abel, surgeon-in-chief to the governor general of British India, who in
1825 published a report of the murder of an orangutan in Sumatra. He had not
seen the killing himself but had obtained the story from others who had been

aboard the British brig *Mary Anne Sophia* when it dispatched a boat to look for water on the western coast of the island, near the town of Trumon. Upon coming ashore, the British party noticed "a gigantic animal of the monkey tribe." They gave chase and cornered the orangutan in a small clump of trees, which they proceeded to cut down, one by one, in order to isolate him. They shot five bullets into him. He vomited blood and appeared ready to collapse, but as soon as they approached, he leaped into another tree. Only after all the trees in the clump had been cut down could they

> drive him to combat his enemies on the ground, against whom he still exhibited surprising strength and agility, although he was at length overpowered by numbers, and destroyed by the thrusts of spears and the blows of stones and other missiles. When nearly in a dying state, he seized a spear made of a supple wood which would have withstood the strength of the stoutest man, and shivered it in pieces; . . . he broke it as if it had been a carrot. It is stated by those who aided in his death, that the human like expression of his countenance, and piteous manner of placing his hands over his wounds, distressed their feelings and almost made them question the nature of the act they were committing.[27]

The shore party brought the orangutan corpse to the *Mary Anne Sophia,* where Captain Cornfoot had it skinned, before eventually delivering it to the Asiatic Society museum in Calcutta. There Abel examined it and, in proper scientific manner, recorded its dimensions. Abel's report was quickly reprinted and widely read, and allusions to the animal he described and the circumstances of its death recur repeatedly during the nineteenth century both in popular science and in creative works.

Reinforcing the trope of orangutan ferocity was an illustration captioned "Orang-Utan attacked by Dyaks," which appeared as the frontispiece to the account by the Scottish naturalist Alfred Russel Wallace of his travels in the East Indies in the 1850s. The orangutans in Wallace's text are not savage monsters. They hoot and howl and throw branches at him, and his Dayak assistants have to treat wounded orangutans carefully to avoid being injured, but there is nothing in Wallace's account to inspire dread. His illustration, however, despite its caption, gives the impression of a group of Dayaks perturbed at the ferocity of an orangutan, rather than the reverse. The American wildlife collector and hunter Hornaday described the first orangutan he shot as having eyes that were "villainously small." Later, he examined the body of an orangutan brought to him by Dayak hunters

The severed head of the Trumon orangutan, 1825
(Abel, "Some Account of an Orang Outang," n.p.)

and, finding evidence of wounds, concluded that the animal must have been "a regular prize-fighter," "a first-class desperado."[28] For *Fight in the Treetops,* although in some respects conscientiously attempting to depict the realities of orangutan behavior in the wild, Hornaday deliberately chose a tableau that would make the orangutan seem dangerous.

In the nineteenth century Borneo became, with central Africa and the Amazon, the exemplar of tropical luxuriance and danger, a place of intense discomfort for Westerners because of its heat and because of the abundance of small predators such as ants, mosquitoes, leeches, and spiders, where the luxuriance of the vegetation made travel challenging, where the possibility of death from fever was never far away. Fictional works set in this forbidding world, such as Rudyard Kipling's "Bertran and Bimi" (1891) and Edgar Rice Burroughs' *Tarzan and "The Foreign Legion"* (1947) cast the orangutan in a ferocious light. Abel's account also appears to be ancestral to a series of fictional writings, explored in chapter 6, that depict the

Wallace's depiction, captioned "Orang-Utan attacked
by Dyaks," gave the impression rather of Dayaks
attacked by a ferocious orangutan, 1869
(Wallace, *The Malay Archipelago*, frontispiece)

ape's combination of physical strength and savagery, especially when abstracted into
Western society. It may be the ultimate source, too, for Frémiet's celebrated sculpture
of an orangutan strangling a Dayak (discussed in chapter 3). It took repeated form
in brief news reports over the decades. In 1927, for instance, newspapers around the
world gleefully reported the story of a mutiny at sea, in which a collection of animals
destined for a menagerie had managed to take charge of a ship, the orangutan burst-
ing into the engine room with a hammer, scaring the crew away and sending the
ship in a dizzy set of maneuvers as he arbitrarily turned the wheel. Most notably, in

the suppression of the mutiny, only the orangutan was shot dead.[29] Frank Buck, too, commented laconically, "The orang's favourite method of doing damage in a fight is to pull the foe to him with his mighty arms and rip his victim to pieces with his teeth."[30] These images, although deriving from the early nineteenth century, meshed easily in the second half of the century with Darwinian interpretations of the natural world as a place of struggle in which only the fittest could survive.[31]

There is only meager, mainly indirect, evidence of a commercial trade in orangutans before the nineteenth century. Southeast Asia had a long history of supplying rare jungle products to China, India, and the Middle East, but we find no hint of a trade in orangutans to these destinations. Even in the eighteenth century, when it became apparent that Westerners might pay generously for curiosities such as infant orangutans or the hands and feet of older ones, the scale of the demand does not seem to have been enough to stimulate any organized hunting of orangutans. When in the 1770s Vosmaer pressed his contacts in the Indies to find him a large orangutan, there was no wholesaler of red apes to whom he could turn. One of his correspondents replied, "This much I can tell you: an orangutan of the size you mention has never been seen here and I doubt that they actually exist. In the past we did have some smaller orangutans, but there are none any more."[32] Another informant told him:

> Concerning the orangutan you have requested: I even went to Kayutangi and asked the local prince, in the politest possible way, whether he might be able to help me obtain a creature of the size that you requested, namely five foot high. He promised to do so, but commented that they are very rare. The oldest Natives tell me that they have never heard of a creature of such a size. But I nonetheless commissioned people right and left to find one.[33]

Both these accounts imply that small orangutans were indeed to be seen, and Jacob Radermacher, leading figure in the Batavian Society for Arts and Sciences, commented in the 1780s that he had seen fifty small orangutans during his time in the Indies. If his account is correct, then these animals were presumably juveniles being traded or kept as pets.[34]

Direct reports of the techniques used to capture orangutans are scarce. An early twentieth-century account of the killing of an orangutan suggests that Dayaks also used the technique described by Abel in 1825:

> One of them finally seeing the creature up a large tree had fetched the others and they had felled the smaller trees all around, leaving the bigger tree with

the Maias standing, so that it could not jump or spring on to the smaller trees and thus escape them again. They then formed themselves into a circle, leaving only two men to fell the tree. It fell, and with it the Maias, which was stunned, and so they were able to capture and kill it.[35]

In other accounts, Dayaks bring orangutans down with darts from their blowpipes or sometimes lure them eventually to the ground with the offer of fruits laced with a sedative.[36] We can imagine that parties of Dayak hunters in the interior of Borneo would occasionally surprise a mature orangutan on the ground or in an isolated tree, shoot it with darts or spear it to death, and hack off hands and head for delivery downriver to a trader with a sideline in curiosities. And sometimes an infant orangutan, caught with its mother in similar circumstances, would be traded downriver and overseas to Java, Singapore, or Hong Kong. Without parental care, only a small proportion of captured infant orangutans could have survived to reach the hands of European seafarers. Indeed, for the whole of the eighteenth century we have reliable accounts of only eight live orangutans reaching the shores of Europe.[37]

The only evidence of something more than an opportunistic trade in orangutans comes from the scanty evidence we have of a trade in fake orangutan body parts from the Indonesian archipelago. In 1771 Jean Allamand, a Leiden naturalist, published a drawing of an unusual detached hand as part of the first additional comments he provided on Buffon's account of the orangutan. Allamand described the hand as having been sent from Batavia to a professor in Rotterdam, who had then passed it on to him. He estimated from its size that it must have belonged to an animal six feet tall, possibly an adult orangutan.[38] In the second set of additions to Buffon in 1783, however, he ruefully notes that the hand had proven to be a forgery.[39] The 1821 Feejee Mermaid, discussed in chapter 3, was another example of the use of orangutan body parts for fakery.

In fact, the sale of such forgeries in the Indonesian archipelago seems to have a long history. In his discussion of "Lesser Java" [Sumatra], the thirteenth-century Venetian traveler Marco Polo reports:

> It must be noted that merchants who claim to have brought back pygmies as a rarity from India are involved in great deception. There is a breed of tiny monkey on the island, with almost human features. They are caught, killed and skinned. Only their facial hair is left. Then they are dried and embalmed with camphor and other substances. They look like men but it is a great imposture. These "men" are fabricated in the way I have described. They exist neither in India nor in any other wild country.[40]

Allamand's giant hand, 1771 (Buffon, *Histoire naturelle,* vol. 14, plate xii)

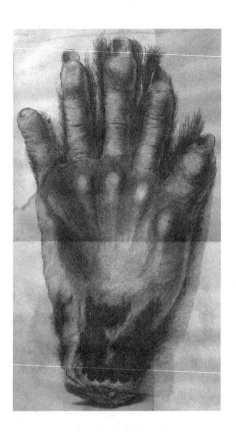

In the absence of extensive records, the safest conclusion we can draw is probably that until the end of the eighteenth century, existing trade and social structures in maritime Southeast Asia were sufficient to supply the modest international demand that existed for living and dead orangutans. From late in the century, however, that demand began to grow, both for scientific and quasi-scientific collections and for public display. The older tradition continued of presenting exotic wild animals to rulers. In 1811, while Britain temporarily occupied the Netherlands Indies during the Napoleonic Wars, the sultan of Sambas gave two living orangutans to the British governor of Java, Stamford Raffles, who later founded the London Zoo. Alongside this traditional form of transaction, however, there emerged a new and ruthless commercial market for the red ape.

The trade in living orangutans was a sordid one, and we have only a few glimpses of its earliest stages. In 1838, the Belgian politician and botanist Baron Dumortier reported that he had obtained fourteen orangutans through the good offices of a Lt. Col. Henrici.[41] Schlegel and Müller paid 80 guilders for a specimen (having beaten the Dayak sellers down from an initial price of 140 guilders).[42] The

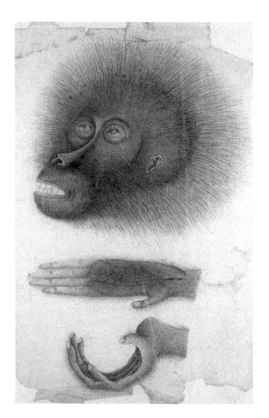

Drawing of an orangutan from the Raffles collection (British Library, Raffles Collection, NHD 49.31)

Hungarian entomologist Vincent Kollár, who collected in Borneo and Sumatra between 1840 and 1848, reported that between 100 and 150 live orangutans were brought to the coast each year, of which not more than twenty-five reached Singapore and not more than five reached Europe. All of them were infants, captured by killing the mother.[43] Although there is no mention of an orangutan, an account by Edward Belcher, captain of the HMS *Samarang*, which sailed through the archipelago in 1843–1846, gives some idea of the existence of animal and bird markets at the time:

> The Straits of Sunda being considered the "Gate of the East," the natives of the villages along this part of the coast of Java, find a ready sale for natural curiosities among the passengers of homeward-bound Indiamen. At Anjer [Anyer], especially, a fair of the most remarkable character is held under the shade of a magnificent Banyan tree, where, for a few dollars, may be purchased longarmed Apes, hideous Baboons, pigmy Musks, Java Finches, graceful Doves, pert Paroquets, satin Grackles, gentle Love-birds, and splendid Peacocks.[44]

In 1929, a German firm arranged the export of several orangutan families to Europe, the last such shipment before the colonial government banned the export of orangutans. The zoo in Düsseldorf purchased one family for 22,000 marks (equivalent to about US$5,000 at the time), and spent a further 2,500 marks converting a large aviary into new quarters for the apes. It was a worthwhile investment, because the zoo received 85,000 marks in income from tickets to see the animals before they died.[45] By this time, attitudes toward the wanton slaughter and callous treatment of wild animals had begun to change. Frank Buck made a name for himself in the 1930s with his book *Bring 'Em Back Alive,* in which he emphasized that he collected live animals for display to the interested public, not dead ones as trophies:

> I am proud of the fact [he later wrote] that in my whole career of dealing with wild creatures I have never willfully or unnecessarily harmed or injured a single one. I have made it my business to bring them back alive, for I have only feelings of kindness for every creature that breathes on this earth.[46]

Yet his account of the acquisition of an unusually large orangutan in Singapore ends, as so many such transactions must have ended, in the death of the animal while en route to America.

In fact, the orangutan was the victim in almost every encounter with human beings. During the nineteenth century, European colonial powers tightened their control of Borneo and Sumatra. The southern three-quarters of Borneo came firmly under Dutch control, while in the north the so-called white raja of Sarawak, James Brooke, created a semi-independent private kingdom from 1841 at the territorial expense of the Sultanate of Brunei. Rubber and tobacco plantations were carved from the jungle in both islands, bringing much larger European populations than ever before. For the first time, orangutans became the prey in hunting carried out only for pleasure. Brooke was the first to write of a hunting trip whose purpose was to locate and kill an orangutan for nothing other than the thrill of slaughter. The animal, he wrote,

> is justly named Satyrus from the ugly face and disgusting callosities. The adult male I killed was seated lazily on a tree, and when approached only took the trouble to interpose the trunk between us, peeping at me and dodging as I dodged. I hit him on the wrist and he was afterwards dispatched. . . . The following is an extract from my journal. . . . "Great was our triumph as we gazed on the huge animal dead at our feet, and proud we were of having shot the first

orang we had seen, and shot him in his native woods, in a Borneo forest, hitherto untrodden by European feet."[47]

Echoing Goldsmith's denigration of orangutans, Brooke went on to berate them for their "lazy and apathetic disposition," ignoring the implications of this judgment for his own prowess as a hunter.

Brooke's callous description of the killing of an orangutan who was "peeping" at him from behind a tree is the earliest of a series of nineteenth-century hunting narratives that are emotionally indifferent to the orangutan victims. This moral detachment is curiously present in those parts of the account of the naturalist Alfred Russel Wallace where he describes the killing of orangutans. Wallace arrived in Sarawak in 1854 and lived there for a little more than a year. He made a living from collecting natural history specimens, especially insects, for museums and private natural history collections in Europe, but he was also an energetic observer of nature. His account of his time in Sarawak includes a chapter in which he describes in meticulous detail the circumstances in which he shot each of sixteen orangutans for dispatch to Europe as skeletons or skins.[48]

The same moral detachment characterizes the Italian naturalist Odoardo Beccari, who lived for three years in the Indies (1865–1868) collecting a vast range of biological specimens. He spent a month seeking orangutan specimens and he describes a succession of dead orangutans, old and young, male and female, brought out of the jungle by local Chinese and Malays keen to collect the bounty he paid. Occasionally he gives a hint of the brutality of the capture: "Towards evening, a Chinese hunter brought me the first orang-utan, but it was so mauled and covered with parang [machete] cuts that I did not skin it." Apart from recording this gruesome procession of men with murdered orangutans, his account discusses at some length the difficulties of preserving the skins so that they were suitable for later museum display, and he records the complaints about the smell made by the Chinese household where he was carrying out his preparations.[49] His writing contains no hint of pity for the apes he killed, or even regret for the waste of those whose skins were too damaged to be worth preserving.

In many of these accounts, emotional detachment from orangutan victims is part of a broader device to emphasize the rugged, yet somehow nonchalant, spirit of the hunter. Hornaday in particular conjures up this tone:

> I would not have exchanged the pleasures of that day, when we had those seven orangs to dissect, for a box at the opera the whole season through. My good friend Helen feels the same way about opera.

It is a pity that men who "don't see how you can do it" could not have been there on that memorable occasion. When we finished, there was a small mountain of orang flesh, a long row of ghastly, grinning skeletons, and big, red-haired skins enough to have carpeted a good-sized room. I forgot to eat and, did not think of sleeping till after midnight. It was the most valuable day's work I ever did, for the specimens we preserved were worth, unmounted, not less than eight hundred dollars.[50]

Hornaday's novel *The Man Who Became a Savage* (1898) takes up the subject of orangutans as the quarry of a great white hunter, and in this book fame is acquired not by the man who discovers some new scientific facts but rather by the man who can kill the most. "We made the orang-utans look pretty sick," his character Gerry Rock comments laconically, "for I shot and skinned thirty of them, mostly big ones, too."[51]

Even in the eighteenth century, however, there were people who were interested in orangutans as more than mere objects for trade or sport. In the 1770s, the first signs of scientific interest in ecology and behavior began to emerge, as naturalists began to appreciate that the nature of an animal was intimately tied to the environment in which it was found. The otherwise unknown Mr. Relian of Batavia wrote presciently in 1770:

It would be a very informative and remarkable performance if one were able to observe these wild people [*wilde menschen*] in their forests without being seen by them, and so could be a witness of their household activities without them suspecting anything.[52]

Clarke Abel, whose baby orangutan had delighted London in 1817 (see chapter 3), offered an entertaining description of the young animal's behavior aboard ship, commenting that his account was useful because "his actions on board ship were less restrained, and therefore more natural, than since his arrival in England."[53] The next to express this curiosity was J. Grant, a British anatomist working in Bengal who, after dissecting a captive male and a female orangutan named Maharajah and Rannee, drew up in 1831 a thoughtful list of research questions, including whether male orangutans were naturally polygamous and whether orangutans took care of their aged relatives.[54]

In the same period, natural scientists became increasingly puzzled by the fact that the animals and plants of tropical Asia, Africa, and the Americas were strikingly different, despite living in apparently similar geographical and climatic conditions. The conviction that similar climates should have similar animals underlay the easy assumption in the seventeenth century that the orangutan and chimpanzee were the same creature. In 1777, the German Eberhard Zimmerman, who gave the ape the inspired scientific name of *Ourangus outangus,* published the first serious work in biogeography, complete with a map showing Borneo, Sumatra, and Central Africa as the habitat of *O. outangus.* Although we now know that Zimmerman's view of orangutan distribution was mistaken, his work signaled a shift in scientific thinking from examining the animal alone to examining it in situ. Even Linnaeus' *Systema Naturae* in its later editions began to add distribution information as an element in the diagnosis of species, though it obdurately refrained from including any information on behavior. The 1770s thus marked the beginning of a change in European research methodology. Part of the impulse for this intrusion was scientific curiosity. One of the clearest examples of this change was the expedition of Captain James Cook in the *Endeavour* from 1769 to 1771. Whereas previous European expeditions had systematically collected geographical knowledge, Cook's voyage was novel for the presence on board of a team of capable and eager scientists, including the botanist Joseph Banks, whose description of an orangutan in Batavia in 1770 appears unfortunately not to have survived.[55]

The first observers to take up the challenge posed by Relian and Grant were

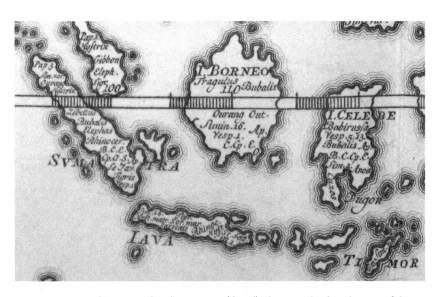

Zimmerman's biogeographical map, 1777 (detail), showing the distribution of the orangutan in Sumatra and Borneo (Zimmerman, *Specimen Zoologiae* [1777], insert)

The orangutan of Schlegel and Müller, 1841 (Schlegel and Müller, "Bijdragen tot de natuurlijke Historie van den Orang-oetan," following p. 55)

Captive orangutan, probably in Batavia, 1905 (Source: KITLV/Royal Netherlands Institute of Southeast Asian and Caribbean Studies, image 81259, used by permission)

the German naturalists Hermann Schlegel and Salomon Müller, sent to the Indies in the 1830s by the director of the Natural History Museum in Leiden. Their report is the first serious field study of the orangutan, providing details of habitat, diet, movement (and the lack of movement that later observers often found frustrating), growth rates, longevity, and reproduction, as well as commenting on the different moods of three captive orangutans and on the acuity of orangutan senses. The two researchers evaluated the evidence of the distribution of the orangutan and concluded that it was restricted to Borneo and a small part of northern Sumatra; they wondered whether there might be more than one species, but concluded that the wide range of forms they had observed simply reflected diversity within a single species. They also fought perhaps the last minor skirmish of the scientific war over the early travelers' tales when they used a footnote to refute a pair of geographers who had unwisely repeated Hamilton's story of orangutans kindling fires.[56]

Although Wallace had been an important perpetrator of the slaughter of orangutans, his natural history account *The Malay Archipelago* was also remarkable in expressing Wallace's willingness to admit affection for a great ape, especially in an age and culture when public expression of emotion by men was rare.

Wallace tells two versions of a story of a baby orangutan orphaned by his murder of her mother, an affectionate account in *The Malay Archipelago* and an ironically emotional one in his later autobiography, where he quotes, apparently verbatim, from one of his letters home to his mother and sister. In the letter, he teases his family by not revealing at the start that the baby is an orangutan: "I must

Wallace's baby orangutan
(Wallace, *The Malay
Archipelago*, 41)

now tell you of the addition to my household of an orphan baby, a curious little halfnigger baby, which I have nursed now more than a month." After describing his ministrations further, he continues teasing: "I am afraid you would call it an ugly baby, for it has dark brown skin and red hair, a very large mouth, but very pretty little hands and feet.... But I must now tell you how I came to take charge of it. Don't be alarmed; I was the cause of its mother's death." He describes how he shot the mother up a tree. "I have preserved her skin and skeleton, and am trying to bring up her only daughter and"—the teasing turns to irony as the reader realizes that this is not a human baby—"hope some day to introduce her to fashionable society at the Zoological Gardens." The account continues with a doting report of the behavior of the little orangutan and concludes with an effusive, but once again teasing, parody of fatherly love for the little creature: "I can safely say, what so many have said before with much less truth, 'There never was such a baby as my baby,' and I am sure nobody ever had such a dear little duck of a darling of a little brown hairy baby before."[57] *The Malay Archipelago* takes the story to its sad conclusion: the little creature died. "I much regretted the loss of my little pet, which I had at one time looked forward to bringing up to years of maturity, and taking home to England. For several months it had afforded me daily amusement by its curious ways and the inimitably ludicrous expression of its little countenance."[58] Half a century after Wallace, Hornaday had two encounters with infant orangutans that started in a similar way, with the killing of their mothers. In one case, the young animal had "the temper of a tiger" and continued to try to bite the hunter, even when it had been trussed. Later he remarked, "The little baby orangutan relieved me of all anxiety on its account by dying," so that he was able to skin and "skeletonize" it, as he put it, along with his other victims.[59] Later, however, Hornaday acquired another infant, whose placid disposition enabled the hunter to see in it the same endearing resemblances to a human child that Wallace had noted. He fed it on his lap, allowed it to sleep with him in bed and wonders, in his memoirs, how so ugly a baby could arouse such affection.[60]

Wallace and Hornaday were the first scientific writers to express fondness for orangutans, and their affection prefigured the much later emergence of a movement dedicated to the welfare of individual red apes. The possibilities opened up by emotional encounters between orangutans and humans, however, were already being explored. With science and religion both unable to provide absolute answers, it fell to literature to explore the possible meanings of the uncertain boundary between humans and apes.

5 ꣑ Imagining Orangutans
Fictions, Fantasies, Futures

During the eighteenth century, the orangutan began to appear in plays and novels, and later in short stories. Most of the early authors had not seen an orangutan, even in captivity, but the writings of travelers such as Beeckman and of scientists such as Tyson were so widely circulated, the illustrations of Bontius and Tulp so often redrawn, and public displays of orangutan-like creatures so common that a knowledge of the term "orangutan" and a sense of what it stood for was widespread within the literate elite of western Europe by the second half of the eighteenth century. The fiction writers who drew upon this public knowledge, with all its inconsistencies, were not primarily concerned with presenting an accurate portrait of the orangutan in terms of the scientific and philosophical discourses of the day. Rather, they saw in the ape's close resemblance to humans an opportunity to explore aspects of the human condition in a new way. They were also in a position to ask awkward questions that scientists often did not and to imagine possibilities beyond the protocols of scientific method.

Until the first half of the nineteenth century, scientists commonly used the term "orangutan" to refer to all great apes. Many fictional great apes are thus composites of imagery and information relating to gorillas and chimpanzees, and sometimes monkeys and other apes, as well as to actual orangutans. There was also scientific uncertainty about the distribution of great apes. It is to be expected, therefore, that fiction writers also use the term loosely and locate their animals in a wide variety of warm, exotic places: India, Southeast Asia, South America, Angola, the Cape, Africa in general, or the South Pacific. Neither Western science nor the Western intellectual world in general had any clear idea of ecology. When Westerners read about or saw pictures or exhibitions of "orangutans," the very nature of foreign exhibition itself meant that the creatures were interpreted within Western epistemological frameworks, having, in Renata Wasserman's formulation, little meaning beyond that conferred on them by the foreign contexts in which they were exhibited in whole or in part.[1] The fictional orangutan, then, is a creature composed less of the skulls, skins, and skeletal remains so prized by the anatomists than of legend, rumor, and political and philosophical debate.

Representation of apes and monkeys in Western writing had a long genealogy, stretching back at least to the Greek classics, and these simian stereotypes necessarily inflect eighteenth-century writing about orangutans. The "orangutans" of earlier fictional works frequently became a composite or selective reference point for subsequent depictions, so that themes developed by one writer found new (and sometimes not so new) expression in later literature. Changes in attitudes to human relations with other primates were thus reciprocally interwoven with changes in modes of writing, pictorial representation, and exhibition. Fictional accounts, moreover, in general retained a readership much longer than did scientific works. Once the distinct character of orangutans was scientifically settled in the 1830s, no one read Tyson or Camper for their science; the works of the scientists who superseded them were in turn eclipsed by later research. By contrast, some fictional works featuring orangutans continue to attract readers and generate insights and impressions today, carrying striking images and narratives from one century to the next.

Apes had traditionally been used in fable and fiction to tell Western societies something about themselves. In some cases simians write or speak in order to impart insights to the West; in others, their actions are used to illuminate generally negative features of Western societies. In this chapter we explore the orangutan commentator, and in the next chapter we look at instances of its proximate entanglements with humans. We have attempted as comprehensive a coverage as possible of orangutans in fiction (except that written for children), emphasizing both the repetition of tropes and the way these were changed, challenged, and inverted over more than two centuries.

The first fictional orangutan to write about human society was a hybrid, the unnamed offspring of a "négresse" and an orangutan, in a work titled *Tintinnabulum naturae* (The Bell of Nature, 1772) by the pseudonymous A. Ardra. The half-orangutan describes himself as a "metaphysician of the woods" and presents a set of reflections on the nature of the universe, proposing a proto-evolutionary model for the emergence of species, based on hybridization and adaptation to different climates.[2] This argument was still a controversial one, and putting it into the mouth of an orangutan-human hybrid was a useful device to express these disturbing ideas in public without attracting religious censure.

Nine years later, in 1781, the prolific French author Nicolas-Edmé Rétif de la Bretonne published a *Lettre d'un singe aux animaux de son espèce* (Letter of an Ape to Others of his Species),[3] purportedly written by "César of Malacca," the off-spring of a "baboon" and a local woman who is herself the product of a human-ape liaison. Abandoned by his mother, raised by a Dutch trader, and eventually passed to a kind mistress, César completed his education with the reading of Rousseau. His eighty-eight-page letter is an unremitting invective against mankind, written as a warning to apes "who have the misfortune to want to imitate humans" (20) and declaring that human society is marked by cruelty, despair, shame, spite, humiliation, and hatred (30). Life in the forest is immeasurably happier than life with people. Attacking religion, law, family structure, education, manners, fashion, hypocrisy, foreign policy, and social inequality, César berates the animal world for its enthusiasm in approaching such a dangerous species. White men will do to apes what they are doing to the blacks who have been enslaved in the French colonies of the Caribbean: "I have said that man is cruel, and singularly cruel to his own kind. . . . [and] to be convinced of this truth, we have only to see how he treats Negroes! It is a cruelty that passes the imagination and which he will exercise towards any of us" (53). The letter concludes with forty pages of notes, in which Rétif de la Bretonne both provides information and references to other works that illuminate César's text and elaborates on matters close to his heart, such as prostitution and pornography, a term Rétif de la Bretonne invented.

Half a century later, the anonymous author of "The Orang Outang" similarly employs the device of the captured simian—a full-blood orangutan—to critique humans, but this time the censure is far more specific, fully imagined as proceeding from an orangutan's particular point of view. The captive orangutan's letter, written from America, is addressed to a friend in Java, and her criticism of Boston society is developed from a comparative orangutan context. Her induction into human society is not simply taken for granted: She apologizes for her tardiness in writing to her friend by remarking on the silence to which "wonder and sorrow" have reduced her in the new land, and she complains of seasickness and the rigors of the journey. Moreover, her letter, although ostensibly directed to her friend in Java, is to be published in the *New-England Magazine* as a riposte to the slanders and calumnies to which she has been subjected by the public in general and scientists in particular, specifically "the pitiful speculations of some ignorant New York-Doctor upon my conformation and personal habits."[4]

Whereas, she notes, orangutans "partake sparingly of the fruits most liberally dispensed . . . by the influences of a delightful climate and the perpetual vegetation of the tropics," man by contrast "cannot make a single meal without some exhibition of his cannibal propensities," greedily demolishing "some other animal as good as he to satiate his raging appetite" (498). This, notes the orangutan, is particularly hypocritical when human philosophers continually recommend moderation but do not practice what they preach. Appetite and accumulation characterize human society. Humans pride themselves on their "gift of speech," forgetting "in their vainglory that the same gift is enjoyed by the parrot and cockatoo." But since their philosophers (contradictorily) trumpet silence as a virtue, "how superior in moral dignity is the Orang-Outang who practises this virtue on principle, and on all occasions, to the man or woman who would rather die outright than hold his or her tongue for half an hour" (498–499). Strangely, too, although man regards his noblest quality as the soul, he spends his life "engrossed by the care of the body" while his "soul" is "treated with so little ceremony that they have nothing, but what they call their own inward consciousness to vouch for its existence" (498), a circular testimony the orangutan scribe finds particularly unconvincing. Moreover, these "soul-and-body, want-multiplying, eternally talking people, have the impudence to call me ugly" (499).

Ardra's human-orangutan hybrid wrote as a philosopher; his half-orangutan identity played no role in his argument. Rétif de la Bretonne's briefly compared the idyllic life of apes in the jungle with the misery of proximity to people, but both he and the Boston orangutan wrote as observers of, rather than participants in, human society. In 1817 the English novelist Thomas Love Peacock, by contrast, created a silent orangutan who participated fully in English society. Sir Oran Haut-ton[5] is gentle, strong, educable, musical, contemplative, and genteel—excellent company, if lacking in conversation. *Melincourt,* the work in which he has become the de facto hero, has remained significant for its political satire and thinly disguised portraits of the Romantic poets and thinkers, as well as for its characterization of Sir Oran, still often described as the most engaging picture of an orangutan in all literature.

Melincourt is a satire, attacking rotten boroughs (parliamentary seats controlled by local landowners), rural poverty, upper-class wealth, marriages in which monetary motives outweigh the meeting of minds, and the introduction of paper money. The central plot of the novel is the making of a match between the chief character, Sylvan Forester, and Anthelia Melincourt, during which many obstacles emerge, some of which are overcome by the intervention of Sir Oran. This storyline provides a thin framework for a series of discussions between Forester and others, in which the taciturn Sir Oran plays no role but of which he is sometimes the subject. The approved opinions in the novel are those of the loquacious hero, Forester (a character based on the mature Percy Bysshe Shelley); his fiancée, Anthelia;

and Mr. Fax, a portrait of T. R. Malthus, celebrated for his powerful prognosis on the future of human populations. Sylvan Forester, a proto-conservationist, has Sir Oran as his constant companion. In *Melincourt,* through Forester, Anthelia, Fax, and the actions (rather than speech) of Sir Oran Haut-ton, Peacock appears to endorse many of Lord Monboddo's ideas, including the humanity of orangutans.

Peacock's portrait is an inspired literary device to castigate human cowardice, greed, folly, and inequality and to lampoon the electoral system of an apparently "civilized" British society; in short, Sir Oran Haut-ton, as "natural man," stands in marked contrast to urban sophistication and corruption. But he is much more than just a traditional novelistic device for satirizing human follies. A different kind of human—but still very much a human—Sir Oran Haut-ton is stronger, more reliable, direct, and honest in his dealings with the upper-class company into which Forester has introduced him than are most of the other characters in the novel. Although he has acquired a baronetcy, plays the flute, rescues the heroine, and is elected to Parliament from the borough of Onevote, he never speaks, a lack for which his mentor and companion Forester amply compensates. When Sir Oran objects to being carried in triumph by the citizens of Onevote after his election, he expresses his displeasure—in what will become a trope of orangutan behavior—by laying about him with a cudgel and thus creating a wholesale riot from which he has to be extricated by Forester. Usually a benign and mannerly presence, however, he sits down to dinner with Sylvan Forester and Sir Telegraph Paxarett:

> 'I think you will like this Madeira?' said Mr Forester. 'Capital!' said Sir Telegraph: 'Sir Oran, shall I have the pleasure of taking wine with you?' Sir Oran-Haut-Ton bowed gracefully to Sir Telegraph Paxarett, and the glasses were tossed off with the usual ceremonies. Sir Oran preserved an inflexible silence during the whole duration of dinner, but showed great proficiency in the dissection of game. . . . Sir Telegraph entered into a rapturous encomium of the heiress of Melincourt which was suddenly cut short by Sir Oran, who, having taken a glass too much, rose suddenly from the table, took a flying leap though the window, and went dancing along the woods like a harlequin.[6]

Sir Oran's gentle strength, apparent moral goodness, and loyalty to Forester make him the real hero of *Melincourt,* rescuing Anthelia from her attempted abductors and becoming instrumental in her liberation from the clutches of her would-be seducer at the end.

Sir Oran Haut-ton was not, as far as we know, based on Peacock's sighting of

Sir Oran Haut-ton and his flute, 1896 (Thomas Love Peacock, *Melincourt, or Sir Oran Haut-ton,* frontispiece)

or any anatomical knowledge about a live or dead orangutan or chimpanzee. As the extensive footnotes to the novel suggest, Sir Oran's origin is not "Angola" (as Forester claims), but lies instead in the works of Linnaeus and Buffon and in Rousseau's *Discours sur l'origine et les fondements de l'inégalité parmi les hommes* (1755). His more immediate and most important progenitor is, however, Lord Monboddo's account of the orangutan in *The Origin and Progress of Language* (1773–1792), a work Peacock footnotes extensively in *Melincourt* and from which his character Sylvan Forester often quotes. Forester also refers to a "learned mythologist" (51) who has been so impressed by Sir Oran that he refuses to call him by any name other than "Pan" because his flute playing and bibulous habits reflect "the natural mode of life of our friend Pan among the woods of Angola" (52). This reference also alludes to the long-standing identification of orangutans with satyrs, of whom Pan was the best known. Sir Oran's "attachment to the bottle" thus links him to the Greek classics more than to the discoveries of Tulp, Tyson, or any of the other contemporary anatomists.

In *Melincourt* the modish Sir Telegraph Paxarett remains skeptical of Sir

Oran's "humanity," especially since the orangutan never utters a single word. Sylvan Forester's riposte includes an extensive quotation drawing on Monboddo's *Antient Metaphysics* (1779–1799) and citing Bontius, Buffon, and Rousseau:

> He is a specimen of the natural and original man—the wild man of the woods. . . . Some presumptuous naturalists have refused his species the honours of humanity: but the most enlightened and illustrious philosophers agree in considering him in his true light as the natural and original man. One French philosopher, indeed has been guilty of an inaccuracy, in considering him as a degenerated man; degenerated he cannot be; as his prodigious physical strength, his uninterrupted health and his amiable simplicity of manners demonstrate (44–45). . . .
>
> As to Buffon, it is astonishing how that great naturalist could have placed him among the *singes* when the very words of his description give him all the characteristics of human nature. (50)

But Peacock was a complex satirist. In spite of his apparent endorsement of many of Monboddo's ideas about orangutans, there is also an element of satire directed *at* them. Sir Oran as "natural man" is certainly used to expose civilized folly, but Sir Oran is himself gently lampooned, particularly in the comic episodes for which he is so fondly remembered. Peacock is thus able to have it both ways: ridiculing a corrupt democratic system by demonstrating that a speechless ape could be elected to the English Parliament and in so doing drawing a definitive line between humans and orangutans (the satire depends on orangutans being inferior to humans in order to demonstrate the folly and hypocrisy of civilized "man"), while at the same time depicting Sir Oran as a convincing equal (and better) of the novel's other (human) characters. Peacock was certainly ambivalent about Monboddo's claims for the humanity of orangutans, but Sir Oran's demeanor and actions in the novel do not suggest that Monboddo or his ideas are the primary satirical target. Sir Oran Haut-ton is thus never an animal simply standing in for a human being (or beings), since he himself is indeed considered human. Whereas Monboddo's claim that orangutans were human—albeit a different kind of human—was not typical of the thinking even of his own times, it did express a possibility that had to be taken seriously, or, by the time Peacock was writing *Melincourt,* at least half seriously.[7] With the disappearance of the intellectual framework that enabled Monboddo and Rousseau to see orangutans as human, so too disappeared the representation of orangutans in fiction as human. They could continue to be ambiguous, liminal creatures, be transparently anthropomorphized, be described as "humanlike" or "almost human," be even better in their behavior and mode of life than humans, but they could never again *be* human.

In the long tradition of Western literature, animals have usually been represented or read as stand-ins for humans or human traits, foibles, and virtues. Speech and conscious thought are precisely the stuff of fiction, and since animals were (and in many ways still are) believed to lack both, most attempts to credit animals with such qualities are either intended as allegorical or, given our traditions of reading, interpreted allegorically. The only exceptions are to be found in works for children (though here the animals are usually anthropomorphized) and in biographical or autobiographical writings that deal with encounters between a particular species or a single animal and a human or humans. Because nonhuman primates share with humans so many anatomical and behavioral traits, the possibility exists that they might participate in human society in ways other animals cannot, even if they remain distinct from it. Sir Oran Haut-ton is a particularly ambitious creation in this respect, because he is accepted in nineteenth-century English society even though he looks unambiguously like an ape and occasionally breaks into distinctly apelike behavior. But the line between humans and animals, increasingly firmly drawn since Monboddo, has concentrated on animals' lack of, or deficiency in, a complex of supposedly human characteristics including speech, consciousness, self-consciousness, and reason.[8]

Frank Challice Constable's *The Curse of Intellect* (1895) is also concerned with critiquing human society and also features a human and his constant orangutan companion, but the relationship is very different from that between Sylvan Forester and Sir Oran Haut-ton. As one might expect from the title, the worth of reason (or intellect) as opposed to instinct is the major subject of the novel, and while *The Curse of Intellect* does contain passages of savage social satire, these are incidental to a much darker study of the nature and meaning of the relationship between Reuben Power and his "Beast." The novel is ostensibly penned by two authors, the human narrator, Colin Clout, an acquaintance of Power's, and by the "orangutan" who refers to himself (as the Europeans he encounters do) as "Power's Beast." The first section of the novel is by Clout, the second by Power's Beast, and the third and final section again by Clout. Power's Beast comes from Borneo and has red hair, but the invitations he and Power extend to what become their increasingly fashionable entertainments are signed by Power and "*Semnopithecus Rubicundus,*" that is, not an orangutan.[9] Like a number of the works discussed earlier, the novel addresses the difference(s) between human (self-)consciousness and reason and unself-conscious, instinctual being-in-the-world; but rather than being primarily satirical or comic, the tone throughout is tragic and melancholic, and there are a number of other interesting differences between this powerful novel and the earlier satires.

Reuben Power disappears from European society for twenty-five years. He has left for Borneo to capture and train a simian in order "to know what a beast like that might think of us."[10] Before this goal can be accomplished, of course, "the beast" must both be taught to speak and, even more significantly, acquire self-consciousness, undergoing a long and painful process of separation from his former life and world. The Beast's only recollection is of "a man and a whip" and a tragic loss of "a life above thought . . . instinctive happiness from reasonable life and the unaffected intercourse of living creatures" (95–96). But it is out of this unself-conscious mode of being that Power must bring him if he is to be the "unbiased" commentator on human society for which the misanthropic Power hopes. Thus, in his own account in the novel, included between Clout's two sections (and published after the deaths of both Reuben Power and the Beast), the Beast writes of a "long sad time of some horrible striving—a striving of something in me, yet foreign to my instinct, to conquer nature" (96). Power had initially captured two simians, and the mark of his success with this particular animal is, significantly, the Beast's murder of his companion, who had "not yet learned to think. Instinct was stronger in him than in me" (97). With the loss of instinct comes not just murder, but memory and imagination, so that even though the Beast comes to realize he is stronger than his erstwhile master and could thus escape the whip, reflection and imagination, which he has now acquired, persuade him otherwise. That "shadow," he notes, had become "stronger than fear of the whip I knew" (99). Articulate speech is mastered following, not preceding, reason, and his reading develops. Power himself is unremittingly cynical about the "glories of man" but initially the Beast is enchanted by the riches that an apparently reasoned, intellectual existence promises over his former natural state. Yet, like Power, he becomes increasingly disillusioned with human society and by what self-consciousness, intellect, and imagination bring with them: not only the horrors of memory and a loss of innocence, but dread of the future. "What has man gained by reason?" the Beast also asks. Nothing but the "[m]isery of the many [and] mean material happiness of the few" (137).

On their return to Europe, the impression gained by Clout and others is that Reuben Power is increasingly subjugated to the Beast rather than the expected reverse. Eventually both retire from society to a country house surrounded by a gloomy park, which contains a room resembling "a grassy break in a thick forest, a tiny glade shut in by trees" (162). The Beast is by now referred to as Power's "alter ego" and comes to feel that Power's (and humanity's) soul has gone into him, while

Power feels that his soul has gone into the Beast. Eventually the Beast releases the unhappy Power by strangling him and then immediately shoots himself. Clout arranges that they be buried together, commenting, "I feel pity for the poor Beast. . . . His life must have been very sad" (177). Humans, Clout reflects, suffer great disillusionment as they grow up, discovering that a world of potentially "noble and unselfish men" is but "a stinking slough of selfish, dirt-bespattered, dirt-bespattering creatures." How much more painful, Clout muses, "for the poor Beast, with power of reflection suddenly born in him . . . to be confronted suddenly with the human beast he is! . . . May he rest in peace!" (177).

Much of the Beast's critique of European society—when Reuben Power brings him home after the twenty-five years of "training" in Borneo—sounds like earlier simian observations. But no earlier orangutan participating in human society was represented as having to give up so much. Rétif de la Bretonne's hybrid and the Boston orangutan were removed from an idyllic forest existence to observe human society at close quarters, but they themselves were not morally compromised; Sir Oran Haut-ton positively reveled in human society. In *The Curse of Intellect,* the distance between human and orangutan mentality is a crucial factor. The futility of the shift that the Beast has made draws the reader to his actions, his loss, the horror of his past, and the sufferings of his present state. Although "the Beast" has mythical and allegorical potential as well as his obvious satirical role, the reader sees his portrait less as an elaboration of "the monkey as critic" motif than as a potential questioning of the terms of the genre itself. To produce a genuine simian critic of human society takes twenty-five years, costs the Beast his unself-conscious mode of being and his life, and spells tragic consequences for all involved.

In the post-Darwinian late nineteenth and early twentieth centuries, ideas of degeneration and in particular a degenerative atavism increasingly haunted English and European writing. That "civilization" might simply be a veneer over a savage inner heart was expressed in a number of ways, for instance in Robert Louis Stevenson's *Dr Jekyll and Mr Hyde* (1886) and most famously in Joseph Conrad's *Heart of Darkness* (1901). But although Reuben Power seems to be to some degree in the power of the Beast, each, in becoming the other's "alter ego," changes the more familiar pattern. It is so-called human civilization and reason that is ultimately damned. The Beast points out that when humans seem to perform unselfish acts (such as a mother sacrificing herself to save her baby), it is a matter of instinct overcoming intellect rather than the reverse. But neither is the instinctive life of "natural man/beast" romanticized. The (significantly named) Shelley Chullaby, a poet, imagines what the Beast's former life would have been like, and his rhapsodic version of "natural life" in the forest is reminiscent of eighteenth-century accounts by disciples of Rousseau and Monboddo. Extravagantly evoked, his vision stands in marked contrast to the Beast's own version of his separation, through the acquisition of a human intellect, from a form of being characterized by its innate continuities with the environment and other jungle inhabitants. This balanced way of life is not in itself idealized, but with its loss comes recognizably human fear, unhappiness, meanness, mendacity, a horror of the past, and unease about the future.

Less a demonic than a genuinely tragic figure, Power's Beast cannot so easily be read as an allegorical equivalent of the atavistic heart of man. Rather he seems to stand as an individual in his own right, challenging the value of precisely those qualities that exile him and his fellows to the beastly side of the human-animal divide. And ironically, neither can he be seen merely as that traditional device, the monkey as critic of human society, the role for which Power initially trained him. Whether or not the Beast is a true orangutan, what remains important is the complexity of his character, the pity the reader is induced to feel for him (and other captured apes), and the strategies by which the author resists his allegorization at a time when racism inclined audiences to read through animal figures to "savages" and the zeitgeist encouraged a psychological interpretation of the relationship between the human protagonist and his animality as a drama of man's inner "savagery" struggling with his "civilized" exterior. While *The Curse of Intellect* has been interpreted in the latter way, it is difficult, given the Beast's writing of his own history and the degree to which his character, training, and independent actions in Europe are elaborated, to consider him as a "mere metaphor" for a human "heart of darkness."[11] Indeed, it is possible to read *The Curse of Intellect* as the reverse of Stevenson's and Conrad's ideological trajectories. Amalgamating the "monkey-critic" tradition with fin de siècle fears of a post-Darwinian atavism and critiquing both, Constable champions instinct over intellect, the so-called savage over contemporary civilization.[12]

In the early twentieth century, Marcel Roland produced a novel that narrowed the daunting gap Constable had identified by creating orangutans who had evolved to become much closer to humans. His science fiction work *Le Presqu'homme* (Almost a Man, 1905), set in twenty-fourth-century Paris, is part of a trilogy about the rise of apedom to supplant humanity. A scientist and traveler, Murlich, has raised a young orangutan and brought him to Europe at the age of thirteen. Gulluliou speaks a "pongo" language that Murlich has been able to master and in which he and the ape communicate. In Paris the orangutan also acquires a limited French vocabulary, and Murlich can thus introduce his "almost man" to the scientific community at a public lecture, but members of the French scientific academy remain skeptical of Murlich's claim that Gulluliou's manners, dress, and even speech actually constitute "humanity." Peculiarly, given that the society has allegedly become thoroughly mechanistic and materialistic, what is demanded of Gulluliou by the academy to prove his "near-humanity" is that he demonstrate capacity for emotion/feeling and memory. He must thus pass what Murlich terms a "cruel" test: he will be shown a photograph of his mother's murder at the hands of the hunters who captured him in the "traditional" way of shooting the adult—otherwise too hard to deal with in her defense of her child—and kidnapping the infant. Gulluliou does not fail this test. In one of many evocative episodes the orangutan recognizes his dead mother and sheds tears, convincing at least some science academy members of his essential (almost) "humanity."[13]

Roland's novella is the first of a trilogy about "future times." The second and third books in the trilogy, *Le Déluge future* (1910) and *La Conquête d'Anthar* (1913), lack the psychological subtlety of the first volume. War breaks out between humans and pongos, and the whole world is flooded. Both sides fight for control of the island of Anthar, which is the last piece of land left above the waves. The story swings through kidnapping, rescue, betrayal, drunkenness, and murder before a mad scientist destroys the island.[14] The sometimes contradictory directions of *Le Presqu'homme* are thus perhaps partially due to its placement in the trilogy. Gulluliou is highly intelligent, has his own language and the potential to learn another, demonstrates human sensibility, and, in the scene where he is taken to visit the twenty-fourth-century parliament, enacts the traditional simian satirical role when the members of the assembly descend to fisticuffs and the educated red ape is tempted to join the fracas, his sublimated instincts briefly catalyzed by the assembly members' "beastly" behavior. In the end, the increasingly pitiful Gulluliou dies a lengthy, "Dickensian" death from tuberculosis, his (human) friends at his bedside after he looks out to the Parisian woods one last time.

Pierre Boulle's science fiction classic *Monkey Planet* (*La planète des singes,* 1963) also recruits accelerated evolution to imagine a reversal of dominance between human and nonhuman apes. The book, like *Melincourt,* not only remains in print in the twenty-first century, but it has been influential in its representations of orangutans, particularly through the iconic films based on Boulle's novel. The first of this film series, *Planet of the Apes* (1968), used human actors in ape costumes to depict the three species who control the "Monkey Planet": chimpanzees, orangutans, and gorillas. Both the novel and the films have been pivotal in the development and continuing popularity of science fiction, particularly in the United States.

On the planet on which Ulysse Mérou and his fellow voyagers land, the dominant structure of twentieth-century Earth has been reversed. Here the apes are the ones who are civilized and socially organized, while the humans are wild, degenerate "animals," deprived of any (human) dignity. The three groups of great

Cover of Roland's
Le Presqu'homme (1908)

apes—chimpanzees, orangutans, and gorillas—wear clothes and between them run the planet. Naked humans roam wild in the forest, lacking any accoutrements of civilization or, in the view of the apes (and Mérou and his companions), any signs of an active intelligence. Their instincts (especially in relation to impending danger) are keen; they can make and use "primitive" tools, and have a limited degree of sociality. In short, they possess those attributes humans generally ascribe to the great apes, while those capacities accorded "humans" on earth are instead possessed, to different degrees, by the three "ape" species. Gorillas, the most militaristic and least intelligent, hunt humans for sport, but also to capture subjects for experiments—behavioral, anatomical, physiological, social, and psychological—which are carried out by the most intelligent group, the chimpanzees, under the management of the more conservative and less innovative orangutans. Humans are subjected to various intelligence tests, almost always about problem solving in relation to the acquisition of food, experiments not dissimilar to Wolfgang Köhler's on chimpanzees and orangutans early in the twentieth century, a telling critique of which is included in Nobel Prize winner J. M. Coetzee's groundbreaking novel *The Lives of Animals* (1999). Experimentation does not stop there in *Monkey Planet*, either. As with Köhler and other behavioral scientists, the chimpanzees and orang-utans of Boulle's fictional world are interested in the sexual habits of their human captives as well as being involved in vivisecting them to learn more about these apparently "wild" ancestors, and thus about their own more evolutionarily "primitive" selves.

Mérou is placed in a cage with the beautiful but apparently vacuous Monkey Planet human Nova, but eventually escapes with the help of two of the chimpanzee scientists, though not before demonstrating intelligence, knowledge, and linguistic skills well surpassing those of the planet's "wild" humans. Nevertheless, it is a long struggle for Mérou to have his capacities recognized and acknowledged. What is genuinely intelligent initiative on his part is persistently interpreted as merely a superior capacity for imitation, and rebellion against his incarceration is taken as further proof of savagery and lack of civilization, while its alternative, apparent acceptance, is read as mindless apathy. The unearthing of ruins of an earlier civilization by one of the two sympathetic chimpanzee scientists suggests that humans may once have ruled their planet, and this discovery of the potential rise and fall of various civilizations alarms the orangutan leaders. They thus become especially fearful of the half-monkey-planet, half-earth-human child born to Nova and Ulysse. Though all three are eventually enabled to escape, they find, on their return to Earth, apparently centuries later, that it too is now dominated by nonhuman apes. Horrified to be greeted on the airfield by a vehicle driven by a gorilla, Ulysse, Nova, and the child re-embark for outer space.

This reversal of dominance in Boulle's novel, its replication of the ways we have treated the nonhuman apes, together with its exploration of the modes by which such dominant attitudes and practices, once naturalized, necessarily become self-fulfilling, self-sustaining, and self-serving, constitutes a stringent critique of our current treatment of other apes (and perhaps "animals") generally. But interpretations of Boulle's work have tended to occlude this more obvious reading, particularly since the 1968 film version of the work is better known than the original novel. Among the most popular films ever made, *Planet of the Apes* and its successors, produced in the United States during the racial tensions of the 1970s and 1980s, ensured that audiences would read a potentially interspecies text in racial terms. The 1970 advertisement for the second film of the series (*Beneath the Planet of the Apes*) asked the question: "Can a planet long endure half human and half ape?" These advertisements, as Eric Greene notes, posed this as

the 'ultimate question' and pronounced the answer 'terrifying'. Filled with historical resonance, the question echoed Abraham Lincoln's prediction made over a century earlier that 'this government cannot endure permanently half-slave and half-free'. . . . The issues of racial conflict and racial oppression at the heart of the crisis which Lincoln addressed are also the central issues of the five films that comprise the *Planet of the Apes* film series.[15]

As Greene points out, "as the series progressed the racial conflict theme increasingly took center stage." In the first two, the racial implications were clearly there, "but not always highlighted." The following three films, however, directly engaged with particular racial incidents and themes.[16] Both *Le Presqu'homme* and *Monkey Planet* are set in a future where evolution has failed to reinforce human superiority to apes. In contrast to the earlier stories, however, the orangutans are not commentators on this process; rather it is their success that shows up human failure.

The orangutan-as-commentator returns, however, in Jenny Diski's novel *Monkey's Uncle* (1994). *Monkey's Uncle* offers us a "looking-glass" entry into another world, and quotations from Lewis Carroll's *Alice in Wonderland* preface each chapter. Diski's chief character, Charlotte FitzRoy, meets Jenny, her "phantasmagoric" orangutan guide, as she falls through her own psychological "rabbit hole" into a state of sane madness. Charlotte has suffered a nervous breakdown occasioned by a combination of circumstances: her disappointment at the collapse of communism and thus of its ideals; the death of her daughter in a car accident; and her increasing fear of having inherited the pathological depression of her assumed ancestor Robert FitzRoy, the captain on Darwin's *Beagle* voyage, who eventually committed suicide.

Three aspects of Charlotte's life related to these crises are structurally interwoven: her everyday life before, during, and after her breakdown; her visions of Robert FitzRoy as he loses his struggle with depression in circumstances where his faith and intellect are at war over the question of evolution; and her participation in an imagined other world where Jenny the orangutan is her friend and guide. Just as Charlotte's sense of actually being in the presence of her ancestor, FitzRoy, has been precipitated by her reading of his biography, so her introduction to and imagining of Jenny are catalyzed by her exchange of looks with Suka, an orangutan in the London Zoo. Although Charlotte knows "anthropomorphism to be a cardinal sin,"[17] Suka's long red hair, hanging from her shoulders and arms and hair pushed back from her face, make her, in Charlotte's eyes

one of those women who live on the street, bundled into all the clothes they possess, topped by an ancient fur coat grabbed at a jumble sale, and worn winter and summer, a protective outer layer, not just for the cold but to keep the world at a distance. Charlotte tried to lose the image, remembering about anthropomorphism, but the picture was reinforced by the orangutan's eyes. It was not just their peculiar humanity. (9)

What Charlotte also perceives is Suka's air of "*resigned suffering,* a package, rather than two separate qualities. Quiet desperation. Hopelessness" (9). Standing outside the cage, Charlotte encounters an elderly couple who, being long-time devotees of the zoo's orangutans, have been accorded the privilege of naming Suka, a Bornean word meaning "delightful," so they tell Charlotte.[18] Considering Suka and remembering the chimpanzee tea parties at the zoo to which she had long ago taken her children, Charlotte slips through reverie into a world of her imagination where the first "person" she encounters is Jenny, an orangutan dressed for tea in a suitable floral frock and a hat decorated with fruit. Jenny's first words are "Jenny, actually," a deliberate correction to what she regards as the misnomer of "Suka." The orangutan regards herself as the current representative of a line of red apes who since the 1830s have been captured and transported to the London Zoo. Since the "first" 1839 captive was called "Jenny," all subsequent orangutans have adopted the name. However, as Jenny explains to Charlotte, that 1839 Jenny was not really the first. She was preceded by another, "Jenny Zero," who arrived in 1830 but was, alas, preserved in spirits after her death on the journey, a death caused by jumping overboard in horror at seeing the ship's butcher hacking at chunks of bloodied meat.[19]

Relaxing in the sea with her "phantasmagoric Jenny," whom Charlotte has persuaded to enjoy the water (and float contemplatively on it), Charlotte wonders what it would be like never to have to think, meaning, of course, like other animals. Jenny is quick to retort, however, that there is no such creature: "You human beings always think that thinking gives you the edge over the natural world. But I have to tell you that there's no species alive in the world which doesn't believe the same thing about themselves" (127).

Although much of the conversation between Charlotte and Jenny (and between the "three men in a boat," Darwin, Marx, and Freud) is comically rendered, all discussion, including that of the "looking glass" world of madness, has a serious underlay. Charlotte's questions, and the specific causes of her breakdown, are at base about the nature and purpose of life, genes and inheritance, ancestors and descendants. But subjected to cross-species scrutiny, these issues are necessarily relativized. As Jenny points out, to a sloth it is the rest of the world that is moving too rapidly, and descent defined by bloodline may be irrelevant. What matters to this fictional Jenny are her Jenny "ancestors," the previous orangutan captives in the London Zoo. And in contrast to Suka's apparent air of suffering, Jenny tells Charlotte that though she and the first Jenny may miss the forest, not having to find one's own food is a very acceptable benefit of zoo life.

In the end, Charlotte is restored to what the human world regards as sanity or normalcy, but although this signals the disappearance of the no-longer-needed phantasm of Jenny, Charlotte does not completely sever the orangutan connection. Instead of remaining in her former home, she takes up residence with the elderly couple she first met at Suka's cage, becoming caretaker of their boarding house when they go to visit Suka, who has since been transferred with her baby to a French zoo. In short, Charlotte abandons her "ancestral" obsessions and conventional family connections to forge new ones—however tentative and tenuous—on the basis of a shared interest in orangutans rather than on bloodlines and chromosomes. Jenny/Suka is a didactic orangutan like César of Malacca two centuries earlier, but her didacticism is of a very different kind: it is ostensibly addressed to a single individual rather than to a broad audience, and the information she imparts is primarily about orangutans, genteel in captivity, not about humans. Jenny's understanding of orangutans, however, becomes a model for a human, a means by which Charlotte can escape the presumed curse of her ancestry and construct a future based on choice.

The orangutan Librarian in Terry Pratchett's *Discworld* series is also an instructor of humans. Accidentally transformed from a human into an orangutan by a misdirected magic spell, the Librarian soon decides that his new form is better suited to his profession than the old one. Not only is he better able to climb to the tops of shelves to retrieve and replace books, but his new strength gives him greater authority in chastising borrowers who damage books or return them late. Although his vocabulary mainly consists of the words "Ook" and "Eek," he is a master of expression and context. In one of the novels he compiles a lexicon of primate language, which runs about eight hundred pages and explains only these two words. Like the orangutan who was displayed in London's Exeter 'Change in 1817, the Librarian is portrayed as highly conscious of his superior status within the animal world and is particularly annoyed to be described as a "monkey."[20]

The figure of the orangutan as guide and teacher of humans (rather than as merely their critical foil) is still more pronounced in Dale Smith's *What the Orangutan Told Alice* (2004), a novel unashamedly didactic in purpose and tone. It shares with Peacock's *Melincourt* lengthy passages designed to teach readers about orangutans and in this case about their current plight, a plight due entirely to human presence and human practices. Peacock's Sir Oran Haut-ton was a nonspeaking example of "natural man," whose presence in the novel implicitly critiqued human urban sophistication, and in case audiences ignored the implications of his behavior, Sylvan Forester was there to point to the comparison directly. Humans and human societies are also stringently criticized in *What the Orangutan Told Alice,* but the crucial difference is that in the 2004 novel, humans are held responsible *for what they have done to orangutans, other animal species, and the environment;* not for, as in the early nineteenth-century novel, their own social inconsistencies and abuses.

The teenage protagonists, Alice and Shane, are visitors to Indonesian Borneo (Kalimantan). Shane is an American exchange student, while Alice has come with her father, who is writing a novel about orangutans. After the two pass into the forest in the company of a male and female gibbon—their passage not so much through as into a "looking glass" that will be held up to them as humans—they meet Anne, a scientist working in orangutan rehabilitation. Alice explains that her father is doing research into orangutans to write "a kind of fantasy thing," a novel:

> 'Fantasy? You mean fiction?' Anne's eyebrows arched above her glasses. 'Who needs fiction when you've got science?' she said rhetorically.
> 'He calls his books "environmental fiction"' explained Alice. 'He thinks novels are a good way to teach kids about endangered species, you know, like orangutans'.
> 'Well, who knows, he could be right' Anne conceded.[21]

Anne is one of the young adults' two major "teachers" in the novel, and they are receptive listeners. Drawn into the forest by their concern for the gibbons, the apt pupils are led toward their second wise teacher, "the Old Man," by a younger orangutan. It is eventually the Old Man orangutan who shows the children the destruction wrought by humans, while during their journey they learn a great deal about orangutan habits and ways of life. Though the animals' capacity for speech communication with humans is thus "fantastic," the novel is unremittingly realist in its assessment of species and environmental prospects for an earth dominated by humans and their greed. Marco the orangutan points out to the children the far greater wisdom of the forest dwellers. "Can you name," asks Marco, "another species on this planet who spoils his own home, his own habitat as humans do?" (138). In the forest also, various plants and animals coexist. Enjoining them to look around, their guide points out that there are no landfills, tire dumps, aluminum cans, Styrofoam cups, or plastic garbage bags. Moreover, "There is no dark cloud of pollution hanging over our heads here. You don't see any animals and plants starving while others are well fed. Here in my world we do not poison our water supplies or make war on each other" (138). The orangutan also shows Shane and Alice the widespread deforestation humans are now causing. Heavy vehicles belching "thunderous grey clouds" hauled the "mutilated corpses of trees along rutted access roads," while

> men with red plastic containers dashed the remaining slash piles of limbs and stumps with gasoline. Then there were orange flames and bellowing, energetic masses of rolling smoke. The flames leaped higher and higher and the leaves of trees still standing at the edge of the clear-cut curled against the heat. Everywhere, there was the sound of panic as birds and animals tried to escape. But there was no shelter for the animals . . . nowhere for the birds to roost. (140)

This destruction is not the end but the beginning of further degradation, caused, as both Anne and the Old Man point out, by human greed, a greed that at its most fundamental level is for self-reproduction. What Shane and Alice witness, the old orangutan tells them, is "against the laws of nature, but not the laws of men. Man only imagines he has morals, but I could put all the morals man has into my pocket" (140). "As long as human beings keep reproducing at their present rate, it's not going to stop."

> After the timber is cleared and the slash is burned off, they'll plant rows and rows of oil palms, or rubber trees, or terrace the land for rice paddies. Of course all the nutrients in the soil will be depleted after two or three years, and after that the crops will fail and factories and houses will spring up in their places. Towns and villages will grow together to become part of a big bustling city. (141)

Anne agrees, and in this novel, neither Westerners nor Southeast Asians are spared blame. "The problem," she says, "is that there are too many people and they are having way too many babies" (74).

Finding humans to have a total lack of respect for the planet, in contrast to other species with whom it should be shared, the Old Man has thought long and hard about human theories of evolution. Since man is emphatically not a part of the natural world, and abuses it so flagrantly, from where did he come? The wise orangutan concludes that humans are the only species never to have evolved and this is because—the irony of our constant search for extraterrestrials does not escape him—we are in fact aliens ourselves. This, he proposes, is why Cro-Magnon man appears so suddenly (after such a lengthy period of Neanderthal stagnation) and with so many tools; there is in fact no "missing link" between "man" and apes. Indeed, the Old Man is as shocked at the thought of his being genetically related to such a species as were the Victorians at the suggestion they had evolved from apes.

Having passed through a "looking glass" from Indonesian village to forest, then, the two teenagers (unlike Lewis Carroll's Alice) have not really passed into a fantasy world inhabited by strange creatures so much as into a reality check on their own impact on the earth, and in particular on the orangutans and the Borneo rain forest. In spite of its underlying basis in fantasy, the novel employs a quasi-realist mode and includes a great deal of current information on orangutans: the places in which they are most likely to be found, their favorite foods, individual and social behaviors, their dwindling numbers, and the reasons for their accelerating loss of habitat. But it also focuses on the treatment of orangutans by humans, the species responsible for their current condition in what is left of their "wild" and in the variously mean forms of their captivity in Indonesia and other countries. *Melincourt*'s reviewers criticized Peacock's didacticism and "speechifying," but Smith's novel returns us to didacticism and even to a degree to Monboddo, who thought his version of the orangutan not only part of but even superior to humankind. When Peacock adopted some of Monboddo's ideas, he did so playfully, only half seriously. Although this novel of the twenty-first century certainly includes lighthearted moments, its attitude toward orangutans (that they are better than humans) is presented just as seriously, if not more so, than in Monboddo's writing. But even if better than humans, orangutans in *What the Orangutan Told Alice* are not in fact humans. And this is where this contemporary work parts company from Monboddo. While Smith's novel obviously employs some anthropomorphic apparatus—the speaking, teaching monkey—it deliberately challenges anthropocentrism and potential dismissals of its own anthropomorphic apparatus by combining its fantasy frame with hard, realist data.

From Ardra's unnamed "metaphysician of the forest" to Smith's "Old Man," fictional orangutans have been represented as taking on human characteristics as a means of articulating the shortcomings of human society. The only major exception, Sir Oran Haut-ton, was able to be both human and orangutan because the scientific and philosophical beliefs of his day allowed him to have characteristics, as an orangutan, which in other eras would come to be regarded as exclusively human. Fiction was neither bound nor driven by changes in science, but the scientific shift from speculating over the possible humanity of orangutans to conviction that they were only one animal species among many was to a degree paralleled in the literary world: voluble, or at least expressive, in the eighteenth and early nineteenth centuries, orangutans lost the capacity to meet humans on equivalent terms except in fantasy worlds where they had evolved well beyond the red apes known to science. In the twentieth century, orangutans could once again take up a conversation with humans, but this time they talked of themselves, lending a specifically orangutan cast to the critique of human society.

6 ✿ Close Encounters and Dangerous Liaisons

The fictional orangutans discussed in chapter 5 offer commentary on Western society, either by directly addressing a human audience or by behaving in ways that highlight human shortcomings. Most of these apes have been removed from their original environment, and even Boulle's advanced apes live in a world that is functionally similar to the human world of the twentieth century rather than the jungle world of their ancestors. Only the Old Man in *What the Orangutan Told Alice* is still in his natural surroundings, but his message to humans relates specifically to what they are doing to that environment in the conduct of their own selfish affairs. Other orangutans in fictional works either remain in their original habitats or bring their jungle wildness into contact with human society, often with catastrophic consequences for both. These orangutans are presented as clever creatures, but for the most part not as near-humans. They may show resourcefulness or glimmerings of human emotion, but there is never a doubt as to their animality. Their primary literary purpose is not to offer comment on the ways of humans or to educate humans about orangutans but rather to dramatize an encounter between Western society and animality or savagery. In these encounters, the orangutan may be an ally, an antagonist, or a liminal figure. Whether the contact is benevolent, savage and dangerous, or implicitly or explicitly sexual, there is a juxtaposition of beings—orangutan and human—who are assumed not to belong naturally together. In some of these instances, the orangutan can be read as the necessarily repressed inner "animal" soul of "civilized" man, the atavistic savagery that must be controlled in order to constitute the essentially human as the very opposite of the animal, the category to which other apes have been definitively exiled.

The most famous benevolent orangutan in late nineteenth-century literature is Joop, of Jules Verne's *The Mysterious Island* (1875). Joop plays a relatively small role in the novel, but one that, as critics have pointed out,[1] is pivotal in displacing the position of Cyrus Smith's former slave (Neb), whose continuing loyalty and servitude might otherwise constitute a political embarrassment. A group of Americans become castaways on the "mysterious" island when the balloon in which they are traveling is brought down in a storm. As Union loyalists, most of them are escapees from incarceration in the South during the American Civil War. They might thus be expected to be pro-emancipation and antislavery; while this does appear to be the case, their strongest motive for fighting for the North is allegiance to the concept of a single country. Neb, a black slave, has formerly been freed by Cyrus Smith's family, but his loyalty to Cyrus remains so strong that, by choice, he continues to accompany him everywhere in the capacity of servant. Neb's persistence in this role risks a potentially awkward compromise of Union principles, but happily the mysterious island on which the party lands is inhabited by a motley collection of wild and domesticated animals, including orangutans. One of the latter, subsequently christened Jupiter (abbreviated as Jup or Joop by different translators[2]), proves readily educable as a servant, able to assist the men under Neb's directions. Joop's easy accommodation to this role would appear to endorse European opinion—advanced since at least the eighteenth century—that orangutans could be made into useful servants. After his initial shock at capture, Joop becomes a tractable and gentle member of the party, conveniently inferior in rank to Neb and usefully strong when defensive measures are required. While the orangutan's presence thus ameliorates any awkwardness in relation to slavery and servitude, the race/species interchange persists in Smith's reassurance of Neb that he will always be his master's "favorite," in the close friendship Neb and Joop share, and in the sailor Pencroff's reference to Joop as a "blackamoor." The new island colonists are also interested in Joop's physiognomy, declaring him to have "a facial angle not perceptibly inferior to that of an Australian Hottentot" and confirming "his place in the ranks of the anthropomorphs."[3] His status in the ape/monkey world, in terms of temperament and intelligence, is also clear:

[He] was an orang-utan and as such possessed neither the ferocity of the baboon, nor the heedlessness of the macaque, nor the slovenliness of the marmoset, nor the impatience of the Barbary ape, nor the fierce instincts of the cynocephalus. This race exhibits many traits that suggest an almost human intelligence. (319)

The most obvious indication of Joop's adeptness in human ways is his facility as a servant, and here he is seen as no exception to other orangutans:

> Put to work in houses they can serve at table, clean the rooms, care for the laundry, wax shoes skilfully, use a knife, fork and spoon, even drink wine . . . much like any other bipedal featherless domestic. It is known that Buffon owned just such an ape, a faithful and zealous servant who remained in his employ for many years. (319)

Joop's "remarkable intelligence" (340) is also manifest in his ability to communicate and sympathize with Neb. Even though Joop does not have the faculty of speech, this can be regarded as a decided asset:

> 'Never a word of recrimination from our fine friend Joop,' Pencroff often said. 'Never indiscreet, never too talkative, never a disrespectful remark.'
>   'What a servant he is, Neb' . . .
>   'My pupil,' Neb answered, 'and soon my equal.'
>   'Your better,' the sailor replied with a laugh. (348)

Joop has his own portrait painted and is declared to be happy with it, from time to time contemplating his picture with a "smug satisfaction" that testifies to his possession of self-consciousness, the trait so frequently used (together with intelligence) to separate human and animal. But an orangutan such as Joop, it would seem, can only exist on a "mysterious island." While all humans survive the final cataclysm, the orangutan does not. Swallowed, as the ground opens up beneath his feet, the faithful (almost human) orangutan shares the fate of the other animals. Given the influence of Verne's works in the European, American, and English literary traditions, it is not surprising that Joop has joined Sir Oran Haut-ton as one of the most important fictional representations of orangutans still in circulation.

Frank Sheridan's *The Young Marooner: or, An American Robinson Crusoe* (1908) also owes a debt to *The Mysterious Island* and an even larger one to Daniel Defoe. Like Crusoe, the youth leaves home without his father's consent to go to sea, is violently ill on his first local voyage south, but recovers to ship on a whaler whose officers are particularly cruel to new crew members. After the boat overturns in a whale hunt, the youth finds himself on the poor creature's back and "rides" the whale till he passes out, returning to consciousness stranded on the shore of an apparently deserted island. There the young marooner-narrator encounters an orangutan whom he is able to attract by cracking a coconut for him. A second

Joop examining his reflection (Jules Verne, *L'Île mystérieuse*, 1916)

coconut particularly pleases the animal, and after reflecting that this is the wrong way round, the marooner decides to lure the animal back to his cave, since he too has heard (from sailors) that orangutans make excellent servants, being "less ferocious than the gorilla and more intelligent than the baboon."[4] "My new companion," he records, "had a well-proportioned frame, a broad chest, head of moderate size, round skull, projecting nose, skin covered with soft glossy hair; in fact, he was a very good specimen of the orang-utang race" (26). "Mr Ape," as the marooner first calls him, is then christened "Spero" (Hope), and chained in case he attacks during the night. This proves quite unnecessary, however, since Spero has unfortunately found the marooner's fermenting palm juice, consequently sleeping very soundly and waking with a dreadful hangover, a condition that seems to confirm the orangutan's humanness:

> 'Aha, Spero!' I said aloud. 'So you have taken too much arack! It is lucky I tied your hands and feet or you might do some mischief.' Spero raised his head and groaned.

It was very human.

'I'm sorry for you, Spero.' I said for I knew his head was aching.

I fetched some spring water from the kitchen . . . and bathed Spero's head.

His little eyes spoke to me his thanks.

I untied the cord.

Again there was an intelligent glance.

I loosened his feet.

He was grateful.

A pocket-hankerchief [*sic*] which I had found in the dead man's pocket was dipped in water and placed on Spero's head. Never did human being appear more grateful. (27/73)

Spero does become an excellent servant, helping to clean rusty tools from a toolbox the marooner—again Robinson Crusoe–like—is fortunate enough to find. The orangutan is particularly taken with polishing the saw, since on completion of the task he is able to see and experiment with his own reflection or that of other objects. These activities would seem to demonstrate Sheridan's belief in orangutan consciousness and self-consciousness and the animal's "almost human" behavior. The young marooner also remembers his father telling him that orangutans could speak but are always careful not to in case they be captured as servants or slaves. (Notwithstanding their silence, however, Joop of *The Mysterious Island* and Spero have been made servants, though not apparently too unwillingly.) When the narrator first encounters Spero both are wielding clubs, but for the single purpose of bringing down coconuts. Sheridan's novel thus employs most of the by-now-familiar, if sometimes contradictory, tropes associated with fictional orangutans: club wielding, attraction to alcohol, "prodigious strength," comparative gentleness, occasional ferocity, imitativeness, and intelligence. Peacock's Forester, Roland's Murlich, Verne's stranded party, and the "Young Marooner" also find them "good company." These characterizations persist into the late twentieth century in the figure of an orangutan barmaid in Paul Carter's allegedly factual *Don't Tell Mum I Work on the Rigs: She Thinks I'm a Piano Player in a Whorehouse* (2005).

In Maurice de Moulins' 1930 adventure novel *Morok, l'orang-outang*, a British planter in Sarawak, Fred Pangborn, captures a young orangutan and, naming it Morok, keeps it as a household pet. As Morok grows, he develops a "poodle-like" love for the Pangborn's beautiful daughter, Doris, and springs to her defense when she is threatened by a malevolent Dutchman, Hans Ruckert. Harold Windham, deputy manager of the estate, kills Fred in a hunting "accident" and soon after poisons Fred's wife Mary using an untraceable powder obtained from a Dayak sorcerer. As the distraught Doris contemplates her mother's body, Morok senses her anguish and recalls the death of his own mother at the time of his capture. He at first seeks to console Doris with the antics that cheered her as a child but quickly realizes they are inappropriate. Morok's recollection of his mother's death is reminiscent of the French academy's test of Gulluliou's possession of emotion and memory and of course reminds us of the cruelty of Indonesian and European orangutan hunts. Realizing that Harold plans to kill her, too, Doris flees into the jungle, where Morok ensures her survival by helping her find shelter and collecting food for her. A giant snake appears (as often happened in the stock jungle combat scenes of early ape melodramas), but Doris saves the orangutan by intervening with her machete. After Morok is captured by Dayak headhunters, Doris breaks her concealment to rescue the chief's child from drowning and thereby earns the right to ask for Morok's release. The Dayaks are suitably impressed that a white woman can handle so dangerous a beast, and Morok repays his debt to Doris by strangling a Dayak who tries to stop her from escaping the jungle village and later by killing the unpleasant Ruckert. At the end of the novel, however, he confirms his status as an innocent child of nature by gamboling in the garden and chasing butterflies.

These stories imply an orangutan intelligence that is closer to that of humans than we generally recognize, but they give no hint of the acuity of judgment that characterizes some of the fictional orangutans described in chapter 5. Orangutans like Joop and Morok are attached to particular humans, but the attachment is uncomplicated: the consequence of being captured, cared for, or rescued by that person. In all cases, the orangutan's devotion to a master or mistress is far greater than that of the human to the orangutan. The apes are thus dispensable, allowed to perish or languish on the island when the castaways depart. In one way or another, the stories all elide the exploitation involved in domestic labor and slavery by presenting service as a matter of pleasure, not compulsion, for the orangutan. These orangutans are above all friendly faces in the jungle, animals marked by a unique resourcefulness and a unique willingness to help humans who have entered an unfamiliar and hostile environment.

Another group of fictional works, by contrast, investigates the consequences of closer emotional ties between orangutans and humans. Charles de Pougens' *Jocko,* published in 1824, is concerned with a male human and a female "monkey" whose species is never revealed but whose size and name, Jocko, strongly suggest an orangutan. Her place of origin the narrator intends to keep secret although we are told it is a tropical island and circumstantial evidence suggests that it is Java. This bizarre story with a strangely uneven tone is accompanied by an appendix titled "Proofs," a section nearly as long as the tale itself. The "proof" includes references to works by philosophers, natural historians, hunters, and scientists about the habits of "pongos." The most widely cited source is Buffon, but Pougens' second source preference, the works of the eighteenth-century Swiss scientist Charles Bonnet, seems to sum up the point he wishes to make with his supporting evidence and in the tale itself: "If the orang-utan is not actually a man . . . he is the most perfect prototype on the earth."[5] As far as is known, Pougens never traveled outside Europe, but he may have seen a chimpanzee or orangutan in England or in his home country. Certainly his textual sources include references to a female orang-utan seen in Paris in 1808, probably the one that was given to Empress Joséphine and christened "Mademoiselle Desbois" by Napoleon.

The male narrator of Pougens' tale encounters a female "Jocko" with whom he gradually develops a friendship. She builds them a hut, which he delights in helping her furnish. She does the cooking, they often dine together, and he teaches her how to clink glasses as they eat at the table he has provided. He only once spends the night in her hut, but in other respects they appear to cohabit: "When we had finished our little snack . . . I would dream and compose poetry. . . . [T]hese fiery pages, those passionate verses that the public so appreciate, . . . I composed them beside a female of the wild and fierce Pongo" (43–44). Although she evinces every sign of loving him, the relationship does not become sexual, however charmed the narrator seems by her delicacy and ready intelligence. They share no language but can communicate, and one day the narrator is astonished to discover she has brought him a present of shells and diamonds. Obsessively overcome with greed, his former romantic attachment to her immediately gives way to avarice, and he presses her to bring him more, even though she obviously incurs physical risks in so doing. Thus his "dear Jocko" is fatally attacked by a snake, dying a pitiful death. The narrator's abrupt comment, "three days later I left for Europe" (67), concludes the tale, reinforcing its peculiar changes in tone and dismissing the proto-romance between ape and human. Intriguingly, Alfred Russel Wallace's account (much later in the century) of his adoption of a baby orangutan whom he has orphaned by killing its mother has an almost novelistic character that echoes Pougens' story. There is an unexpected meeting of boy and girl, with the chastity of the relationship ensured this time by the boy's taking a paternal, rather than amatory, role. The sweetness of the friendship is overshadowed by an uncertain future because introduction to "fashionable society at the Zoological Gardens,"[6] which Wallace envisages as the future for his girl, means eventual separation and a degrading, uncomfortable life for her. Conveniently, however, she dies before Wallace's seriocomic future scenario can be enacted.

Building on Pougens' initiative was a derivative short story by the Russian author Antonii Pogorel'skii titled "Journey in a Stagecoach" (1829). Employing the old device of reporting a story heard from a stranger, this time in a stagecoach traveling between Moscow and St. Petersburg, Pogorel'skii tells of an infant, Fritz van der K., the child of Dutch settlers in Borneo, who is kidnapped by orangutans and brought up as part of orangutan society. He loses the capacity to speak but develops a close attachment to his orangutan foster mother, whom he calls Toutou. He is devoted to her and tells his listener in the stagecoach: "in the course of my life I met but few women whose character was both so mild and kind, whose amiability was so natural and whose gaiety was never impaired by anything. Alas! The moral perfection of this kind creature enhances my terrible guilt towards her."[7] After years in the jungle, Fritz stumbles on his family home and is restored to human society, but three years later Toutou comes to visit him. His old affection for her returns and he meets her every night. In time, Fritz becomes engaged to the charming Amalia, but she becomes suspicious of his nighttime meetings and is horrified when he reveals he is meeting an ape, insisting that he choose between them. Dejected and preoccupied, he goes into the forest for his regular tryst with Toutou. When she tries to cheer him up, he becomes angry and shoots her, then faints. When he regains consciousness, Toutou is beside him, bleeding to death. She licks his hand in friendship and he tries to bandage her wounds, but it is too late. She dies, and like Pougens' narrator, Fritz leaves for Europe the next day. French literature was immensely popular with the Russian elite and Pogorel'skii would certainly have been familiar with Pougens' tale. Pogorel'skii mutes the sexual element that is strongly implicit in the original story by making Toutou the foster mother of Fritz, rather than a potential lover, but Amalia's jealousy is presented in terms of rivalry in love, even if she misunderstands the circumstances. Whereas the affection of Pougens' narrator for his Jocko is made ambivalent by his greed and by his lack of direct complicity in the orangutan's death, Fritz directly betrays and kills his own foster mother and thus carries an unassuagable guilt for the rest of his life.[8]

In *L'Enfant du Bois,* published in 1865 and translated into English in 1868 as *The Wild Man of the Woods,* Élie Berthet takes up the trope of orangutans kidnapping humans and raising them within the orangutan forest community. Richard Palmer, his English wife, and young son Edward live near the colonial town of New Drontheim. Wandering off to the forest, Edward is menaced by a tiger but is saved by someone his servant Maria describes as "very big, very strong. He be one of those men 'who do not talk'."[9] Their rescuer has used a "formidable club, broken from a casuarina or iron-wood tree" to deliver the tiger such a forceful deathblow that the club itself had "been broken by the violence" of it (59). That he owes his son's life to the orangutan does not endear the creature to Palmer, but the orangutan seems fascinated by the child and remains in proximity to the settlement, where he is believed to have killed an evil local known as "[Opium] Smoker." Smoker's death provokes Palmer to offer "ten pagodas for the skin of the orang," although he cared nothing for Smoker and, it appears, is motivated to profit from the animal's rarity: "of all the creatures in the world, there is not one so little known in Europe" (140). Edward disapproves: "Smoker was ugly and wicked—and then the man who does not talk killed the tiger; and I will not have him killed" (144). He is distressed to learn that his former rescuer has been cornered in a natural declivity from which the hunters intend to starve him out. The naturalist, Van Stetten, has offered them a greater reward if the orangutan is captured alive. Setting off to feed his friend, Edward is recognized by the trapped ape, who "all at once assumed an expression so sad, so gentle and at the same time so supplicating that it was impossible not to be touched by it" (184–185). Edward supplies his friend with a rope, but when the animal gains the top of the cliff, Edward, forgetting the exchanges of good offices that have passed between them, is "frightened out of his senses" on suddenly seeing this "strange hideous creature of gigantic stature sprung up before him like an evil spirit let loose" (189).

Now menaced by those guarding the culvert, the orangutan seizes Edward, presses him "to his shaggy breast," and climbs the nearest large tree. Pursued by Edward's father, bounty hunters, and officials of the settlement, the animal is nonetheless careful of the boy:

> It was strange; but one would have said that the orang was aware of the suffering and fatigue of the strange creature he had stolen. He took pains to save him from too violent shocks, and very cleverly pushed aside the branches that might have hurt him. For a minute even, seeing that he was some way in advance of his enemies he stopped for an instant to let him take breath and rocked him gently in his long arms like a nurse. (193–194)

This apparently considerate treatment of the human child by his abductor stands in marked contrast to the human treatment of an orangutan youngster when, after five years of living as an ape with his adopted family, Edward is discovered. In order to find, rescue, and avenge, Palmer has spent long periods in these intervening years learning to be a "savage" in the forest, but he rarely sees an orangutan, and only then at a distance. When Palmer does at last locate the orangutan colony, he is impressed by the construction of their "huts" as he watches his now adolescent son, "a robust, sunburnt youth" (291–292), play with his younger orangutan step-brother, a little creature quite besotted by his human sibling.

The attitudes toward orangutans expressed by the novel's third-person narrative voice often seem surprisingly contradictory. Apparently observing the family from a distance, the narrator notes that "there was something sedate and thoughtful in their way of moving, not commonly observed in any other kind of monkey. The little one looked lively and almost intelligent and seemed as if he showed signs of a certain capacity for education" (297). Yet when either Palmer or indeed the narrative voice focuses on the human Edward, now "as beautiful as a Greek Apollo" (344), the lament is that he had "sunk into a mere savage . . . fallen into a state of degradation . . . the slave of these hideous apes" (297). And there is a particularly stark (and unremarked) contrast between the orangutan's relatively gentle abduction of the boy and the hunting party's forcible recapture of a now reluctant Edward. While the treatment of Edward's orangutan family is utterly pitiless, it is recounted in such a way as to evoke pity in readers. As Edward is being led away, cries of despair are heard from the young orangutan, and Edward, communicating with him and his family through growls and guttural sounds, refuses to go further. He is quickly trussed by his father's party and, now rendered "completely powerless," begins "to utter cries so frightful, so different" from the sounds that "terror or anger sometimes wring from human beings, that those who heard them never forgot them. He rolled on the ground in a perfect frenzy, biting everything within his reach" (358). Edward is now, of course, unable to release himself, and the young orangutan makes a brave but futile endeavor to help him. Though mortally wounded by Edward's "rescuers," the youngster attempts to take Edward in his arms and, covering him with kisses, tries to "carry him away to the huts" (359). Again the narrative voice is equivocal in its description of the scene. As the dying orangutan brandishes a club in his last bid to retrieve his stepbrother, the rescue party refrains from shooting him because "either they were afraid of hurting each other in the confusion, or perhaps they were filled with pity at the sign of the poor creature's devotion to young Palmer" (360). But the Malays in the party are described as feeling none of this pity; they attack the young orangutan with swords, krises, and bayonets. Finally, overpowered by numbers and covered in blood, he throws down his club and ceases all resistance. In a passage that strongly echoes Abel's 1825 account of the actual killing of an orangutan in Sumatra, Berthet describes his death:

> By his cries, tears and supplicating gestures, he seemed to be imploring his enemies to have pity on him. His dying look was strangely intelligent. He

placed his hands on his gaping wounds; he uttered groans so sad, so expressive, that no human language could equal them. All the spectators of this miserable scene were deeply touched, and Palmer turned away his head with horror. One of the Lascars, perhaps with the intention of putting an end to the sufferings of Edward's defender, gave him a blow with a bayonet; but he had miscalculated the strength that the orang still possessed, for the latter, seizing the weapon, broke it as he had broken the *kris,* while his dying look seemed to reproach the assailant for this piece of unnecessary cruelty. (361)

Although it is thus clear that Edward does not yearn to be restored to "civilization" and that the author deliberately evokes pity for the orangutan sibling in the description of his cruel death, the reluctant Edward is reassured by his father—without irony—that "You need not be afraid of those savage beasts any more" (150). Yet all the "savagery" in these later scenes is perpetrated by the humans, not the orangutans. Edward's subsequent re-entry into "civilization" is marked by his re-acquisition of human (as opposed to orangutan) language.

In spite of the equivocal attitudes to orangutans expressed both through Palmer (and other characters) and in the predominant third-person narrative voice, the certainty of the superiority of European civilization and its values remains inviolable. Perhaps it is this inviolability that generates, at least for the modern reader, the apparent contradictions in attitude toward both orangutans and those who, from the European point of view, share some of their animality: the "savages" in the text. The huts the orangutans build are better than those of "the aboriginal inhabitants of the South Sea" (289). Orangutans (often described as "brutes") demonstrate compassion and caring, while the Malays, "loving nothing, respecting nothing" (245), do not. Filial ties are clearly important to both Europeans and orangutans, but not, apparently, to the Malays. To rescue his son, Palmer must become "a savage," and Edward, in his five years with the orangutans, has also become "a savage." Yet the text's insistence on orangutan intelligence, compassion, and even reason, in contrast to the instinctual behavior of the "savages" and the consideration orangutans evince for humans as well as their fellows, often contrasts with that of the Malays in the novel. Such contradictions are not unique to this text, to fiction, or to attitudes about orangutans in general. As much as anything, they expose the radical inconsistency of human responses to animals: instinctive empathy with their suffering censored by conventional and scientific approaches to the nonhuman.

Like Monboddo and Peacock earlier in the century, Berthet was interested in feral children and had published a work about one ten years earlier titled *La bête du Gevaudau.* But writing more than half a century later than Monboddo, Rousseau, and Peacock, he does not regard either the orangutan or the feral child as representatives of "natural man." In *The Wild Man of the Woods,* nature and civilization are diametrically opposed; the orangutan is the wild (and potentially savage) kidnapper of a colonial's young son and the father's greatest fear aside from the child's murder or death from exposure is that having been raised by orangutans, he will be altogether lost to "civilization." Here, in spite of much of the red ape's humanlike behavior as described in the novel itself, the distinction between human and orangutan represents the obdurate and apparently impermeable line of division between "man" and "animal." Even though the young Edward Palmer may seem to cross this line in a purely literal sense in the course of the novel, this passage in no way disturbs the basic philosophical dichotomy on which the work is premised.

In Albert Robida's delightful farce *Le roi des singes* (The King of the Monkeys, 1879), a child is raised by what are sometimes called orangutans, although they have tails. Bizarre and frenetically picaresque, the work consists of the loosely connected adventures of one Saturnin Farandoul and is itself part of a much longer work titled *Voyages tres extraordinaires de Saturnin Farandoul dans les 5 ou 6 parties du monde, et dans tous les pays connus et meme inconnus de Monsieur Jules Verne.*[10] A melange of genres—adventure novel, comedy, satire, fairy tale, nationalist allegory— the plot is risibly episodic, beginning with Farandoul's early childhood wherein, after a shipwreck, he is washed ashore as a baby, Moses-like, and found in a basket by a family of "monkeys," who are the sole inhabitants of a terra incognita roughly situated in the area of "the Polynesian isles of Pomotou" (17).[11] The monkeys are of much the same size as adult humans and are reddish in color. Although they consider Farandoul odd because he lacks a tail, his adopted family and the society at large accepts the child as a monkey/orangutan and raises him. Yet his human traits make him incompetent in this Rousseauesque paradise in which "The simian race was the pinnacle of the evolutionary scale, dominating by its intelligence the entire natural order of the island. Man was unknown here, never having repressed it with his barbarity, nor perverted it with his example" (23).

But fretting at being regarded "with an offensive air of commiseration" (27), Farandoul quits his loved and loving orangutan/monkey parents and sets "sail" on a palm trunk.[12] From this inadequate vessel he is rescued by *La Belle Locadie* under Captain Lastic, who happily adopts him. The new "monkeys," however, do not seem to be "imprinted with the least moral quality" (30), unlike those he had left. Undeterred, Farandoul, who speaks "no European language" on his rescue, quickly masters French, English, Spanish, Malay, Chinese, and Breton, "all at the same time" (32). Together he and his human companions sail the seas for a number of years, yet Farandoul never forgets his adoptive parents. On being shown "engravings of monkeys in an account of ocean voyages," he "covers them with tender kisses" (31), and when he sees a stuffed monkey at the Liverpool Museum of Natural History, he is unable to contain his sorrow and anger, throwing himself "upon the terrified curators with such fury that they had only been torn from his hands in a considerably damaged state" (32–33).

Farandoul later returns to his home island courtesy of Verne's ubiquitous *deus ex machina*, Captain Nemo, to recruit club-wielding orangutans for a battle against Malay pirates. The monkeys depart on the captured pirate vessel while the indefatigable picaro, Farandoul, now delightedly outfitted in one of Nemo's full diving suits, explores the undersea world. An undersea romance with a Malay princess (Mysora) follows, but in fleeing her father to be with Farandoul she is unfortunately swallowed by a whale. Farandoul gives chase in a *Moby Dick*–like quest, eventually arriving in Victoria, Australia, where the whale beaches itself. He is, however, moments too late to save the day as the celebrated scientist Crocknuft has claimed both the whale and its contents (Mysora) and is exhibiting them at the Melbourne Aquarium. Farandoul's *cri de coeur* for his lost love proves no match for the legal intricacies of English property rights, so war is declared: Saturnin Farandoul versus Crocknuft, together with "the State of Melbourne," Australia, and England (74). The main body of Farandoul's army "whirling heavy clubs" (79) comes from Borneo and New Guinea, the elite officers and troops drawn from the island where Farandoul was raised. A local Australian reporter (with the suggestive name of "Dick Broken") now becomes the reporter/narrator of the battle. Farandoul and his quadrumane armies are initially triumphant and are given "good press" by Broken, who reports that "the monkeys, so terrible in battle, now seem very amiable," "full of concern for our well-being," and in fact "rather good chaps" (80). In place of "Australia," Farandoul institutes the "Oceanian Empire of Farandoulia," with "bimane" (human) and "quadrumane" (ape/monkey) citizens equal before the law (85). He enjoins the bimanes of Melbourne: "resume the course of your everyday labours in peace, under the protection of the quadrumane armies" and "Live henceforth in peace with your formerly disinherited brothers, the noble and generous monkeys" (85).

In spite of his apparently resounding victory, Farandoul is cheated of his beloved Mysora when Crocknuft blows up the aquarium after he dreams of being stuffed for display in a new quadrumane museum of natural history. Meanwhile, the situation is also deteriorating in the new Oceanian Empire of Farandoulia. The billeting of quadrumane troops on the Melbourne bimanes has proved unpopular, and the army is weakened from within by an ambitious general. The coup de grace is delivered with the arrival of the British Fleet, which employs the apparently traditional tactic of substituting kirsch for the quadrumane armies' water. The "monkeys of Borneo," mounted on kangaroos and "armed with heavy ironwood clubs" (113), offer strong resistance, but the kirsch does its evil alcoholic work and perfidious Albion is once again in control. Forced into surrender, the quadrumane armies are rescued by Verne's ubiquitous Captain Nemo. Shipwrecked en route to France, Farandoul has just enough money left to buy back his monkey/orangutan foster father from a sordid captivity and settle him "comfortably" in his own apartment at the Botanical Gardens before sailing off for new adventures in America. Comedic and parodic though it is, *Le roi des singe*'s equal rights for bimanes and quadrumanes prefigures—however adventitiously—the belated push for such rights since the late twentieth century.

Pougens' unconsummated romance between a human and orangutan led to tragedy for the simian; Pogorel'skii and Berthet portrayed growing up with orangutans as a prelude to misery caused by human action. How much more intense, then, would be the consequences of a real sexual encounter? Power's Beast in *The Curse of Intellect* is conscious of, though he dismisses, the possibility of sexual liaison with human women. He admires and is grateful to Lady Champerowne's daughter Kitty, but he concedes that the process of his sad fall into man's estate has "put beyond reach for [him] all question of love" so that he "never felt more than a passing titillation of vulgar passion" (117). These hints of sexual awareness on the part of male apes echo the strong trope of orangutan lustfulness initiated by Aelian and sustained over the centuries by Bloemaart and others, including Thomas Jefferson, who had made the notorious comment in the late eighteenth century that male orangutans naturally lusted after African women in the way that African men lusted after European women.[13] While the assumed animalistic sexual attraction of native men to white women was a recurrent source of anxiety in colonial societies,[14] there is little evidence that orangutans were feared as violators of white women in colonial Southeast Asia. When orangutans kidnapped white women, as in Mayne Reid's *The Castaways* (1870), O. C. Vane's "To the Rescue" (1875),[15] and Edgar Rice Burroughs' *Tarzan and "The Foreign Legion"* (1947), they did so in wonder at a precious object rather than with lustful intentions.

Gustave Flaubert's 1837 short story with the Latin title "Quidquid volueris" ("Whatever You Want"), written when he was still a teenager, concerns an animal termed an "orang-utan" (though allegedly from South America), who has been captured by the unpleasant (anti)romantic hero, Paul. Because Paul's sexual advances to a young negress have been rejected, he vengefully locks her in a cage with the captured animal. The result, as Paul had hoped, is her rape by the orangutan, who then escapes, and the subsequent birth of Djalioh, whom Paul eventually brings back to France. Now a young adult, Djalioh's "large white teeth showed between thick lips like a Negro's or a monkey's" and thus there was a "wild and animalistic aspect about him, which made him look less human than strangely bestial."[16] Though Djalioh can speak, he says very little at all in company, but he is described as "feeling deeply," even being pathologically "oversensitive." He desires the beautiful but vacuous Adèle, Paul's betrothed, but is initially capable of suppressing his strong feelings for her and continues to live with the couple after they are married. By the time Djalioh can no longer contain his love/lust for Adèle, she and Paul have had a child. Apparently reverting to the bestial half of his inheritance, Djalioh murders the child before raping and killing Adèle, thereafter dashing out his own brains in the fireplace.

This violent tragedy, the final result of the cross-species rape, is, in contrast to Roland's novella, entirely devoid of pity for either humans or the crossbred Djalioh. Not untypically for the author, the narrative evinces a deep cynicism about human nature, men, women, romance, and class. The horrific fate of Adèle and her child is ameliorated for the reader by the tale's self-conscious acknowledgment of the artifice of all storytelling, by the general unpleasantness or vacuousness of most of the characters, and by the cynical tone and distancing structure of the work. In contrast to Gulluliou's slow death from tuberculosis while in European exile in Roland's *Le Presqu'homme* (1905), Djalioh's demise is sudden and violent, the inevitable conclusion to his own lustful and murderous frenzy. Djalioh, however, is only half-orangutan and half-human, and his bitter cruelty seems to arise from frustration over his not being able to possess Adèle, who actively dislikes him, whereas he feels an obsessive love for her. Djalioh's liminal, apparently mixed-race status seems more important as a difficulty for him than his part-animal origins, but his extreme violence is less a product of his mixed parentage than it is of sexual frustration.

Roland's *Le Presqu'homme* is mentioned in the previous chapter as a novel that summons an accelerated evolutionary process to bring orangutans back into the intellectual vicinity of humans, but it also toys with the issue of sexual relations between humans and orangutans when Gulluliou falls in love with Murlich's cousin, Alix, with whom the two are lodging in Paris. On two occasions the orangutan's desire for Alix (a symptom of his "beastly" nature) almost overcomes his gentlemanly humanity. The unapproachable Alix has continually refused the advances of another suitor, her longtime friend, Maximin, and the novel presents her "virginity" as if it were the essential key to womanly independence. While Maximin only attempts a kiss after he is emboldened by the success of his play, it is Gulluliou who most threatens Alix's chastity, twice almost raping her, before "the beast" that she perceives "rising in his eyes" is suddenly quelled by his education as a man, and, on the second occasion, by the weakness induced by his fatal tubercular condition, an illness that has also increased his ardor for Alix.[17] Although Gulluliou's love for Alix and his attempted rape of her are frustrated, the story raises the early twentieth-century interest in the subject of interspecies sex, if only to abandon it. Gulluliou's sexual desire for the aloof Alix is pitiable rather than threatening.

Pre- and post-Darwinian evolutionary philosophy and biology exerted a significant influence on Roland's representation of orangutans and to a lesser degree on Flaubert's. In the background of all these stories, moreover, is the issue of race. Nineteenth-century Western society was marked by both complacent certainties and deep and often contradictory uncertainties over the nature and significance of racial differences. Depending on circumstances, which included especially class, nationality, geographical location, and the specific character of the colonial power, interracial sexual liaisons could be viewed as a profound threat to racial integrity, as a perquisite of power, as a path to hybrid vigor, as a sign of private weakness, as a source of moral contamination and degradation, or even as a normal expression of heterosexual love. In writing of actual or potential sexual encounters between humans and orangutans, Pougens, Roland, and Flaubert not only evoked ancient anxieties about human-animal sexual relations but also reminded their readers of the meaning and implications of interracial sex. Yet none of the three works is a simple racial allegory; rather, they evince interest in and exploration of the sexuality of nonhuman primates and interspecies (as much as interracial) conjugality. In all three texts, the fictional orangutans, whether female as in "Jocko" or male as in Roland's and Flaubert's works, are represented as highly emotional and having outward expressions that can, unequivocally, be understood by humans even where no "common" language is possible.

But it was Felicien Champsaur's *Ouha, Roi des Singes* (Ouha, King of the Monkeys, 1922) that took the issues of human-orangutan sexuality and species miscegenation to an extreme. In this novel, racism, negrophobia, and the dangers of the degeneration of European civilization through black-white human misalliances are most evidently connected. Champsaur was writing in the context of widely known experiments by a Russian Jewish surgeon, Serge Voronoff, with the implantation of monkey glands as a means of rejuvenating and renewing the sexual potency of men and, later, women. *Ouha, Roi des Singes* is dedicated to Voronoff, and actually set in Borneo, a Borneo imagined by Champsaur to be predominantly populated by orangutans and negroes.

Harry Smith and his daughter Mabel escape the ennui of postwar Europe by visiting Borneo and inviting a friend, Dr. Abraham Goldry, to join them. Goldry is fascinated by orangutans, whom he considers the ancestors of humans. Ouha, orangutan king of his own group, has captured a "négresse" who has been his not entirely unwilling mistress. The narrator comments that male orangutans have always carried off human women[18] and that "assimilation was easy with an indigenous people of ape-like aspect."[19] Ouha is himself captured by the Smiths but, desiring the white woman Mabel, succeeds in carrying her off. This proves not quite as easy as did his abduction of the négresse: Mabel initially resists Ouha's sexual advances, but eventually succumbs to his powerful and virile sexuality. This experience proves not degrading, but rejuvenating for Mabel. When she is "rescued" from Ouha she rejects her former suitor on the grounds that she has now discovered "needs" he will be unable to satisfy, but which clearly Ouha could—in fact, to which, we assume, he has introduced her. As Brett Berliner points out, *Ouha* addresses "profound social anxieties, including sexual ones, concerning blacks and whites, and, especially the modern white woman."[20] *Ouha* also prefigures a small number of twentieth-century works that present male great apes as lovers who are, in some respects at least, more satisfactory than human males.[21]

The frisson of cross-species sex dramatized in *Ouha* recurs in Boulle's *Monkey Planet* in the unconsummated but erotically tinged relationship that Ulysse Mérou has with a female ape scientist. Their developing rapport suggests that intellectual and emotional communion between humans and apes is possible, whereas Mérou's sexual liaison with the beautiful human woman, Nova, is a purely physical matter—he can neither initially speak nor emotionally communicate with her—the way sex with an orangutan would have been envisaged on his home planet.

The savage orangutan began to appear in fiction from the late 1820s, in sharp contrast to previous representations, in which orangutans are for the most part gentle, friendly, and well-disposed to humans, even if like Sir Oran Haut-ton they are willing to wield a club in a good cause. In the few years that had elapsed between the writing of Pougens' 1824 story and Pogorel'skii's tale in 1829, Clarke Abel had published his influential account of the apparently savage orangutan on the shore at Trumon in Sumatra. Even though the foster mother orangutan in "Journey in a Stagecoach" treats her human ward kindly, Pogorel'skii writes of orangutans in general that they are "the fiercest and most terrible enemies of the Europeans":

Endowed with incredible instincts, they seem to carry out their attacks according to a well thought-out plan. The strongest among them, armed with
thick clubs, are the main line of attack, while an innumerable host of other
apes throw stones at the enemies from all sides, and they aim so well that
no stone is wasted! Sometimes these apes, having hidden themselves among
the farthest branches of the immensely tall trees, let people pass by their
hideouts and then come down, as quick as arrows, jump onto the shoulders
of the passers-by, biting their heads and ripping out their eyes with their
sharp claws.[22]

Pogorel'skii appears to have been the first writer of fiction to explore the implications of the new trope, but he was not the last. The English author and bookseller
William Hone repeated Abel's account in his popular *Table Book* of 1827,[23] where
it was read by the aging Walter Scott. In Scott's late novel, *Count Robert of Paris*
(1829), a leading minor character is an orangutan of prodigious strength called
Sylvan, who intervenes at crucial points, including an incident in which he com-
mits murder and escapes out the window.

Scott's story in turn appears to have been part of the inspiration for Edgar
Allan Poe's "Murders in the Rue Morgue" (1841), which remains one of the most
influential of all fictional portraits of an orangutan, even though Poe may never
have seen an actual orangutan, let alone in its own habitat.[24] Poe presents us with
a murderous and savage beast capable only of inept imitation of humans, definitely
not man but animal. His "Murders in the Rue Morgue" is credited with being the
first modern detective story, progenitor of an entire genre, and it continues to attract
the attention of critics and writers in the form of rewritings of, re-engagements
with, and homage to the original story. One contemporary example, Luis Fernando Verissimo's *Borges and the Eternal Orangutans* (2005), will be considered
later in the chapter.

Strong, imitative, and murderous, Poe's animal has neither the gentility nor
the intelligence of Peacock's Sir Oran. Captured and brought to Paris by a sailor
who intends to sell him, the orangutan escapes the mariner's custody and enters
an apartment in the Rue Morgue that happens to be occupied by two women.
There, it is assumed, he attempts to imitate the actions of the sailor whom he has
observed shaving, seizing one of the women and attempting to remove her hair.
Naturally this action terrifies both occupants, and in the ensuing struggle the
orangutan, still in possession of the razor, horribly murders the two women, escaping his pursuing owner by exiting through a high window. Left with such a chaotic
and grisly scene of violence, the seeming physical impossibility of the perpetrator's

exit, and the apparent lack of motive, the police are baffled. But Poe's proto-detective, Auguste Dupin, and his companion, the narrator, eventually solve the mystery. Dupin finds, through application of cause-and-effect reasoning plus, significantly, an imaginative genius that remains characteristic of his literary descendants such as Sherlock Holmes, that the perpetrator was not human—the assumption upon which the police have of course been proceeding—but rather an orangutan.

In complete contrast to the didactic apes discussed in chapter 5 and to the servants, hybrids, and sexual partners examined earlier in this chapter, Poe's orangutan is pure animal with no hint of humanity or of attachment to humans. Its sailor owner has brought the ape to Paris only for the purpose of selling it; there is no hint of a dimension of companionship. The orangutan's behavior is wholly imitative and it has no moral sense, with the consequence that it is not punished at the end of the story, but simply led off to captivity in a zoo. Thus the murderous orangutan of Poe's story is not only completely different in character from Pea-

Arthur Rackham's 1935 illustration of the orangutan in "Murders in the Rue Morgue" (The British Library Board, 12625.t.14, p.258, used by permission)

cock's Sir Oran Haut-ton, but occupies a radically different ontological position in relation to man. Sir Oran is never an animal figure simply standing in for a human being (or beings), since he himself is considered human, if of a different kind from the other characters. By contrast, the unnamed orangutan in Poe's tale, though categorized as beast, has often been read by Poe's contemporaries and present-day critics as *representing* something other than a wild animal; precisely because his identity is not complicated by any truly human element, he can stand as a metaphor for some aspect of human society.[25] In this case, his behavior can be seen to offer a warning against slave emancipation in the south of the United States, or to represent, in its motiveless murder, a world without purpose or design. Southern planters and, as we noted earlier, West Indian whites like Long likened slaves to apes, and the new anatomy had, sometimes intentionally and sometimes, as in the case of Camper, unintentionally, bolstered such racist sentiments and proslavery positions. The campaign to abolish slavery threatened to dissolve the borders between master and slave. The extreme violence of the Haitian revolution (1791–1804) in response to the cruelties of French colonialism gave substance to this disquiet among many contemporaries of Poe. Anxiety about freeing such potentially "wild beasts"—however apparently docile and domesticated—was at its height. Poe seems to have been reluctant to address this issue directly in his stories, but we know from biographical sources he was deeply concerned about the possible consequences of abolition and used his writing to warn against it.[26] A second reading of Poe's orangutan arises from the general anxiety provoked by proto-evolutionary ideas in circulation well before Darwin's 1859 *Origin of Species*. The replacement of the notion of God's divine plan with the random carelessness of nature seemed to many to usher in the reign of blind chance, perhaps expressed, as in the case of "The Murders in the Rue Morgue," as motiveless murder.[27] Dupin's solving of the case offers human rationality as the "solution" to a godless potential chaos, ushering in, as Lawrence Frank argues, the succeeding popularity of the detective genre.[28]

A German tale, "The Young Englishman" ("Der Affe als Mensch: Der junge Engländer," 1827), by Wilhelm Hauff helps pinpoint the distinction between Poe's use of the red ape as metaphor and a revelation of a different order when an orangutan, tutored to aspire to the ranks of the human, is eventually revealed as a beast. Hauff draws, in the manner of "The Pied Piper of Hamelin," on the European folk tale tradition of the duping of the smug inhabitants of a small town. A rich stranger arrives in Gruenwiesel, buys a local mansion, and lives in seclusion, precipitating a great deal of gossip about his origins and intentions. Observing traveling performers from his window, he is amused by a circus act involving a man and an orangutan. Secretly following the troupe from the town, the stranger purchases the animal and returns to Gruenwiesel, with the orangutan now disguised as a young man. He then proceeds, using "brutal" methods, to train the creature he describes to the townsfolk as his "English nephew" to fit him to appear in local society. The eccentric behavior of the nephew (who, unlike Sir Oran Haut-ton is often kept literally and always figuratively on a tight leash by his mentor) is attributed to his foreign origins. The townsfolk gradually come to accept and then laud his wild and unmannerly escapades until, left to his own devices at a public performance, the "English nephew" creates mayhem and his real identity as an orangutan becomes all too clear. The burghers of Gruenwiesel have been deliberately tricked by the stranger into believing an ape to be a man and are humiliated at having so zealously courted his favors. Hauff's orangutan was never, as Sir Oran Haut-ton was in *Melincourt,* intended to be considered a human. Indeed, the whole moral of the tale in the trick perpetrated by the stranger actually depends, as in Poe's story, on the imitator not being a human after all; yet the revelation of his pretense cannot easily be read as a solution to the contemporary ontological dilemma of what distinguishes humans from apes.

Nineteenth-century adventure literature enthusiastically took up the trope of the orangutan as a fearsome monster. An early novel in this genre set in Borneo, James Greenwood's *The Adventures of Reuben Davidger* (1865), has its protagonist in flight with a companion through the jungles when they are pursued by an orangutan with "hideous green eyes and leathern lips."[29] Just as they despair of escaping, however, "the hideous man-monkey, as though he thought it an excellent joke, swung himself into a tree and was off and away, chattering and barking with all his might."[30] The jungle is not so savage after all. In a similar vein, Mayne Reid's 1870 adventure novel for boys titled *The Castaways* tells the story of Captain Robert Redwood, who has landed in Borneo with his two children after being set adrift in a small boat during a typhoon. The narrative is the vehicle for a series of confrontations with strange animals and plants such as a hammerhead shark, a boa constrictor, and a poisonous *upas* tree, but its climax is an encounter with an orangutan, or "red gorilla." The ape first triumphs in a spectacular battle with a crocodile, then kidnaps Redwood's daughter, Helen, carrying her through the trees to his nest, where he presents her to his mate and her child:

[T]he mother at intervals [seized] her hairy offspring, and grotesquely [caressed] it; then letting it go free to dance fantastically around the recumbent form of the unconscious captive child. This it did, amusing itself by now and then tearing off a strip of the girl's dress, either with its claws or teeth.[31]

Redwood wants to shoot the animals, but the ship's Malay pilot, Saloo, kills them first with a poisoned dart from his blowpipe. With the girl safe at last, the party reaches the British settlement at Labuan. No longer an unexpected friendly face in the jungle, the orangutan has become the most formidable of the hazards to Europeans in an alien environment.

Rudyard Kipling's "Bertran and Bimi" (1891) extends the fictionalization of simian violence with an orangutan who combines superhuman strength and an almost human intelligence with an evil disposition. The short story, described by one critic as the "most repellent" Kipling ever wrote,[32] uses a German narrator, Hans Breitmann, to relay the tale in an exaggerated accent. He tells of a naturalist, Bertran, and his devoted pet orangutan, Bimi, who sleeps in a bed, smokes cigars, and understands English. When Bertran plans to marry, Breitmann warns him

Helen and the Orangutan, 1870 (Reid, *The Castaways* [1870 ed.], frontispiece)

that Bimi will become jealous: "If I was you, Bertran, I would gif my wife for wedding-present der stuff figure of Bimi."[33] Bertran dismisses the concerns, but returns one day to find that Bimi has ransacked his house and savagely killed the woman. Bertran then lures Bimi back to the house with professions of forgiveness, gets him drunk, and strangles him. The orangutan is so strong, however, that Bertran also dies in the fight. The story differs from Poe's in the crucial sense that the orangutan's intelligence gives the animal's violence intentionality and guilt, apparently justifying Bertran's revenge.

In another adventure tale, *Breath of the Jungle* (1915) by James Francis Dwyer, a professional wildlife collector has managed to assemble ten magnificent orangutans for shipment from Borneo to Singapore, but the vessel on which they embark, the *Papuan Queen*, runs into bad weather. The captain eventually orders everyone to abandon ship, and a kind-hearted crew member releases the orangutans so that they will not drown in their cages. Reluctant to abandon his prize specimens, the collector climbs back aboard the ship and spends the next fourteen days both steering it against the wind to keep it from capsizing and fending off highly organized attacks by the orangutans, who understand that he is the cause of their problems.

By the second half of the twentieth century, the trope of the violent orangutan had largely disappeared from creative literature. Orangutans were instead portrayed not just as subject to the cruelty of individual humans as in Berthet's *L'Enfant du Bois*, but more generally as victims of human rapacity. In James Hall's 1995 novel *Gone Wild*, the murdered are not the unfortunate humans of Poe's tale, but the critically endangered red apes of Borneo, though one woman is brutally killed at the outset by the orangutan poachers she captures on film. Though not the initiator of it, *Gone Wild* is a pioneering work in another new genre, the "eco-warrior" narrative. Most "eco-warrior" fictions are at least partially didactic in tone and method, qualities *Gone Wild* shares with Dale Smith's *What the Orangutan Told Alice* and indeed with Peacock's *Melincourt*. In this respect, Hall's text is deliberately crafted to educate readers about orangutans and to establish what the protagonist, Allison Farleigh, sees as the relative moral order of humans and these "animals."

In the course of the novel, Allison succeeds in tracking down the murder culprits, thereby exposing professional smugglers of orangutans (and other endangered species) as well as uncovering a plot to corral the last of the lines of all endangered animals in one single park for, of course, the massive profits such a monopoly would potentially produce. The story opens with Allison and her two daughters assisting with a rehabilitation study that involves an annual count of orangutans:

Pongo pygmaeus. People of the forest, the orange apes, Wild Men of Borneo. Their long reddish hair standing out from their bodies as if it were charged with static electricity. Bulging stomachs and eyes full of melancholy wisdom. They spent their lives in the forest canopy, dangling from limber branches hundreds of feet up, or bending saplings like pole vaulters to reach down for pieces of fruit, jambu-air, longsat, nangka.

Same number of teeth as man; blood pressure, body temperature the same. Ninety-eight percent of their genetic material identical to humans. And as far as she was concerned, whatever comprised that other two percent was pretty damned wonderful. The orangutans Allison had spent time with were a hell of a lot more intelligent and certainly more trustworthy than most of the humans she'd run into.[34]

As she carries out her work in the reserve, Allison and the animals are being stalked by three men. Winslow, her elder daughter, manages to photograph the killers before she herself is killed and her film stolen. At first Allison's chief suspect for the two murders is a well-known United States animal smuggler who has grown "rich and respectable" by buying and selling apes and is now "the golf partner of federal judges and congressmen" (65). The police in Sarawak do little to solve Winslow's murder or catch the smugglers, while back in the United States Allison seems to have even less chance of bringing the criminals to justice. Considered "trivial shit," animal smuggling is not high on the authorities' agendas:

From running red lights to holding up a 7-Eleven, hey, the cops didn't have time for the dinky stuff, man, they're too busy bull-dozing the bodies off Biscayne Boulevard, unloading the freighters full of heroin.
    With all the heavy-duty crime going on ... nobody cared if the White brothers were trafficking in a few of the lower life forms.... The only ones who got in an uproar about it were a bunch of lonely old ladies like Allison Farleigh and her gang of ape-kissers. (115)

Although apparently narrated in the third person, this passage is being filtered through the consciousness of the White brothers, hired American animal kidnappers, and told from their perspective. "Voice" shifts in this way throughout the novel, apparently entering the perspectives of both human and orangutan characters. The White brothers, however, are only secondary smugglers, acting on behalf of a much larger and more ambitious villain, Patrick Bendari Sagawan, nephew of the Sultan of Brunei. Patrick has been given the task of diversifying the state's assets for a potentially less oil-dependent twenty-first century, and his plans—unknown to the Sultan—are horrific. (The text notes that oil-rich Brunei, unlike much of the rest of Borneo, has kept its forests by not mortgaging itself to the oil palm industry. Its forests, though, contain no orangutans, for they were hunted to extinction by the end of the nineteenth century.) Patrick is behind the project to build the largest zoo on earth to house pairs of endangered animals. In case they are not quite endangered enough, he embarks on a killing spree, transporting some animals to Asia for his zoo and annihilating the remainder.

Unlike Charlotte (and apparently the "Jennys") of Diski's *Monkey's Uncle*, Allison Farleigh has always had a particular hatred for zoos. She is aware of the significance of forest destruction in Borneo, but her own campaigns focus on poachers, traders, and their buyers: pet keepers, experimental scientists, zoos. "Cities are the problem," she says. "That's where it all begins. . . . City people want some reminder, some connection to the world they've lost" (266–267). Although the "pet" traders Allison confronts frequently sell orangutan skins as well, "there wasn't as much money in the skin trade as in the pet trade. A live orangutan was worth more than the sum of its parts. But hunters couldn't always capture them alive. Often both mother and child were killed in the fall. For every live one captured, half a dozen died" (86).

The mother orangutan, who is murdered on the same day as Allison's daughter, had previously been a captive in the United States before being returned to the Borneo reserve where she had lived for fifteen years. During that time she had had one son, with a white blaze on his forehead. Shortly before his brutal capture by the poachers, the young orangutan—moderately used to people because of his mother's sporadic visits to the rehabilitation center—had had a close encounter with Allison, who, like Winslow, witnesses the mother orangutan's killing. Just as we follow Allison's investigation into the poaching and her daughter's death—the fate of this particular young orangutan being the key to both—we also trace his journey until his reunion with Allison and presumed reshipment back to Borneo. First the orangutan is stored in cramped conditions with other sick, dying, and/ or dead animals, then he is sold to a movie star as a cute accessory to his wedding. Almost immediately a college professor buys him for inane experiments (to prove that torturing the animals raises their blood pressure) but is thwarted by a junior scientist who kidnaps the orangutan and by a fortunate coincidence takes him to one of the members of Allison's rescue network.

Much of the material concerning the orangutan attempts to reproduce his perceptions of events and his current situations.

He no longer knew when to sleep and when to rise. . . . He looked out of his cage, up at the light through the windows. The sun was no longer where he was, shining only through those few bright squares. Outside, inside. Something new. Something he was learning about. The sun outside, the orangutan inside. (61)

As well as passages such as this one that shift perspectives from humans to the orangutans (there is another called "Broom" whom Allison has previously rescued from appalling conditions in a roadside zoo), *Gone Wild* offers facts about orangutans, not just through Allison but via lectures delivered by her helper, Thorn, at an animal retreat called Parrot Jungle. Thorn is depicted facing a group of cynical schoolchildren and their supercilious teacher. As well as dealing with apathy, obdurate misinformation, and heckling, Thorn is confronted by an angry girl who has heard that male orangutans actually rape females of their species. (This is also an issue for the teenagers in *What the Orangutan Told Alice*.) Outrage at this information, like the pronoun slippage, tends to suggest a breach, or at least a chink, in the wall of the species boundary, since the sphere of human morality (and in this case its apparent violation) is being extended to orangutans. And if animals can be considered morally culpable, they are close to inclusion in the ranks of the human.

The young orangutan's mother, even when released into the Borneo preserve, had retained a memory of her time in the United States:

From its months in the research centre in Arizona, the mother orangutan had retained a half dozen of the words the Professor taught it. At the time of her death, she could still sign the words *tree, nest, fruit, sad, sleepy,* and her favourite word, *shit.*

The orangutan with the silver patch had seen his mother use all six of these signs in the four years it had spent with her. And now, as the young orangutan hung upside down in its four-by-four cage and looked round at the stark warehouse, it made the sign for *fruit.* Because it was hungry and bored, it made this sign again and again. (132)

In this passage, the shift in pronoun usage seems to reflect a continuing dilemma. Science has long taught us not to anthropomorphize, asserting that it is inaccurate. Nevertheless, anthropomorphization can also draw animals closer to us and thus include them in "our" circle of similar beings with moral rights. Pronoun usage reveals an author's perspective on the status of orangutans vis-à-vis the human, as Alfred Russel Wallace's letter about the baby orangutan (discussed in chapter 4) demonstrates.[35] An animal is usually still referred to as an "it" to distinguish it from humans.

The "eco-warrior" theme is a widespread and popular phenomenon, even recently taking the form of a *nouvelle manga*.[36] The two-volume manga *Eco-Warriors: Orang-Utan* (2010), by Richard Marazano and Chris Lamquet, is set in the Indonesian jungle, where the protagonist, Chris, is engaged in orangutan rehabilitation but finds his work thwarted on a number of fronts: by a large pharmaceutical firm that has conducted gruesome medical experiments on orangutans; by violent but unnamed rebels; and by commercial interests that clear the rain forest to plant oil palms, destroying the orangutans in a holocaust of napalm. In the real world of the twenty-first century, as *Gone Wild* and *What the Orangutan Told Alice* also confirm, we may in the end return to an orangutan who can rarely if ever be seen, who can never again exist independently of us, who can "live" on only as he was born, an imagined creature, the stuff of story. The stories about orangutans, and certainly of orangutans in the wild, might well outlive the species itself.

Luis Fernando Verissimo's intriguing novel *Borges and the Eternal Orang-utans* (2005), takes us back to the classical and medieval past and into the contemporary world. Verissimo's book is not, however, concerned with biological orangutans or their possible extinction, but rather with a fully imaginary "orang-utan," one who has been conjured even before the animals themselves were seen or named by Europeans. This novel's orangutan is a philosophical and literary conceit, one apparently traceable to Egyptian and Kabbalistic sources, to the Renaissance scientist and necromancer John Dee, to Edgar Allan Poe, to H. P. Lovecraft, and on to a fictionalized version of the famous Argentinean writer Jorge Luis Borges. Verissimo employs the device of the hypothetical orangutan as representative of the author of the entire corpus of human writing to today and beyond. The eternal orangutan, perpetually writing, includes the potential for planetary annihilation in the accidental inscribing of names or words that are theologically prohibited from speech and inscription, such as names of gods or devils. This "eternal orangutan" is purely the product of human story, having no existence outside the real and imaginary literary and theological traditions that both produce and prohibit it. The term "orangutan," insofar as can be deciphered from the melange of facts and fiction that complicate this wonderfully sophisticated detective story, is used (*post hoc* of course) to refer to a kind of blind chance with an ancient lineage. The "orangutan" is a never-ending writer who (re)writes all the texts of the known world and more, and who might or might not stumble on the secret name of God or the powers of darkness, the naming of which, in writing, accidentally precipitates the end of the world. Every writer is potentially such an "orangutan" and the metaphor has some affinities with the common saying in English that if a monkey or chimpanzee were given a typewriter he or she would eventually write *Hamlet*. The novel is thus about writing, about the telling of stories, especially detective stories, and its most obvious subject is Poe, particularly his "Murders in the Rue Morgue." Through this story the term "orangutan" gains its second set of references in the novel, but one which reinforces the first, since the murderer of a man in an apparently locked room turns out to be the unreliable narrator of the novel itself.

Reminding readers of Poe's orangutan throughout—and thus throwing into doubt the status of the "tail" of the narrative as well—nothing could be further, in its use of the red ape figure, from Smith's and Hall's two near-contemporary novels about orangutans. In Verissimo's enjoyable postmodern romp through Western (and some Eastern) literary esoterica, only the *hypothesis* (even further removed from the real animal than metaphor) remains eternal. In *What the Orangutan Told Alice* and *Gone Wild,* by contrast, the genuine animal is threatened with extinction in the wild. While, as this chapter has posited, literature with both its affective and imaginative purchase can indeed influence environmental and animal conservation through the reform of human attitudes and practices, it would be tragic if the orangutan remained "eternal" only within literary memory.

7 🜊 Monkey Business
Orangutans on Stage and Screen

On March 16, 1825, a young French "character dancer" known simply as Mazurier created a buzz among Paris audiences with his beguiling simian antics in a ballet-drama that was to become one of the most influential stage pieces of its time: *Jocko, ou le singe du Brésil.*[1] Although set in Brazil, well away from the known habitat of any great ape, the drama appears to have drawn substantially on existing representations of orangutans. Within months, entrepreneurs had negotiated to bring the production to London to showcase Mazurier's singular talents as Jocko, the trickster ape of the play's title, and plans were afoot for a New York version. *Jocko* parodies and *singeries* (the act of "aping") rapidly appeared in a variety of French venues, often attributed to fictitious authors such as Sapajou[2] or Monsieur Monkey. Acknowledged imitations of the ballet-drama were soon staged at the Royal Theatre in Brussels (1826) and the Stuttgart Opera (1827), among other places.[3] One *Jocko* play, *Jack, l'orang-outang* (c. 1826), inspired the name of the first orangutan kept at the Jardin des Plantes menagerie; others offered generations of performers in Europe and North America opportunities to impress spectators with intriguing masquerades and virtuoso displays of athleticism. Always responsive to controversial issues, nineteenth-century popular theater had found the vehicle—the "man-monkey" role—through which to engage with growing public interest in the nature and origins of humankind.

Despite its misleading name, the man-monkey role was enacted by adults in costumes without tails and was understood as a generic marker for apes, as distinct from monkeys, which were normally played with tails by children. Like the literary fictions of the era, this form of theater cared little for specificity when it came to representing orangutans or other primates, and even less for realistic plots. Yet, being an embodied art form, stage performance could never escape entirely into the imaginary realm as literature could; it was confined, but also energized, by the insistent presence of the actor. Whereas real orangutans recruited to perform in circuses and zoos would later come to evoke humanness (see chapter 8), the challenge for human actors in man-monkey dramas was to simulate the ape as animal. The most successful works in this genre excited audiences with the frisson of species similarity through skillfully executed performances, even if their plots were generally unequivocal about where the lines between humans and animals should be drawn. Spurred by Darwinian debates on evolution, the man-monkey character remained popular until nearly the end of the nineteenth century, by which time trained orangutan acts had begun to eclipse the spectacle of human ape imitation. In addition, film was emerging as a new visual technology through which to advance apparently more realistic orangutan dramas.

Jocko was not by any means the first performance to harness the theatrical potential of simian roles. Matthew Cohen lists Javanese *wayang wong* among what appear to be old and widespread genres of acrobatic drama in Indonesia featuring men and women costumed as monkeys. Such enactments, he notes, ennoble the animal as the Hindu monkey god Hanuman.[4] By the early nineteenth century, the human-ape performer likewise had a long history in Western fairground entertainments, but largely as a grotesque clown. This character type and the related figure of the ape-as-fool lent themselves well to the popular theater forms of the time and could be readily adapted to probe human fears and foibles in topical narratives. A small number of plays had also taken up the figure of the orangutan as a servant, usually in an exotic location, following enduring debates on whether such apes could be put to useful work. One early drama in this mode, *Der Orang Utang, oder der Tigerfiest* (The Orangutan, or the Tiger Party), by Karl Friedrich Hensler, premiered at the Theater in der Leopoldstadt in Vienna on the same night in September 1791 as Mozart's *The Magic Flute*. The play features two Spaniards cast away on the coast of Peru and, like Mozart's opera, opens with a scene in which a leading character is threatened by a giant snake. He is saved by an orangutan called Danti, who also brings food to sustain the pair. Similarly, the plot of French actor Édouard Bignon's 1806 play *L'orang-outang, ou les amans du desert* (The Orangutan, or the Lovers in the Desert), turns on the intervention of a resourceful orangutan. In this case, a pirate marooned on an island rescues the animal from a snare, after which it devotes itself to his service, making a shelter, carrying the fishing equipment, and fighting valorously alongside him against ferocious "savages." The play was not widely performed, but it contributed to a narrative tradition of orangutans as helpful aides. A counterpart character, the devious and lustful ape, whose genealogy reached back to the satyr-apes of medieval folklore discussed in chapter 1, had also been incorporated into Western stage dramas, offering a ready visual portrait of a villain. Not long before *Jocko* first played in London, the Theatre Royal English Opera staged one such sketch with an orangutan as its chief menace. *Monkey Island, or, Harlequin and the Loadstone Rock* (1824) presents a young man and woman shipwrecked on an island run by scheming monkeys and baboons that dress and behave like humans. Their ruler, an "ourang outang," captures the woman and attempts to rape her, but she escapes. After an extended acrobatic combat between humans and apes, the sketch ends when a fairy intervenes to save the day, transforming the players into stock pantomime characters: the humans morph into the lovers Harlequin and Columbine, and the apes become Clown and the deceitful, lecherous Pantaloon.[5]

None of these works presented new ideas about orangutans in particular or apes in general; the simian roles were primarily ciphers for human character types that were already the stock-in-trade of popular theatricals. Similarly, *Jocko*'s plot is at best derivative, echoing not only elements of the Hensler and Bignon dramas discussed above but also the fate of the hut-building orangutan in Pougens' novella *Jocko*, published just months before. The play's ape protagonist is a pupil and devoted servant of a Portuguese trader, Fernandez, who has saved him from mortal combat with a venomous snake. Jocko lives in his own jungle hut and spends much of his time thwarting or teasing slaves and workers on the plantation. Toward the drama's end, he rescues Fernandez' son from a shipwreck, then saves him a second time from certain death by snatching him away from another giant snake. As he is about to return the boy to his father, Jocko is shot by a sailor who mistakes his intentions. He lives just long enough to struggle to his hut to retrieve some coveted diamonds for his master, then dies, to the grief of all assembled. What was remarkable about Mazurier's rendition of the ape role was that he performed it so precisely and so evocatively in gestural terms that the human actor seemed to disappear beneath the animal-hide costume and grimacing mask. One London reviewer remarked that the dancer had surely "changed his humanity": "he is not a man playing monkey tricks; but the very monkey of the forest."[6] Another critiqued the script as "badly written and clumsily contrived" but judged Mazurier to be so "perfect" in his characterization that "his spirit must have formerly inhabited an apish form."[7] Animated by the highest art in this momentary synthesis of man and beast, the Jocko character could be ennobled as if he were human, ironically sealing his fate as a tragic hero since the standard resolution for a comedy of human proportions—a wedding—could never be allowed. As one spectator remarked to

MAZURIER
Rôle de Jocko.

Jocko performance (Houghton Library,
Harvard Theatre Collection, used by permission)

a young lady who wept at the finale: "What could be done, Miss—you know it was impossible to marry him!"[8]

In establishing a performance idiom for the noble orangutan, Mazurier reached across the species line between humans and animals to suggest their physical, behavioral, and moral affinities. Yet this flirtation with the idea of species equivalence was also a theatrical conceit designed to breathe new life into the well-worn trope of the trickster servant whose subversive energies helped to drive genres such as farce, melodrama, and pantomime. The slapstick routines that had hitherto been stage business now became "monkey business" in countless reprises, imitations, and adaptations of the *Jocko* script. Most of them ended happily rather than in Jocko's death, though the tragic version circulated until at least the 1860s and was animated by many of the acknowledged man-monkey masters. Such actors confounded audiences with their simian impersonations just as orangutans did when they copied humans. Staged renditions of ape characters were, nonetheless, highly codified in performance terms. An 1836 print showing poses by German acrobat Eduard Klischnigg, who made a career of performing Jocko, suggests something of the "combination of balletic grace, acrobatic spectacle, mimed pathos and grotesque comedy" required of the role.[9]

The man-monkey dramas reached the height of their popularity in Europe in the 1840s to 1850s, by which time juvenile orangutans such as Jack (in Paris) and Jenny (at the London Zoo) had intrigued visitors with their intelligence and mannerisms, fueling speculation that apes might be closely linked to humans. Some man-monkey roles involved simple imitative sequences; others sensational feats or elaborate ruses. An early *Jocko* derivation, *The Dumb Savoyard and His Monkey* (1825), pairs an ape and a mute boy, both stowaways on a ferry boat, in a quest to save an imprisoned soldier from execution, along with his wife and their child. Following in the pattern of the *pas de deux* between Mazurier's Jocko and the Portuguese trader's son, which is "choreographed as a sequence of mirror images,"[10]

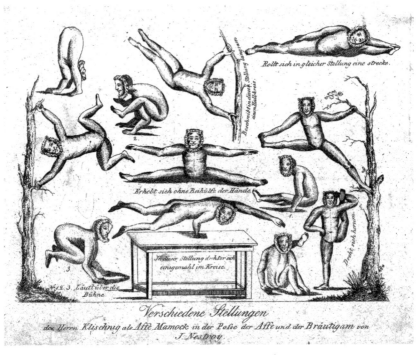

Monkey postures by Eduard Klischnigg (Österreichisches Theatermuseum, Vienna, used by permission)

Theatre, Inverness.

SEVENTH NIGHT OF

Mons. GOUFFE'S

ENGAGEMENT,

THE CELEBRATED CONTORTIONIST AND MAN-MONKEY,

Whose Wonderful Delineations are Nightly received with shouts of applause, and universal marks of general approbation.

This present Evening, TUESDAY, July 19, 1842,

Will be represented the Historical Drama (in Two Acts) of

CHARLES XII.

OR, THE SIEGE OF STRALSUND.

A COMIC SONG BY MR WATKINS.

After which, the favourite Interlude of The

A DEAD SHOT,

OR, MY MAIDEN DUEL.

A SAILOR'S HORNPIPE, IN CHARACTER, BY MR THOMPSON.

To conclude with, for the last time, and by particular desire, the interesting and peculiarly constructed Drama of

Brazilian Ape.

OR   **THE**

MAN-   MONKEY.

The Part of Jocko (the Island Ape) by Mons. GOUFFE.

Programme of the different Incidents in the Drama:—

GROTESQUE APPEARANCE OF THE BRAZILIAN APE.

BAMBOO TREE AND THE ROPE,

CONCLUDING WITH HANGING HIMSELF BY THE NECK !!!

Playbill for Monsieur Gouffe starring in *The Brazilian Ape* (1842)

the structural homology between the two dumb characters is played out through mimicry. The ape also dresses up in a women's cloak and jewels at one point, though for no apparent reason. At the other extreme, at least in terms of novelty, a contortionist man-monkey actor by the name of Monsieur Gouffe, who had become known in the British Isles for "hanging himself by the neck," used the grotesque trick in *The Brazilian Ape* (1842) to portray Jocko's "dreadful demise."[11] In their own ways, each of these performances aligned the ape with abject humans, though the playbill for Gouffe's show depicts an ordinary looking orangutan.

At the narrative level, the man-monkey plays fell roughly into two main categories. The most common category staged the figure of the good ape (following Jocko) to metaphorize the good servant or "savage" that had become a standard device in colonial melodramas.[12] Like the early literary fictions about benevolent orangutan helpers, these plays were typically set in the "new world" and used the device of the shipwrecked seafarer to provide the context for European engagement with an imagined tropics where novel animal species—and nonwhite humans—could be encountered. In theater, however, the ape character was less remarkable as a tractable servant than as a canny trickster able to outwit the mutineers, pirates, and hostile natives whose villainy threatened the moral order. He (never she) invariably "saved the day," usually after all sorts of mischievous antics, by learning how to mimic human behavior sufficiently to hide the treasure, rescue the damsel, dupe the guards, untie the hostages, and even dispose of the villain. *Jack Robinson and His Monkey,* which was frequently revived after its London premiere in 1829, amply demonstrates the genre with its portrait of a British sailor marooned for three years with the daughter of his drowned master on a richly vegetated island in the "Indian Sea." They recruit the ape Mushapeg as a servant of sorts, though the animal is rarely obedient and in fact constantly provokes Jack's wrath with such tricks as burning his stockings and sucking the yolks from turtle eggs he has collected to cook. When a Spanish galleon founders on the "infernal" rock that brought Jack's vessel to grief, Mushapeg finally proves his worth, helping to rescue the commander's son and conquer the mutinous crew that caused the accident. Amid the fracas, the ape throws a pile of plates at the villains in what is applauded as "a good broadside," before knocking their leader flat, then shooting him.[13] Moments later, to the strains of "Rule Britannia," the heroes set sail for England with Mushapeg perched on the ship's masthead.

The chief mutineer in *Jack Robinson and His Monkey* is an African, but his race is a simple visual index of his status as villain and does not have any particular consequences for the plot or characterization. As the popular stage increased its engagement with European and American colonialism over the course of the nineteenth century, however, the man-monkey dramas began to incorporate explicitly racial themes, though not in any consistent way. One popular 1840s play, which circulated under several titles, including *Bibboo, or the Shipwreck* and *The Ourang Outang, or the Indian Maid and the Shipwrecked Mariner,* pits a marooned sailor and the "sagacious ape" who befriends him against a tribe of hostile Indians inhabiting an island off the coast of South America. Bibboo, dubbed "Master of the Zoological Emperor Ourang Outang,"[14] shows his humanlike intelligence in elaborate pantomime sequences. As well as saving the sailor from the island's savage headhunters, the orangutan proves himself more faithful than the beautiful and alluring "nut-brown" Indian girl who, Pocahontas-like, defies her tribe to aid the interlopers. Toward the play's end, the sailor's wife arrives with their child, who engages the orangutan as his nurse, proclaiming "one monkey is as good as another."[15] They are all captured by the Indians, but Bibboo strangles the chief, at which point mariners arrive for a closing tableau, waving the British flag in triumph.

By contrast, the black revolutionaries who function as Europe's Others in *Phillip Quarl, or the Mariner and His Monkey* (c. 1847) come to be admired for their principled stance when French villainy undermines any coalition of white races during a slave rebellion in the "Caribbe Isles." The interracial romances at the heart of this tale are eventually allowed to flourish, yet the text's apparent liberalism is by no means unswerving. Significantly, the mariner's orangutan, Beaufiddle, is the flexible foil against which some of the black characters are defined—or not—as moral humans. The cook, for instance, is rebuked as a cannibal when she offers to roast the orangutan for dinner, and she is also cast as a coward when she refuses to swim a rope to the storm-battered ship, leaving Beaufiddle to take her place even though he is terrified of the water. The noble black hero, by contrast, appears to be (just) fully human in the island's species hierarchy, since he is warned against idolizing a mere "brute" as he thanks Beaufiddle for rescuing his lover from the wreck. The issue of race also shadowed the man-monkey plays by dint of their regular programming in variety bills featuring minstrel shows, ethnological performances or melodramas about slave life. Such evenings could take surprising turns: on one occasion, man-monkey actor Harvey Teasdale deliberately leaped (in full monkey suit) into a group of "Bosjesmans" who had been brought to Liverpool to perform their customs. The audience was apparently delighted at the ruckus that ensued and the venue's manager organized a reprise of the "act" the following evening, after first collecting the bewildered Africans' weapons.[16] An enterprising playwright quickly scripted a new drama called *Monkey of the Plantation* (1847) for Teasdale, while local newspapers likened the Bosjesmans to orangutans and other primates.[17]

These patriotic dramas, constantly adapted for different audiences as they shuttled between Europe and North America, spoke to the geopolitical concerns of their day by meshing fears and fantasies about exotic animals, peoples, and places with issues such as national rivalries and colonial insurrections. In this narrative mélange, in which racialized Others were often excluded from humanity's embrace and cast as animals of a lower kind, the orangutan/ape could temporarily cross the species boundary to align himself with the human family as long as his trickery served the heroes' causes. When overt species comparisons entered the action-driven plots, they mostly served to draw servants, children, rogues, and women into the ape's comic circle or to stress the animal's exceptional imitative capacities.

The second main type of man-monkey play was much more interested in species debates. It drew inspiration from the display of live orangutans in menageries and exhibition halls and was quick to exploit the farcical potential of discourses (both popular and scientific) about their perceived similarities to humans. In these dramas, usually set within provincial Europe, a man masquerades as an ape in a convoluted plot to either win a woman's affection or trick a rival into withdrawing his interest. After bizarre twists to the action that are designed to accommodate the same kind of physicalized stage business as the ape-servant dramas, identities are revealed, young lovers thwart the plans of parents, guardians, or lascivious old men, and the species boundary is eventually reaffirmed, at least at the play's narrative level.

A paradigmatic example in this genre is *The Ourang Outang and His Double; or The Run Away Monkey*, a one-act farce that first appeared as *The Mayor and His Monkey* in the late 1830s. The action is set in a village in France, where a menagerie proprietor has bribed the mayor, Pierre, for permission to exhibit his "caravan of outrageous beasts, lions, monkeys and cockatoos" on a Friday in Lent.[18] Pierre wants to marry a friend's daughter, but she spurns his advances in favor of a university student who is rumored to be dabbling in sorcery. Undeterred in his quest and convinced that "many girls are fascinated by monkey tricks," the older man decides to go to the annual village masquerade as an orangutan in an effort to win her attentions: "I've got a complete monkey-dress within," he tells his friend. "I'll just slip it on in an instant, and show you how I shall look the character to the life" (12). The student overhears the plan and dupes Pierre into believing he is now cursed to inhabit the simian shape for a year; if he speaks to reveal his human identity, the transformation will be permanent. In this guise, the mayor gets mistaken for a menagerie escapee, the wondrous ape Pongo, who can "take his hat off to the ladies, dance on a rope, or smoke a cigar with the air of a prince" (13). Pierre is beaten and taken to the caravan to perform as the ape-cum-gentleman while the real orangutan wreaks havoc in the village with a long sequence of tricks that includes peeling carrots for soup, putting a baby in an oven while he hides in its cradle, shaving an unsuspecting customer whose servant has gone to get a better razor, and dressing himself in women's clothes to parade before a mirror. He then wanders back to the menagerie, arriving just in time to see his human double being exhibited. Pierre, lacking the orangutan's willingness to perform tricks, breaks his silence to call for help, at which point Pongo confronts him, they look at each other in disgust, and the menagerie owner exclaims: "They are both so like, that I hardly know which belongs to me" (18). The orangutan returns to his cage, the student breaks the so-called spell, and Pierre, chastened, relinquishes his claim to the girl before articulating the play's lesson: "Come, strike up the music . . . but, before I go, allow me to communicate a curious zoological fact which I have discovered— viz., that a man isn't a monkey" (18).

The Ourang Outang and His Double lampoons the theory of transformism, which supposed the possibility of species transmutation in ways that challenged the orthodoxy of a divinely created (and hence immutable) natural order.[19] The mayor is ridiculed not just for thinking his orangutan suit would attract the girl but also for believing the simian guise could somehow become real. His credulity prompts a wry remark from a neighbor: "Love turns men into many ridiculous shapes. . . . What a horrible transformation! Yet I can still recognise a strong likeness to your former self" (14). While this ludicrous human-to-ape masquerade casts transformism as pure theatrical invention, Pongo's rampage through the village presents the play's "real" orangutan as a more ambiguous figure. At one level, he is a simian double reflecting human folly; at another, he becomes the potentially dangerous beast against which humanity can be defined. The specter of the beast is encapsulated in a sketch printed with the script in 1842 and probably used on playbills; it shows Pongo as the iconic ape-barber brandishing a cutthroat razor, like the murderous orangutan in Poe's tale (published just a year earlier), as he prepares to shave his human client.

Live performance added another layer of ambiguity to such farces. Whereas the ape-servant dramas simply presented the man-monkey figure as the realization of an exceptional animal character with no overt emphasis on the meta-theatricality of the role, the ape masquerade plays self-consciously highlighted the process of role playing. A key mechanism in their dramaturgy was the on-stage dupe (such as the mayor) whose gullibility served to instruct the audience on the importance of differentiating orangutans (and other apes) from humans. Yet both were played by human actors in ape costumes as the masquerade unfolded, introducing a certain irony to the idea that it is (or should be) a simple matter to distinguish the man/performer (as mimic) from the ape/character (as object of mimesis). In this manner, a skillful performance could stage the fixity *and* the mutability of species borders, sometimes simultaneously. Moreover, the imaginative morphing of physicality between humans and other primates was not always confined to the world of the play. In his memoirs, Teasdale recalls a moment in the 1850s when some spectators, greatly impressed with his performance, laid a wager at a public house over whether he "was not a real monkey" and called for him to settle the dispute by "ocular demonstration."[20]

Several years earlier, in 1847, Hervey Leech, who had played numerous ape roles on the antebellum American stage, was likewise subjected to ocular scrutiny to confirm his species status, but in very different circumstances. Leech, who performed under the name of Hervio Nano, had a normal upper body but his legs were dwarfed to the extent that his hands touched the ground when he walked along. His peculiar physique was suggestive enough to prompt P. T. Barnum to strike a deal to exhibit the actor in London's Egyptian Hall disguised as an intermediary species that explicitly evoked the idea of a human-orangutan hybrid but, cleverly, left viewers to decide on the nature of what they saw.[21] The advertisements were typically sensationalist:

> Is it an Animal? Is it Human? Is it an Extraordinary Freak of Nature? Or is it a legitimate member of Nature's works? Or is it the long-sought for Link between Man and the Ourang Outang, which Naturalists have for years decided does exist, but which has hitherto been undiscovered. The Exhibitors of this indescribable Person or Animal do not pretend to assert what it is. They have named it WILD MAN OF THE PRAIRIES; or, 'WHAT IS IT' because this is the universal exclamation of all who have seen it. Its features, hands, and the upper portions of its body are to all appearances human; the lower part of its body, the hind legs, and haunches, are decidedly animal![22]

Leech's masquerade ended less than thirty minutes after it started when a friend (or perhaps a rival showman; accounts are inconsistent) recognized the actor under his suit of hair and stained face and hands.[23] Exposed, the ruse created an uproar but scarcely diminished public interest in the idea of a "missing link" between humans and apes, even if not an evolutionary one in the way Darwin was soon to imagine. The commercial potential of this sort of hybrid or "nondescript" (a term used at the time for species awaiting classification) became fully apparent during the next decade as crowds flocked to see Julia Pastrana, a hirsute woman from an Indian tribe in Mexico who was exhibited first on Broadway in 1855, then in major cities across Europe until she died in childbirth in 1860 in Moscow. Pastrana had an extreme case of congenital hypertrichosis but a seemingly normal intelligence and could speak three languages. After a New York medical doctor declared her to be "a hybrid between human and orangutan," her impresario (whom she later married) was quick to foster rumors about her parentage, insinuating that her mother had conceived her in a mountain region inhabited only by animals. He presented Julia variously as a bear woman, a baboon lady, or a "hybrid wherein the nature of woman predominates over the ourang-outangs."[24]

The orangutan's association with freakishness intensified in popular entertainments as the nineteenth century progressed, particularly on the American circuits. Within months of the publication of Darwin's *The Origin of Species* in 1859, Barnum seized the opportunity to capitalize on science's evolutionary hypotheses with a reprise of the scandalous "What Is It?" exhibit, this time at his own museum in New York, where he cast a dark-skinned performer as the "missing link" between man and beast. A pamphlet claimed the "marvellous creature" on display had been discovered "in a perfectly nude state" climbing among trees in Gambia. It had a very "awkward walk" and "arms too long in proportion to its height." The placement of the forehead and ears "constitute[d] the perfect head and skull of the Orang Outang while the lower part of the face [was] that of the native African."[25] Possibly alluding to Camper's measurements of facial angles among humans and apes (discussed in chapter 2), the blurb took a quasi-scientific approach to cast the taxonomic issue of species difference indirectly in racial terms. But again, Barnum called the figure he was presenting a nondescript and advertised it by posing questions rather than giving answers: "What Is It?" "Is it a lower order of Man? Or is it a higher order of Monkey? None can tell."[26] James Cook reads this imprecision as a calculated strategy on Barnum's part, one designed both to exploit the proven theatrical appeal of liminal figures and to navigate volatile issues surrounding racial definition and social order in a nation lumbering toward civil war.[27] The exhibit was a resounding success, possibly because its contradictions seemed to mock as well as affirm racial typologies. The performer, a microcephalic African American by the name of William Henry Johnson (also called Zip), continued to present variations of the "What Is It?" character for more than half a century afterwards and became one of Barnum's best-known (and best-paid) "living curiosities." The unspeakable notion of human-orangutan sexual relations was subtly embedded in the exhibition of such hybrids and, for that matter, in the image of the man-monkey suitor in the ape masquerade plays. If "What Is It?" seemed to combine characteristics of different species (and races), it also raised questions about the nature of their intercourse, which were given new impetus with the powerful theory of simian ancestry put forward in Darwin's 1871 *Descent of Man*. It was a prescient move on Barnum's part to shift the figure of the "freak," whose monstrosity had long been linked to cross-species fertilization, to that of a human-ape intermediary who seemed to hold the *scientific* (rather than merely prurient) secret of the origins of humankind.

In terms of narrative dramas, the man-monkey role offered a ready mechanism for staging explicit intermediaries between orangutans and humans, but the standard plots had become tired by the 1870s and were not refashioned as extensively as might have been expected given public interest in evolution. A French operetta staged at the Folies-Bergère in 1875, *L'Homme est un singe perfectionné*, appears to be one of the few later-century man-monkey dramas to offer a new theatrical spin on species relations, though it borrows substantially from earlier texts in its use of the ape-masquerade as a parodic device. In this farce, a savant decides to gradually transform an ape into a human with the aid of a depilatory liquid, *eau simiesque,* together with lessons in manners and morals. He intends to create a "new" man to marry his daughter, but she already has a suitor. To thwart the savant's plan, the suitor dresses as an ape and, after pretending to shed his hair as a result of the experiment, is approved as a son-in-law. Meanwhile, a real gorilla, which has likewise been treated with the liquid, turns into a woman (not a man as anticipated) and runs off with an acrobat, much to the savant's disappointment, as he finds her curiously attractive.[28] Here, the taboo against interspecies sex is scarcely evaded by the female gorilla's unlikely transformation, and there is no unmasking scene to recalibrate the moral order of the play's fictional world.

As Darwin's theories filtered into the public realm after 1859, ape parodies and masquerades must have seemed blunt tools with which to explore the complex field of species relations, even on the popular stage. The figure of the man-monkey had become clichéd amid proliferating images of fanciful simian hybrids, including a cartoon of Darwin himself as a "venerable orang outang." Social Darwinism, one

Darwin as a "Venerable Orang-Outang" (*The Hornet,* March 22, 1871)

A VENERABLE ORANG-OUTANG.
A CONTRIBUTION TO UNNATURAL HISTORY.

of the chief influences on theater proper at the turn of the twentieth century, demanded more nuanced human dramas, and both actors and audiences were beginning to stir against entertainments that invoked racial stereotypes. In 1904, for example, a New York producer was brought before the court for attempting to force a red-bearded Irishman in his troupe to perform a wild-man character made up as a huge ape.[29] By this time, zoo and circus orangutans had been taught to beguile audiences with comic performances as quasi-humans, leaving the character of the stage orangutan, as animated by human actors, to exit quietly to the wings. Buster Keaton's silent film *The Playhouse* offered a brief backward glimpse of the man-monkey genre in 1921, but as a humorous tribute to theatrical forms whose mass appeal had waned with the coming of cinema. Using this new medium's facility for simultaneous doubling, Keaton presents himself as a variety-stage actor who plays all the roles in the show, including the part of a circus orangutan he has accidentally set free from a cage. The comedy turns on Keaton's extraordinary ability to imitate an ape imitating a man as he dines at the table, smokes a cigar, and then jumps, unscripted, into the auditorium, making a woman faint. More subtly, the sequence also gains force from a cameo appearance of the real orangutan whose act has been sidelined by Keaton's sly imposture.

Inadvertently, this film pointed to a question that would resurface periodically as cinema drew the orangutan character into new narrative genres: could the real ape actually play the part? While point-of-view filming could enhance character construction, and footage could be shot, edited, spliced, or otherwise manipulated to shape representation in ways never available to the stage, orangutan actors could not be expected to develop the same physical or emotional repertoires as humans. And they were not always biddable before the camera, as a 1915 report of one film shoot involving "an orang outang of enormous proportions" reveals: "An elaborate stage setting was arranged, but when Chang 'walked on' he promptly pulled up a drawing room carpet and rolled in it, turned a big settee into a pushcart and chased George K. Larkan, the hero, right round the studio with a cane."[30] Not surprisingly, some filmmakers chose to use actors in ape suits for orangutan characters, but this convention had its own constraints. Whereas theater had reveled in the human skill visible in renditions of the man-monkey figure, film developed as an essentially realist medium that was expected to efface signs of artifice unless it was conducive to the chosen genre. This demand for verisimilitude effectively limited the range of orangutan representation in cinema and television until about the 1980s, by which time special effects and costuming techniques were sufficiently refined to screen complicated dramatic sequences with animal actors or to bear the scrutiny of close-up shots in the case of the humans in ape suits. By and large, earlier films cast real orangutans as quarry, or pets, in exotic jungle adventures, while human actors scored the part of the monstrous ape unleashed on society in horror and related genres.

The early jungle-based adventures featuring orangutans were in many ways a ready fit for film, which was invented in the 1880s as an aid to examining the physiology of animal motion.[31] Moving pictures of the animal world not only supplied scientists with research data but also promised the general Western populace visual (and emotional) access to "authentic" forms of nature apparently untouched by the ravages of industrial modernity. On the screen, spectators could watch with amazement as scenes conjured in travel narratives and habitat dioramas came alive, often with the added frisson of showing humans and wild animals in close proximity. Jungle quest fantasies shot at least partly on location in tropical regions began to appear in cinemas soon after Edgar Rice Burroughs published *Tarzan of the Apes* in 1914. While the novel situates its orphan boy who is raised by apes specifically in equatorial Africa, its thematizing of the jungle as a place where masculine virtues were honed, tested, and restored could be easily adapted to human-ape encounters elsewhere, as adventure stories set in the East Indies had already demonstrated.

Drama was an essential ingredient in the jungle narrative formula, and filmmakers sometimes provoked aggressive behaviors by setting up fights between animals to enliven their plots. *Rango,* a silent film shot in Sumatra in 1931 with a native cast, features an orangutan slain in battle with a tiger, leaving the ape's father to mourn in the trees.[32] Footage of such scenes could be recycled through different films, thus making certain images iconic. A 1930 feature titled *Ourang,* shot in Borneo, was apparently never brought to completion, but sections of it subsequently found their way into other works, including *The Beast of Borneo* (1934), which the expedition leader Harry Garson directed. Advance advertising for the unfinished *Ourang* promises "virgin jungle color" in "an absorbing tale of love and sacrifice in which a white derelict and a native girl find the only road to happiness stemming the stampede [*sic*] of the fierce ourang-utans."[33] Accompanying the blurb is a sketch showing an orangutan carrying off a protesting woman while two smaller apes bring up the rear. By the time Garson assembled his film three years later, the ape-abducts-woman fantasy had been realized on screen in a definitive version in the gigantic gorilla-like figure created with pioneering special effects for the original *King Kong* (1933). *The Beast of Borneo* wisely took a different tack, maximizing its footage of real apes, around which it crafted a melodramatic story of an American expedition leader pitted against a mad Russian scientist in a quest to capture an adult orangutan.

Posters for the film, perhaps building on *King Kong*'s success, show a scantily clad woman with a ferocious ape sketched in the background and a question alluding explicitly to Serge Voronoff's monkey gland experiments, which had attracted considerable interest in the 1920s: "Can gorilla glands bring back passionate youth?" The "Beast of Borneo," however, is quickly established as a flanged male orangutan in the opening footage. He has just escaped capture, but his infant son is scooped up by Ward (the American) as an appealing pet who also becomes a vehicle for depicting this species' intelligence. Borneo Joe, as the youngster is called, accompanies Ward when he leads Borodoff (the scientist) and his attractive assistant, Alma, in search of their quarry. After three weeks trekking in the "green hell of the jungle" they eventually capture an adult orangutan by luring him down from the trees with fruit soaked in bottles of gin, but Ward lets the animal free when it becomes clear that Borodoff intends to harm it. The beast eventually repays this kindness, saving Ward and Alma by mauling Borodoff to death as he is about to shoot them. Meanwhile, Borneo Joe has narrowly escaped becoming mad science's next primate victim by undoing the knots binding him to a makeshift operating table where he is about to be vivisected. Although the film deals in stereotyped images of orangutans and jungle life, it invites empathy for the apes and shows they are able to distinguish between "good" and "bad" humans, if only intuitively. As for the depiction of orangutan savagery, alert viewers might notice that when the beast attacks humans it takes on the distinct look of an actor in an ape suit.

The most enduring representations of orangutans in the jungle adventure mode are found in the stories, cartoons, and films of Frank Buck. In 1932, he adapted his bestselling book *Bring 'Em Back Alive* into a blockbuster film of the same name featuring the capture of a massive orangutan alongside footage of crocodiles, tapirs, and a tiger locked in battle with a python. The orangutan sequence was included in a radio show spin-off from the film and later reappeared in *Jungle Cavalcade* (1941), a compilation of Buck's earlier screen hits. Among other enterprises, Buck created a simulation of his jungle camp for more than two million visitors at the 1934 Chicago World's Fair, where he incongruously exhibited his pet orangutan in a sailor's suit, smoking a pipe.[34] What made these ventures so successful was that Buck put himself in situ as the dauntless hero, combining suspenseful narration with realist photography to convey firsthand the perils of his encounters with the jungle's denizens. Describing leeches, insects, and steaming heat as all part of the quest, he conjured a sensory world in which spectators could imagine the "Giant Jungle Man" as something more formidable than the hapless orangutan presented on screen. This mythical giant was later realized in a comic mode (played by an actor) in the Abbott and Costello film *Africa Screams* (1949), in which an odd assortment of adventurers, including Buck as a big-game hunter, find themselves in the Congo to search for the "orangutan gargantuan." The creature appears with impeccable timing to thwart diamond-smuggling villains and save the chief coward, Stanley Livingstone, from murderous natives.

After the 1940s, orangutans rarely played central roles in adventure films, though they can be glimpsed in some memorable vignettes, including a famously erotic scene in the 1981 remake of *Tarzan*, starring Bo Derek and Miles O'Keeffe. Here an orangutan (whose appearance in the jungles of Africa is never explained) and two chimpanzees witness Jane's sexual awakening as she caresses Tarzan after he falls unconscious, exhausted from his struggle to free her from a massive python.[35] Whereas the chimpanzees express concern for their injured master, the orangutan becomes a voyeur of sorts in several crosscut shots that show him/ her watching Jane's movements closely before breaking into a lascivious grin. This brief visual parallel between orangutan and human desires is then subsumed into a narrative of species difference: Jane pronounces that Tarzan is "just a man," not the legendary "great white ape" she had feared, upon which he rises from his stupor to lead her to a nest in the treetops where they continue their foreplay in private. (In Burroughs' novel, by contrast, Jane has a much ruder introduction to male sexual desire as she is abducted by the ape Terkoz and almost raped.) The orangutan and chimpanzees essentially disappear from the film at this point, though the long credits sequence at the end rolls out over background images of Jane and Tarzan, apparently in a moment of postcoital bliss, having a romp on a jungle beach with the orangutan. At first, in a reversal of the earlier scene, Tarzan merely watches as Jane, topless, tumbles about with the animal, which at one point sits astride her. Then Tarzan joins the game, vying for Jane's attention. Many of his actions are choreographed to mirror the orangutan's, and there is a brow-raising shot of both man and ape chasing Jane up a dune as the camera lingers on her shapely buttocks. While the scene titillates (perhaps even disturbs) with its suggestiveness, the orangutan soon becomes the odd "man" out, not a character or rival as such but an exotic presence enhancing the film's general appeal to primitivism.[36]

Interspecies desire and miscegenation between orangutans and humans figured, if at all, as peripheral specters in the adventure films, but such tropes became thematically and structurally fundamental to representations of great apes in early horror cinema. As a popular genre, the ape horror film began with the 1932 adaptation of "Murders in the Rue Morgue" for Universal Studios, although works such as *Balaoo: The Demon Baboon* (1913) are clearly antecedents. Once again, art drew its inspiration partly from evolutionary science in shaping what was to become a classic trope: the anthropoid monster. Around the turn of the twentieth century, there was a fresh burst of interest in the possibility of creating human-ape hybrids. An eccentric Dutch teacher, H. M. Bernelot Moens, made serious plans for formal experiments to fertilize female orangutans, gibbons, chimpanzees, and gorillas with human sperm, as well as human females with ape sperm. To avoid the moral problems associated with bestiality, he proposed to use artificial insemination techniques.[37] Film took license to give such imagined hybrids visual form and, in some cases, complex emotional lives. It also seized on the figure of the "mad scientist" as the monster's creator, thereby probing the issue of moral responsibility in the quest for scientific knowledge.[38]

Although Lugosi titled his film *Edgar Allan Poe's Murders in the Rue Morgue,* it diverges markedly from the source text. The chief human protagonist is not Poe's detective, Dupin, who figures as a secondary character in the adaptation, but the menacing Dr. Mirakle, a carnival barker with a passion for evolutionary genetics. Mirakle exhibits his sagacious ape, Erik, at the town fair as part of a diatribe trumpeting evolution's marvels, but his clandestine project is to create a human-ape hybrid as proof of Darwin's theory by injecting a woman with Erik's blood. Erik helps to kidnap potentially suitable victims—only a virgin will suffice—and he is instrumental in their murders when the experiments fail. Finally, he rebels against his master to save Camille, a woman he desires for himself, killing the mad scientist before escaping across the rooftops with the maiden slung across his shoulders. The forces of law and order arrive just in time to give chase until the dramatic finale when Erik, fatally wounded, plummets into the river Seine, leaving the terrified Camille to fall into Dupin's arms. Paul Woolf reads this Hollywood trailblazer in its American context as navigating a path between conservatives and modernists in the wake of the controversial Scopes trial, which found a Tennessee high school instructor guilty of breaching state laws by teaching the theory of evolution.[39] The film gives no verdict on the origin of the human species, though it toys with the figure of the ape rapist as a possible progenitor of novel hybrids. Visual contrasts between the monstrous dark ape and its white female victims add a racial subtext that would not have been at odds with Poe's post-emancipation story.

Most subsequent ape horror films likewise subsumed the various great ape species into a composite cinematic vehicle by which to probe human fears and desires, particularly in relation to female sexuality. With a few notable exceptions, the generic ape suit, enhanced by special effects, sufficed to create the semblance of the beasts and monsters central to the genre, but the animal most often invoked was the gorilla, not the orangutan.[40] One film that does specifically draw on the orangutan in representing its human-ape hybrid (sans costume) is *Dr. Renault's Secret* (1942), featuring character actor J. Carroll Naish, who specialized in ethnic roles. Naish plays a mysterious émigré from Java, Noel, who is both ridiculed and feared by the inhabitants of the small French village in which the action is set. He speaks only with effort and has a vaguely simian look, a heightened sense of smell, and an air of melancholy, but he is devoted to the beautiful niece of Dr. Renault, the scientist with whom he lives. Noel is learning the ways of civilization with Madelon's help, although she remains unaware that her uncle's protégé is in fact his experiment: with plastic surgery, operations to the brain, nerve grafting, and glandular injections, Dr. Renault has transformed Noel from an orangutan into a man. Only his moral capacities remain in doubt. The experiment begins to unravel when Madelon's American fiancé arrives, making Noel painfully aware of his own inadequacies as a human. He is accused (wrongfully) of attempting to murder this interloper, prompting Dr. Renault to confine him to his cage. Noel escapes to the local fairground and, with the atavistic savagery of the beast, kills two villagers who ridicule his attempts to dance. Inevitably, the saga ends with the death of the unfortunate hybrid, but not before he redeems some of his humanity by saving Madelon from the clutches of the other villain of the piece, an ex-convict who abducts her, possibly for a ransom. The film gives emotional substance to its less-than-human, more-than-animal protagonist, while the other characters remain thinly developed. Like the monkey critics discussed in chapter 5, Noel is also endowed with a voice to censure science's hubris: "Why you no left me like I was? I remember good things. You take me away, make me like a man. Why? To make me happy? No, to make you big man!"

The ape monster film had largely exhausted its currency by the 1960s, though periodic remakes of classics such as *King Kong* still drew massive profits at the box office. With this development, the actor-animated ape had only limited uses as a screen character, and interest soon turned to trained orangutans, chimps, and gorillas to play primate parts, mostly in comedy. In a few cases, real apes were recruited to star in horror films, with unsettling effects. The chief orangutan-centered example in this mode, *Link* (1986), unfolds in a rambling manor in Scotland, where a university professor keeps great apes for research into primate intelligence. As his name suggests, Link, the orangutan, is closest to humans in both behavior and mental capacity. Retired from the circus, he acts as Professor Phillips' inscrutable butler, seemingly polite but slightly sinister from the start in his starched vest and trousers. The two remaining inmates are chimpanzees, a caged adult and an infant called Imp. They are implicated in the orangutan's vengeful campaign against humans for a lifetime of subjugation, but the film focuses primarily on Link's stalking of Jane, the newly arrived student assistant. Jane only gradually realizes her predicament because she underestimates the orangutan's intelligence even as he plans his moves, sets up traps, and plays psychological games to allay her suspicions. She also fails to understand the depth of his desire for her affection. One provocative scene shows her undressing to take a bath when the orangutan intrudes, staring at her nakedness. She holds his gaze for a moment, then casually covers up as the camera cuts from his face to hers before zooming out to capture the pair briefly in the one frame as an erotic couple. Even though Link's expression gives nothing away, the scene works to amplify his emotional complexity so that he registers as more than a predatory animal. We also see the hurt that catalyzes his violent rampage when Jane excludes him from her games with Imp. Eventually, the horror (for both human and animal) ends with Jane's escape while Link perishes in an explosion after she pipes gas into his room and persuades him to light up a cigar, a trick he learned in the circus. Within its genre, the film's power to disturb despite the hackneyed plot rests primarily in its realistic rendition of the orangutan. As one reviewer remarks, "at no point do you think 'ah look, he's just playing' or 'it's a bloke in a suit!' "[41]

Cinema's turn toward casting real orangutans in narratives that highlighted their intelligence, even if to dramatize anthropocentric perspectives, drew impetus from decades of behavioral studies and cognitive testing in ape research (see chapter 10). The underlying thrust of a great deal of this research was to investigate the animality of humans rather than the humanity of apes. Desmond Morris' bestselling popular science work *The Naked Ape* (1967) is an exemplar of the trend insofar as it annexes large areas of human behavior (from its conventional placement within the dynamics of human culture) to evolutionary imperatives and to instinct. Identifying humanlike behavior in great apes tended to confirm the simian affinities of humans, not the humanness of apes, though fictional elaborations of the unfolding research sometimes took license to imagine versions of ape societies that probed species similarities—and differences—from ape as well as human perspectives.

Released just a year after Morris' book, the first screen version of *Planet of the Apes* (1968) signaled film's nascent interest in the creative twists that could be applied to serious comparisons between human and ape intelligence as well as to interactions among the species, although director Franklin J. Schaffner explored such themes via costumed actors and not actual animals. The film presented all its ape characters in masks that conveyed generic simian features—color-coded to distinguish between orangutans, gorillas, and chimpanzees—but also admitted human characteristics so that the effect was to position the apes as not only uncanny mirrors for humanity (in ways lending themselves to the racialized interpretations of the film discussed in chapter 5) but also as simians whose revolt against humans could be cast as an understandable response to oppression and exploitation. Tim Burton's 2001 remake achieved much greater visual realism with its depictions of the various ape groups, winning accolades for costume design, but it generally has been judged a lesser film. Although the orangutan characters are more individuated, they also seem more villainous, diminishing the narrative's critique of anthropocentrism and interspecies ethics despite its explicit engagement with these issues.[42]

For the most part, the 1970s sequels that make up the original five-film series of *Planet of the Apes* continued to feature chimpanzee characters as the main nonhuman protagonists, while orangutans reappeared as allies and bureaucrats in conflict-driven plots exploring issues of the day, including rights, nuclear war, experimentation on animals, and, more elliptically, species hierarchies. Dr. Zaius, the orangutan science minister, figures prominently in the second film, *Beneath the Planet of the Apes* (1970), but is not drawn with any great depth. A clever tactician, he leads the successful simian conquest of the Forbidden Zone, having seen through the fire mirages created telepathically by its irradiated human inhabitants in an effort to keep the apes at bay. He also acts as judge and jury in the closing scene, refusing the last human appeal for clemency before the doomsday bomb ends life on Earth.

Orangutan mask from the 2001 *Planet of the Apes* remake
(Profiles in History, used by permission)

Interestingly, the 2011 prequel to the series, *Rise of the Planet of the Apes,* departs from the earlier pattern of simian representation by showing the orangutan character as the sole ape with natural intelligence, though he plays only a secondary role in what is essentially a backstory telling how the apes came to be the dominant species on the planet. Set in San Francisco and explicitly shaped as a critique of biotechnical science, the film focuses on the chimpanzee Caesar, whose extraordinary mental powers derive from artificial means: an experimental gene-repair drug administered to his mother to test its potential as a cure for Alzheimer's disease. The drug's intelligence-enhancing effects pass from mother to son in utero before she dies in an attempted breakout from the laboratory. Caesar is raised by a scientist in a suburban household, where he excels at sign language and complex cognitive tasks, until officials abruptly move him to a crowded ape shelter after he attacks an aggressive neighbor to defend his human family. In the shelter, Maurice, a wise old orangutan, becomes Caesar's friend and teacher. He is the one fellow ape who speaks the chimpanzee's language, having learned to sign in the circus, and who understands the painful choice Caesar will have to make between humanity and apehood. "Careful, human no like smart ape," Maurice warns at their first acquaintance. His natural intelligence is all the more remarkable beside the brute instincts of the shelter's chimpanzees and gorillas, but such distinctions dissolve after Caesar supplies the entire ape contingent, including Maurice, with the gene therapy. They are then armed with the intelligence and will to escape and, it is intimated, to take control of their own futures. The therapy contains a virus, harmless to apes but fatal to humans, that has leaked into the atmosphere during the experiments and that will soon decimate the human population. In the final moments of the film, Maurice holds out his hand to Caesar in a supplicating gesture, indicating that he recognizes the chimpanzee's authority as leader of the yet-to-be formulated ape civilization. The film's ape characters are expertly played by human actors with the help of sophisticated special effects that collapse the visible distinctions between those animating the characters and the particular species rendered. The orangutan, animated here by a woman (Karin Konoval), is named after the late Maurice Evans, who was cast as Dr. Zaius in the earlier *Planet of the Apes* films.

Back in 1978, when *Every Which Way but Loose* launched a spate of comedies featuring real orangutans in major roles, there was a much greater difference between the ways in which apes and costumed humans could register on screen. Like the man-monkey dramas staged more than a century earlier, these films turned on the ingenuity of an unusual actor who added verve and humor to well-worn scenarios by leaping across the shrinking species gap between apes and humans. That leap gained comic resonance from incongruity between the character and the action, so in this case the "act" worked best if the performer in question was recognizably an orangutan placed in human contexts. Whereas the adventure genre had required exotic jungle settings and the horror films preferred atmospheric country houses or villages, the comedies drew force from situating orangutans in urban or semi-urban societies. There, they could parody social and sexual mores, trick people into compromising situations, undermine species hierarchies, and generally create mayhem, all to hilarious effect. The actual repertoire of the orangutan actor was not expansive but could be made to seem so through clever narrativization, aided by judicious filming and editing. In addition, narrative film featuring real orangutans lent itself well to depictions of interspecies intimacy, a topic of great interest to audiences now better informed about human-ape affinities. This trope was typically manifested through a "buddy friendship" between an orangutan and a man or male child, marking a shift from the erotic pairing of woman and ape that characterized the horror genre. Thus the orangutan was again recruited to stories about (human) masculinity, albeit in a different mode from the bravado of the jungle quest films.

The combination of a rough-and-ready bare-knuckles fighter, Philo Beddoe, and his orangutan sidekick, Clyde, in a freewheeling search for love and a little spare income proved to be a winning formula in this comic turn. *Every Which Way but Loose* and its 1980 sequel, *Any Which Way You Can,* not only drew handsome profits at the box office but also enhanced the popularity of screen star Clint Eastwood, despite being panned by critics. As Beddoe and Clyde maneuver their way through pub brawls and drinking sessions, police blockades, sleazy hotel rooms, and numerous showdowns with the Black Widow motorcycle gang, the films express both a vision of egalitarian mateship and a fantasy of kinship between species. Unlike the women in Beddoe's life, Clyde is uncomplicated and trustworthy, even if inclined to minor mischief. His discretion makes him an ideal confidant and his strength proves convenient when man-muscle looks insufficient to get Beddoe out of a tight corner, but in fact Clyde ducks most of the fights, even covering his eyes and grimacing. His secret weapon against macho adversaries is a sudden, sloppy kiss on the lips, a tactic repeated in numerous orangutan comedies. Such gestures work to temper the violence of Beddoe's world and even skew its versions of masculinity.

Beddoe himself never treats Clyde as anything but an intelligent ape who is equal to—or even better than—a person, but these films' representations of species equivalence can only ever be provisional since their comedy inheres in an emphasis on the imperfect symmetry between ape and human. A roadside scene in the second film, for example, is played for laughs when Clyde stands with Beddoe and a friend to urinate while the camera rests on their naked male backsides, all lined up in a slightly odd row. The running visual discourse of species likeness with a critical touch of difference also extends to male sexuality. Both films show Beddoe taking Clyde to the zoo for a rendezvous with his "girl" Bonnie so he can share his human friends' carnal pleasures. In *Any Which Way You Can*, this sex sequence unfolds as a series of images of the orangutans romping in a hotel bed intercut with scenes between Beddoe and Lynne next door and, in another adjacent room, snaps of an old couple who are inspired by the noises to reignite their desires. In this instance, although the narrative works to animalize the humans, Bonnie and Clyde still seem somewhat gauche—and curiously chaste—in their interactions.

Paul Smith notes that Clyde's role in these films is as an "alternative human" who embodies what is normally elided in constructions of the heroic male figure.[43] Hollywood's next popular installment in the comic genre, *Going Ape* (1981), followed suit in positioning its three apes in human circles but made only limited attempts to plumb the intelligence and intimacy themes. The farcical action revolves around three ex-circus orangutans who are bequeathed to a young man, along with five million dollars that he can only access if he keeps all three animals alive and safe from the clutches of scheming zoo officials and sundry opportunists. While the orangutans continually thwart their would-be captors, they are weakly individuated as characters and there is little in their "monkey business" to position them as anything other than clever buffoons. The opening credits specifically state that the film's stars are "Bobby Berosini's orangutans" and indeed its plot seems expediently built around the routines developed for the famous trainer's live act in Las Vegas at the time (see chapter 8). Among other trained orangutan actors whose talents inspired specific screen works is C. J., the eponymous star of a short-running 1983 television sitcom titled *Mr. Smith,* which features a primate genius with an IQ of 256. Having mastered the intricacies of law, medicine, and nuclear physics, he is recruited to work as a special consultant to a government think tank in Washington. His periodic demand for human rights, which is a running gag in the series, points toward philosophical debates about ape rights and the status of animals in human society, but the drama's core action rests on family contretemps.

The pairing of children with orangutans opened up a different perspective on the issue of comparative intelligence, while fitting neatly into a well-established screen genre showing children and certain animals as steadfast companions. *A Summer to Remember* (1984) focuses on a growing friendship between an orangutan, Casey, and a boy, Toby, who has become mute after a bout of meningitis. In

Mr. Smith, television series, 1983 (NBC Productions, used by permission)

their own ways, both are "lost" figures. Casey has been separated from her owner after a van accident, and Toby is caught in the maelstrom of a speech-centered world from which he feels alienated. After initially being terrified of the "gorilla" who turns up in their treehouse, Toby and his sister are astonished to discover that Casey is not only benign but also able to communicate in American sign language, which, she tells them, she has learned from a human. This shared conceptual language becomes the basis for reassessments of species difference—on the part of both the children and the orangutan—particularly when Casey asks Toby why he does not talk. The action includes various comic gags, mostly at the expense of adults, folded into a serious climax when Toby finally speaks again to save the orangutan from being shot. This moment marks the beginning of the boy's reintegration into conventional family structures and of Casey's return to the primatologist "mother" who raised her. Although the orangutan is positioned as a child of sorts, the film does take seriously its capacity for emotional and intellectual intimacies. Casey's facility with language goes well beyond the simplified gestural communication that is typically used in such genres (and Eastwood's adult "buddy" films) to convey animals' bonds with humans or with other animals. As an officially recognized language that deals in abstract symbols rather than gestures, signing conveys a more complex interspecies dialogue, albeit within a romanticized framework in this story.

A Summer to Remember possibly draws on the case of Chantek, an infant male orangutan taught sign language as part of an experiment begun in 1978 by Lyn Miles, an anthropologist at the University of Tennessee. Raised in constant human company but isolated from other members of his species, Chantek acquired a vocabulary of about 150 signs, which he used not just to respond to his keepers but also to initiate conversations. As well as understanding that things, actions, and living beings had characteristics that could be described with language, he asked his teachers for new signs to identify unknown objects, spontaneously executed some signs with his feet, and even invented five signs of his own, including "eye-drink" (contact lens solution) and "no-teeth" (to indicate he would not bite in rough play). Chantek's response to verbal language learning was equally adept. Miles reports that he developed sign-speech correspondence without intentional training and that he appeared to understand hundreds, probably thousands, of English words. He also used signs to deceive his keepers, claiming on one occasion that he had eaten an eraser when in fact he had hidden it in his room.[44] At eight years old, however, he escaped from his compound and broke into a university office. To prevent a repeat of the offense, he was transferred to the Yerkes Primate Research Center, where his enculturation to humans was regarded as a mistake and efforts were made to "put the animal back in him."[45] Reflecting on her experiment, Miles notes that she undoubtedly experienced Chantek as a person and that he had met the definition of this category of being, not only through his problem-solving abilities and mastery of abstract language but also in terms of "invention, will, consciousness and conscience."[46] Films representing orangutans as close companions of humans tend to stress such attributes in their ape characters, thereby implicitly according them personhood, if sometimes contingently.

The 1990s brought comedies engaging in a more sustained way with the issue of orangutan captivity in urban societies, if sometimes obliquely. *Dunston Checks In* (1996), for example, suggests the harshness involved in ape training regimes, even while the film itself harnesses the quirky energies of the simian actor. As Dunston wreaks havoc in a luxury hotel when he escapes from his owner, who uses him to burgle the rooms of rich guests, we catch glimpses of the animal's cruel treatment at the hands of corrupt or narrow-minded humans. Most of the scenes are played for laughs but their resonances are unmistakable. At one point, Dunston knocks for attention from within a large crate that transports him to the crime site; at another, he makes rude gestures to his owner but carefully keeps out of reach to escape being beaten with a cane. There are also moments that hint at the orangutan's displacement from his jungle habitat, again within the film's farcical plot structures. "*Pongo pygmaeus!* You've got an orangutan problem," the wildlife control officer proclaims after flicking through mug shots to identify the exotic ape pest plaguing the hotel. Later, he stalks Dunston through a jungle greenhouse, only to find the orangutan hanging from the barrel of his stun gun. Such gags are interspersed with scenes pairing Dunston with the hotel manager's young son, either in mirrored actions or moments of shared trust such as when the boy and his teenage brother remove a large glass splinter from the ape's bleeding hand. The story ends with a repatriation spoof of sorts: Dunston relocates with his new human friends to Bali, where a final shot shows him sitting comfortably in the trees with two other orangutans—possibly his wife and child—about to drop a coconut on a hapless guest. In a similar mode, the closing scene of *Babe: Pig in the City* (1998) pictures the orangutan, Thelonius, rescued from the depredations of an urban menagerie where he had earned his keep performing in an ape act. Acculturation to human ways is not so easily reversed in this instance, however; while the chimpanzees in his troupe take to the trees on arrival at Farmer Hoggett's rural haven, Thelonius finds his new niche helping humans with the laundry.

By the end of the twentieth century, the question posed by Buster Keaton's comic orangutan act—"Could the real ape play the part?"—had been answered with a resounding "yes!" But there was no consensus as to whether that counted as self-conscious performance despite many reports of the particular intelligence orangutans had applied to their work.[47] In 2004, the American Academy of Television Arts and Sciences ruled that BamBam, an orangutan starring as Nurse Precious in the long-running soap opera *Passions,* could not be nominated for an Emmy Award on the grounds that "a line of distinction" had to be drawn "between animal *characters* that aren't capable of speaking parts and human *actors* whose personal interpretation in character portrayal creates nuance and audience engagement."[48] Around the same time, the show attracted protests on two different fronts. The Center for Nursing Advocacy objected to the implication that a nurse's job could be done by "trained monkeys,"[49] while PETA (People for the Ethical Treatment of Animals) campaigned for BamBam's retirement as part of the organization's larger effort to stop cruelty to animals in the entertainment industry. No Reel Apes, an alliance of artists and scientists launched by primatologist Jane Goodall in 2005 with the specific aim of eliminating ape actors, helped to consolidate a decade-long trend away from using orangutans in screen fiction. By this

Poster for *Creation* (2009) directed by Jon Amiel (Recorded Picture Company© 2009, used by permission)

time, the red ape could be seen much more readily in television documentaries concerned with species conservation (see chapter 9).

The judgment that BamBam could *be* a character but not *create* one suggests the extent to which anthropocentric perspectives continue to influence what is understood of orangutan behavior and cognition. Documentary maker Nicolas Philibert makes a point of exploring this issue in *Nénette,* his minimalist 2010 portrait of an elderly female orangutan at the Jardin des Plantes zoo in Paris. The film works specifically to subvert the expectation that Nénette's life and character will be captured on screen in any comprehensive way, inviting viewers to ponder instead whether the casual voyeurism promoted by zoos merits a lifetime of captivity for the animal. This theme is set up in the opening frame when noisy zoo visitors are reflected in a close-up shot of Nénette's pupil so that her eye/look becomes a mirror and a mask rather than a means of interaction with humans.[50] Subsequently, the camera focuses exclusively on Nénette and her cage mates, while the soundscape encompasses comments and stories from various sources—curious child and adult onlookers, zookeepers, art school students, and finally Buffon's *Histoire naturelle*—so as to emphasize ways in which human viewers construct the object of their gaze. The overall effect is to suggest the unknowability of the orangutan even while we glimpse her intelligence. In this respect, *Nénette* rejects the simple fantasy of connection—encapsulated by the iconic image of human and ape reaching toward each other with pointed fingers almost touching—that has inspired theater and screen productions for almost two hundred years, from Mazurier's early performances as Jocko in the 1820s to the 2009 film biography of Darwin, *Creation*.[51] What Philibert offers instead is a provocation to consider the ethics of human encounters with the orangutan, particularly in the entertainment industries.

8 ❧ Zoo Stories
Becoming Animals, Unbecoming Humans

Ideas of human responsibility for animals, along with debates over the ethics of their captivity, began to shape the orangutan's appearances in zoos more than a century ago, albeit in a more limited range of ways than is evident in fiction or film. By the early 1900s, innovative zoo directors were attempting to present exotic animals in more humane, quasi-natural contexts rather than in bare cages. This mode of presentation lent itself well to incorporating educational and scientific goals as part of the Western zoo mandate, leaving institutions that did not establish themselves in such terms to continue as mere animal show grounds.[1] The First and Second World Wars interrupted the momentum but not the trajectory of this "new" zoo movement, which has expanded since the 1950s, aided by a spirit of transnational cooperation in developing zookeeping techniques and in breeding and exchanging animals. From the 1960s, educational agendas began to inform zoological practices more overtly, resulting in displays aimed to position humans as incidental observers of natural behavior rather than as spectators anticipating a show. More recent exhibits of rare species in zoos and animal parks have foregrounded survival projects, inviting visitors to become involved in species conservation, usually via financial contributions. The landscape immersion concept, integrating animals, plants and human visitors in carefully controlled walk-through enclosures, has also emerged as an alternative display strategy. In some zoos, virtual tours and interfaces add a sense of close proximity to exotic inhabitants, allowing zoo-goers to "experience" them in apparently more ethical ways.

Despite these efforts to reinvent themselves, most zoos continue to function primarily as entertainment ventures that graft research, education, and conservation onto a "recreational rootstock."[2] Such institutions have played contradictory, and sometimes controversial, roles in presenting and representing the orangutan as animal, in part because they necessarily privilege spectacle-based human-animal encounters even when exhibits strive to emphasize natural ecologies. Orangutans are not uniquely affected by this dilemma, but their recognized ability as "natural" actors has favored their recruitment in staged zoo entertainments, particularly as anthropomorphized mascots or zoo icons. Such performances have strengthened public perceptions of the red ape's affinity with humans, while simultaneously showcasing its animality. Although twenty-first-century zoos may work to disassociate themselves from that history, its legacy is abundantly evident in Internet clips, media reports, and zoo memorabilia. Yet the roles played by orangutans within zoo cultures and related ventures such as circuses add up to more than a simple narrative of clever—or vulgar—performances. The complexity of human-orangutan bonds has been a particular focus of zoo stories, along with accounts of the ape's intelligence. Ideological and quasi-scientific agendas have been enfolded into both training practices and shows. Not least, orangutan performances have also galvanized animal welfare debates and high-profile legal battles.

The principles of naturalistic presentation that revolutionized zoos at the beginning of the twentieth century were not easily transposed to exhibits of zoo apes housed in escape-proof enclosures in temperate climates. Yet orangutan performances were part and parcel of a larger aim behind zoo modernization: what Nigel Rothfels calls the project of "managing eloquence," or directing spectator attention away from imagining the experiences of animals in captivity.[3] The German architect of the changes, Carl Hagenbeck Jr., had extensive experience as an animal handler and by the late 1870s had built up Europe's largest live animal dealership, along with a crowded entrepôt-cum-zoo in Hamburg, which contained numerous valuable species. He also became famous for his so-called anthropological-zoological exhibitions, massive undertakings that juxtaposed native peoples with trained animals in shows attracting up to 100,000 spectators a day in some venues.[4] Hagenbeck's innovations in zoo design were driven by an apparently genuine desire to improve the lot of captive animals. But he was also a master at theatrics and used his skill in this area to develop cage-free exhibits via the zoo "panorama," a composite structure of tiered enclosures "laid out as stages" separated from each other and distanced from the public by concealed moats.[5] Within this mise en scène, Hagenbeck could present a mixture of species, even predators and their prey, as part of one unified dramatic image to communicate a harmonious, if highly aestheticized, version of the natural world. He patented his design in Germany in 1896 and opened his own Tierpark (animal park) in the Hamburg suburb of Stellingen in 1907, with vast Arctic and African panoramas as initial features. Although some of his contemporaries scoffed at the theatricality of his exhibits, nobody could deny their sheer beauty or their appeal to urban crowds. The Tierpark's modes of display were quickly adapted across a range of zoological institutions worldwide and remain influential today in efforts to present geographical, habitat, or behavioral groups.[6]

Two Borneo orangutans, Jacob and Rosa, were among the initial exhibits in Hagenbeck's popular park. Hagenbeck believed that the "naturally high" intellectual capacities of anthropoid apes only fully developed by "constant association with human beings."[7] He devised training regimes to enhance human behaviors acquired by the apes in captivity and presented the results of this acculturation to the public at the zoo's primate tea parties. His 1912 memoir, *Beasts and Men,* describes how the chimpanzee played waiter at these events, "bringing the food with great pomp and ceremony," while the other apes, also dressed for the event, sat patiently awaiting their several courses.[8] Hagenbeck also recalls that the three apes often collaborated to have fun at the expense of visitors: Jacob and Rosa would hold out their hands from the cage to greet guests, luring them close enough for the chimp to snatch their hats. After a time, Hagenbeck tired of paying for new hats and built a barrier around the cage to keep spectators at a distance, eventually replacing it with a glass wall that prevented them from giving the animals unwholesome food.[9]

At one level, the apes' antics as ready entertainers and mimics contrast starkly with the images of ostensibly wild nature presented in the Tierpark's panorama exhibits. Yet the tea parties were consistent with Hagenbeck's philosophy of zookeeping as well as with his theatrical bent. These performances subtly incorporated the apes into his model of the modern zoo as a sanctuary where wild animals could follow (and display) their natural propensities—in this case, for acquiring complex physical and social skills—in a safe environment presided over by benevolent humans. Hagenbeck's memoir explicitly casts the Tierpark as a unique institution for "the friendly intercourse of great gatherings between beasts and men." His exhibits, he argued, rendered it "not a place of captivity, but a happy and contented home."[10] Several photographs of the apes around a dinner table and one picturing a Somali child arm in arm with an orangutan illustrate the chapter, as if to support the author's claim. One shot shows the two orangutans, unclothed, in what looks like a training session, their skill at eating with utensils not yet perfected. Hagenbeck's approach to animal training, which he termed *zahme Dressur* (literally, "gentle dressage," sometimes known as "gentling" in English), was an important element in creating this homelike environment. The method eschewed punishment, hinging instead on the trainer's recognition of each animal's character and intelligence, so that those responsive to instruction could be encouraged rather than intimidated into performing. The human-animal intimacy fostered by this training also ameliorated the effects of captivity.

Hagenbeck was not the first twentieth-century zoo director to set up primate performances as visitor attractions. In 1901, William Hornaday, in his new post as inaugural director of New York's recently established Bronx Zoo, launched its immensely popular orangutan dinner parties. These shows likewise focused on the apes' abilities to dine in a human fashion but included sequences showing them dressing themselves and mastering circus-style acts such as riding a velocipede. In setting up the orangutans' training regime, Hornaday also adopted the "gentling" approach, though he was less inclined than Hagenbeck to gloss ape performances as scientifically oriented. The zoo's mandate as a public institution for exhibiting animals offered sufficient context to tone down (but still exploit) the fairground roots of such entertainment. Hornaday was looking to build up the zoo's profile and knew that orangutan performers were proven crowd pleasers, but still rare enough to attract media coverage. The first primate parties involved four orangutans and sparked a rush of reports in the *New York Times* with regular updates published from the first performances in the July until the show's star, Rajah, suddenly died three months later from septic dysentery. The rest of the troupe quickly succumbed to the illness, which was eventually traced to a protozoan parasite brought into the zoo by giant Galapagos tortoises.

Initially, the dinner parties were presented to the public as a lesson-in-progress as the orangutans were still in training:

> The first attempt [at teaching them table etiquette] proved disastrous, for Rajah promptly kicked over the table. Yesterday, however, Rajah's deportment was remarkably creditable, to the great delight of the children. He made proper use of his spoon, although he committed the impropriety of smacking his lips after each mouthful. The fork he was able to use after a little coaching, but he dropped the knife quite suddenly, however, because on making a perfectly useless swing upward with it he caught the tip of his nose. For a minute or more he seemed utterly disgusted with the new fangled way of getting food into his snout, and there was trouble for his teachers.[11]

Within less than a month, however, Rajah was graduated as a "full-fledged 'gentleman' in the presence of several thousand interested spectators." The journalist covering the event noted that "[o]nly once did he disgrace himself" and then "only for an instant" when he put his elbow on the table while turning to see who had

Primate tea party, 1906 (Bronx Zoo, used by permission)

spoken to him.[12] Shows were held daily and occasionally included Hornaday himself in cameo appearances. In one of these, he delivered Rajah a letter from his "brother"; the orangutan inspected it slowly, then tore it to shreds.[13] A few days later, Rajah captured media attention with another impromptu performance when, amid screams from visitors, he entered the crocodile house brandishing an unloaded pistol he had found in the chief trainer's office.[14] During its brief run, the orangutan act increased the zoo's average monthly attendance by an estimated 100,000 visitors, and its sudden cessation elicited much publicly expressed disappointment.[15] Hornaday was quick to arrange a reprise of the show with new actors, including a chimpanzee, and kept it in the repertoire for much of his three-decade tenure as the zoo's director.

A 1906 guide to the zoo presented the shows as "entirely germane to educational purposes" because they "illustrate[d] the mentality of animals and their wonderful likeness to man, far more forcibly than the best printed statements."[16] In the *Zoological Society Bulletin* staff members detailed the orangutans' humane training methods and argued that spectators recognized the performances were driven not by "dumb obedience" but considerable "memory and reasoning power."[17] Some reporters registered the animals' efforts as performers, noting their exceptional skills or that they seemed tired after their exertions or uncomfortable in human garb; others recorded the dinner party act as merely a grotesque banquet.[18] Children were especially delighted with the "lessons," though one group called Rajah "Nigger Nigger" when he greedily devoured a watermelon.[19] The shows brought live (juvenile) orangutans into public view as never before while also seeming to entrench their image as comic entertainers. Nevertheless, these comedies may be more complex as sites of cross-species interaction than is often allowed.[20] While they worked metaphorically to harness the orangutans' transgressions of human rules for amusement (or disgust), effectively presenting both a safe parody of white middle-class manners and an affirmation of their social purchase,[21] the dinner parties also communicated something of the labor, both human and animal, behind the acts. At a simpler level, the humor and vitality of the performances helped to conceal the more sober realities of the red ape's fate in captivity, particularly its susceptibility to fatal illnesses such as tuberculosis and pneumonia. Although survival rates had improved considerably since early zoo exhibits in the nineteenth century, the average span of an orangutan's life at the Bronx park was initially only four years.[22]

Hornaday's own publications present primate training as akin to a theater project requiring a sensible—and sensitive—director. In *The Minds and Manners of Wild Animals* (1922), he critiques animal psychologists for writing about behavior and calls for a recognition of "old-fashioned, commonsense *animal intelligence*" encompassing not just mental capacities but also the moral qualities that shape temperament and disposition: amiability, patience, courage, and obedience.[23] In his view, "[t]he orangutan's sanguine temperament is far more comforting to a trainer than the harum-scarum nervous vivacity of the chimpanzee," even though the chimp was the more showy animal.[24] He stresses the need to sequence the performances with a variety of props, to spend hours in daily rehearsals so the action might reach the precision of a ballet, and, above all, to recruit interested actors to the specific roles. One section compares Rajah's enjoyment of his work and sense of responsibility to the audience with the attitude of another zoo star, Dohong, who cared little for imitation but was an "original thinker and reasoner, with a genius for invention." Dohong found dinner parties "tiresomely dull" and "frivolous" but entertained visitors with his use of a lever and "gymnastic appliances" such as a trapeze and horizontal bars.[25] To Hornaday, the training regimes were justified not only on the grounds of education and health (they provided activity for the animals and mitigated boredom) but also as a *logical* part of zookeeping:

> We hold that it is no more 'cruelty' for an ape or a dog to work in training quarters or on the stage than it is for men, women and young people to work as acrobats, or actors, or to engage in honest toil eight hours per day. Who gave

any warm-blooded animal that consumes food and requires shelter the right to live without work? *No one!*[26]

Hornaday's incorporation of orangutan "labor" into a vision of the zoo as a working institution differs from Hagenbeck's idea of the ark or refuge, but likewise attempts to situate animal performances as part of a larger social ecology.

A subsequent Hornaday book, structured as a series of interviews with wild animals, has an adult orangutan character telling his erstwhile trainer that the shows have been the highlight of his zoo years:

> Don't you remember how I learned in four lessons how to ride a tricycle? I could go anywhere on a level. And then those roller skates. How I loved to dash around Baird Court with them! You don't let me do it now, and I think it's because you are afraid I would get gay, and either bite a scared visitor, or run away. Honestly, I would do none of those things. I'm not looking for trouble as men are.

The orangutan goes on to reminisce about the dinner parties then ends with another sly dig at humans: "I could have smoked, just as correctly as any clubman or flapper in this town. I like acting. It is so easy to amuse idle people."[27] The idea that red apes are not only natural but also *knowing* actors, and that humans are the real butt of the jokes embedded in their performances, would later come to animate and rationalize more nakedly commercial shows such as the controversial orangutan act staged by Bobby Berosini in Las Vegas in the late 1970s and 1980s. In the meantime and until more intense ethical scrutiny began to attend animal entertainments, primate tea parties, mostly featuring chimps, caught on and flourished at the London Zoo (from 1926 to 1972) and in various other locations.

As readily anthropomorphized animals, orangutans have also appeared in many zoos among a cast of individual "characters," usually drawn from large mammal species, with particular abilities to draw crowds. The role of zoo character, as the name implies, is an inherently performative one, especially for apes, though in recent decades such animals have appeared less often as actors than as ventriloquized interlocutors who are meant to teach children about specific exhibits in entertaining formats. In the early stages of the public zoo movement in the nineteenth century, as sketched in chapter 3, orangutans became instant attractions because of their novelty but tended not to achieve character status, at least in their own time. The role requires a degree of longevity so that, aided by media coverage, an animal's specific personality registers more significantly for staff members and visitors than do its general traits as a member of a given species. With the gradual increase in the numbers of captive orangutans, zoos looked to present them in ways that capitalized on their individual differences and talents, often emphasizing gender characteristics. Mollie, one early character, came to the Melbourne Zoo in 1901, where the staff initially tried to teach her to ride a bicycle and put on gloves. She showed scant interest in these activities but readily developed a less refined act for her public performance: drinking large amounts of beer (to no apparent ill effect) and smoking cigarettes. By the time she died after some twenty years at the zoo, Mollie had acquired the reputation of a coarse, irascible clown. She was taxidermied, placed in a glass case, and returned to her cage to form an exhibit lauded by some regular visitors as being "just like old Mollie back again." Several newspapers, alert to public interest in local characters, printed the orangutan's photograph on the tenth anniversary of her death.[28]

Orangutan characters were, and remain, especially prominent in zoos in the United States, where they were assigned ambassadorial functions that slotted easily into an established repertoire of popular animal tropes, strongly influenced by developments in film, particularly animation, from the 1920s. Among other institutions, the St. Louis Zoo in Missouri boasted an early orangutan character, now profiled on the primate page of its website as "A Piece of Zoo History":

> An orangutan named Sam is known at the Zoo as one of its first 'ambassadors.' The charismatic ape would ride a large three-wheeler around the Zoo, stop at the ice cream stand, shake hands with visitors and even enjoyed puffing on a pipe. When the Primate House opened in 1924, Sam attended dressed in white tie and tails. He was part of the reception line and sat at the head table next to Mayor Henry Kiel for dinner.[29]

By contrast, some zoos capitalized on more "natural" performances to produce orangutan ambassadors, thereby distancing such roles from the tawdry excesses of the circus arena. In its early years of operation, the San Diego Zoo produced a character called Maggie, who arrived as an infant in a boat from Singapore in 1928 and quickly endeared herself to crowds. The zoo's manager, Belle Benchley, claims in her bestselling memoir *My Friends the Apes* that Maggie had "never actually been trained"; rather, her acting was developed by "encouraging her to enlarge upon funny little things she did from choice."[30] When the orangutan showed herself adept at human behaviors, she was brought to eat at a table, and children joined her for publicity purposes or as a special treat. She apparently loved to dress up and regularly took a ride on the zoo's bus, once amusing fellow passengers when she donned a discarded pair of pink silk underpants used as a cleaning rag. Benchley recounted such stories as stock-in-trade anecdotes when she took her charge on excursions for media shoots or to children's parties. Her maternal investment in Maggie's characterization as a lovable, if sometimes willful, daughter is never far from the surface of her book, such that the orangutan's life is narrated as a tale of growing up naturally within reasonable parental constraints rather than unnaturally in captivity.

Jiggs, another San Diego Zoo orangutan, is profiled in Benchley's memoirs as the "greatest" of all zoo characters,[31] an accolade she seems to win for her extraordinary empathy when the primate keeper, Henry, suddenly dies. A trained actor, Jiggs came to the zoo as an adult in the early 1930s, after being retired by a Hollywood film studio because she was unmanageable. Although uncooperative at first, she was apparently a "natural born clown" who "loved an audience." With time, she came to trust Henry enough to perform, often improvising on routines by taking his hat and sunglasses or sharing his pipe.[32] When Henry died, the orangutan was distraught and her act was stopped, but not before one final, apparently voluntary, performance for a crowd that had gathered by her cage not knowing the show was cancelled. Benchley recalls this act as a moving tribute to Henry by a consummate professional:

> Jiggs was sitting there with her black oriental face pressed hard against the wire and staring sadly into the crowd. She looked into my eyes, as though she sensed that I, too, was in trouble. Suddenly she turned and, with that gallantry that is the tradition back of the footlights, went mechanically through every part of her performance that she could without Henry's help—even acts she hated most, including chinning her weight to the swing bar ten times.
>
> Henry had always let her stop at that point if she had done well. Today, she looked over at me for approval or a treat, and then dropped down, in a dejected huddle, on the floor of her cage. She waited there an instant, seeming to gather a little courage, and came to the wire to receive the applause which was her due. I found it impossible to explain to visitors what they had seen and turned silently away. Jiggs had done her part better than I could do mine for she had gone on with the show regardless of her grief.[33]

This account, while embellished for theatrical effect, embraces the idea that orangutans may choose to perform to communicate abstract concepts or emotions rather than simply mimicking human acts, a mode of intelligence since observed by scientists.[34] At another level, Benchley's detailed, intimate narratives reveal the indel-

Jiggs dressed in Henry's cap and glasses, 1942 (Benchley, *My Friends, the Apes*)

ibly human basis on which orangutan characters are premised. Not only are these special zoo figures constructed largely through their interaction with humans, rather than with other animals, but their very status as characters necessarily relies on human traits. In essence, they must register as "both animal and not-animal, human and not-human"[35] for their characterization to work. This dual status accounts for a large part of their fascination to readers and zoo visitors.

An emphasis on interesting behaviors, as distinct from theatricalized performances, became a key factor in forging new kinds of orangutan characters during the expansion of Western zoos after the Second World War, as shows in many venues began to shift focus to demonstrate the "natural" animal. Stories of animal ingenuity, a theme popularized in nature documentaries at the time, conveyed the feats of orangutan captives, advancing an educational agenda more readily than the figure of the animal actor ever seemed able to do. In *The Parrot's Lament* (2000) and *The Octopus and the Orangutan* (2002), Eugene Linden brings together many such stories, spanning several decades to assess the mental capacities of red apes. These books assemble remarkable portraits of zoo characters who were inveterate traders, practical jokers, cleaners, game players, and above all cunning escape artists.

Frustrated with what he sees as the limited ability of scientific experiments to account for nonhuman intelligence, Linden approaches the issue by evaluating animals' "thinking" responses to the challenges of captivity. His starting point is Fu Manchu, an orangutan whose repeated escapes from an enclosure in the Omaha Zoo in 1968 accorded him the status of popular legend. After verifying aspects of the story with the zoo staff, Linden detects multiple levels of ingenuity (involving forward planning, creative tool use, and deception) in Fu Manchu's "Houdini" act. The orangutan identified a potential escape route in advance, though it was by no means straightforward, waited until keepers were absent, then tripped a latch with a piece of wire slid up between a door and a metal flange that first had to be pulled back, and finally hid the wire in his cheek to use in subsequent escapades. Fu Manchu also might have traded with another ape specifically to get this tool in the first place.[36] The story prompts Linden to examine numerous other orangutan escape artists, including the San Diego Zoo's notorious Ken Allen, whose serial "jailbreaks" in the 1980s cost the institution vast sums of money in security improvements, but incidentally brought in new audiences and fan club support.[37]

Linden's research also shows, alongside their attempts to outwit keepers, orangutans' ability to make intelligent assessments about human trustworthiness. As one example, he cites a protective mother who chose to bring her sick infant close enough for treatment after being shown a syringe, even though she knew from experience that the injection would hurt.[38] What Linden gleans from his in-depth engagement with the vast archive of zoo stories is "a new window on animal intelligence," understood as "the kind of mental feats they perform when dealing with captivity and the dominant species on the planet—humanity."[39] His emphasis on the interspecies context of captive animal behavior allows him to make a cogent case for the acuity of orangutan intelligence even though extensive tool use has not been documented among red apes in the wild. Whether such stories registered (in terms akin to Linden's) among the broader public as complex responses to the challenges of zoo life remains doubtful. Certainly media reports stress intrigue and ingenuity in escape narratives while glossing over the fact of captivity. Jonathan Burt's argument that zoos elicit spectators' interest in proportion to their perceived vitality of animals as "events" helps to explain the enduring purchase of the fugitive amid a cast of other orangutan characters: "the significance and attractiveness" of the zoo animal was that it "was not simply an object but also an event. From an entertainment point of view, the more dynamic the event, the greater the interest."[40]

Orangutans also played a prominent part in more official zoo events as they were brought into the media spotlight to greet important visitors or inaugurate new facilities. In 1950, the Bronx Zoo opened its Great Apes House with a young orangutan, Andy, as the ribbon cutter, shown in one newspaper photograph clasped awkwardly in the arms of the New York Zoological Society's president. The habitat-attuned "house" consisted of several green and yellow tiled enclosures with drinking fountains and exercise bars, and an outdoor compound separated from visitors by a large moat. Apparently, Andy kissed the boss lovingly and "did his act well enough" to earn applause from the trustees.[41] The same year, he served as the host in *Andy's Animal Alphabet,* a short film introduction to the zoo's exhibits, which now circulates via YouTube. Among numerous other official guests at the Apes House in its first decade of operation were Prince Rainier III of Monaco in 1955 and Fidel Castro in 1960.

Such events said little about the anthropomorphized animals recruited to endorse innovations in zookeeping or enact a lighthearted form of cultural diplomacy. By the 1970s, however, official zoo visits in various parts of the world began to encompass conservation issues, albeit with vested interests. The 1972 British royal tour to Malaysia, for example, included a much-publicized stop at Zoo Negara in Kuala Lumpur, where Prince Phillip opened a captive breeding enclosure designed to supply world zoos with orangutans, thereby helping to stop illegal acquisitions. Widely reported in the international press, the event drew attention to the endangered status of the red ape, though some journalists presented it as an amusing variation on the "meet the locals" excursion that was the standard fare of royal visits to the colonies. One London newspaper wryly noted that the duke managed to combine big-game hunting with membership in the World Wildlife Fund (which was one of the breeding program's partners) and that the orangutan, Minah, had been put "on the wagon" without water for hours in advance of the visit to avoid embarrassing scenarios.[42] Today, conservation agendas have become de rigeur for captive orangutans' official encounters with public figures, many of whom have rallied to support zoo-sponsored survival campaigns.

The presentation of orangutans as quasi-human characters—or as focal points in human-centered events—tends to perpetuate the theatrical tropes that most modern zoos wish to disavow. While overt animal shows have been expunged from zoo repertoires in favor of contextual zoology (information about animals and their environments), ape exhibits in the modes described still readily conjure up their circus predecessors. This link is not only historical but also structural. Recognizable character types and simple events have long been the building blocks of primate circus routines, whether or not the feats performed are difficult for particular animals to execute. In both the zoo enclosure and the circus tent, voicelessness (in human terms) limits animals' self-representation, but the idea of a knowable animal subject is presented to spectators, with different degrees of subtlety. Where zoos and circuses differ most is in their respective framing strategies: zoos increasingly take pains to display natural animals in postures apparently free of human intervention, whereas circuses deliberately foreground human efforts to produce unnatural animal acts. In turn, such acts are meant to communicate human attributes and themes. Big-cat shows, for example, are "little plays" about the ringmaster's courage, while horse acts show the trainer's "love and control" of a domestic animal.[43] In ape performances, which typically dramatize cross-species kinship, physical "tricks" are less important than crude costumes and narratives that place the animal as a human in some kind of social context. The most appealing acts in this vein occur when the training elicits "a behavior from the animal that, within the constructed situation, subtly creates the impression that the animal has humanlike motivations, emotions and reasoning."[44] Orangutans have lent themselves well to this kind of impersonation because of their expressive faces and gestural language as well as their morphological similarity to humans. Like the silent mime, the orangutan on the circus stage "speaks" volumes but always leaves room for interpretation.

Trained orangutan actors (as distinct from menagerie curiosities) began to attract notice in the Western circus arena in the early twentieth century as the gentling system of animal handling spread internationally, boosting the precision and range of circus performances, particularly in the United States. The itinerant life of the "big-top" entertainer, however, suited the red ape even less than the settled routine of the zoo exhibit. A 1913 circus industry article advocating humane treatment of animals notes that most orangutan performers lived only about two years in captivity.[45] From the 1940s, better results in captive breeding ensured a larger pool of trainable recruits, though circus orangutans never became as numerous as their chimpanzee counterparts. The reasons for this went beyond differences in accessibility and cost. (In the 1960s, for instance, orangutans were normally priced at thousands of dollars whereas chimps could be bought for about $500.[46]) While the chimpanzee's talent for mimicry was readily harnessed to the human-animal kinship theme promoted by circuses, the red ape's ability to solve problems through insight, one of its most human traits, could not be demonstrated easily in a single, time-limited performance routine. Moreover, orangutans had a reputation for obstinacy that could sabotage a show, especially in segments based on cooperation with other animals. Yet for entrepreneurs willing to invest in an orangutan, the potential to impress audiences was considerable. Because of its novelty, the red ape often stood out among the standard array of exotic animals recruited to the ring. Orangutan acts could include performing in crude costume dramas such as weddings, riding horses or elephants, or manipulating simple contraptions, but the most enduring routines featured close interactions with trainers, often in scenes structured by comic exaggeration and misbehavior. Cast as a special, gentle clown, orangutans were also trained to communicate directly with audiences via a handshake or some gesture that breached the physical gap between performing animal and human spectator, however momentarily.

Variations of this kinship scenario were staged in other ways, not the least of which involved tame animal visits to public places before the actual show. One circus enthusiast's memory of a childhood encounter with an orangutan at a pub in New Zealand suggests the lasting impressions these events could generate:

> All of a sudden there was a lot of noise and excitement and the crowd moved apart to make way for a pram, one of those old white cane ones that used to be so fashionable in the 1950s. In the pram sat the hairy orang-utan, clapping his long hairy arms, waving his legs excitedly as he eagerly anticipated the first drop of his favourite drink. No doubt this stunt was an arrangement between the hotel owners and the circus crew, as it obviously attracted quite a patronage the night the circus train arrived.

The narrator was too young to be allowed into the bar to see the exotic visitor close up, but the event made her annual circus pilgrimage much more exciting:

> We'd watch the entertainment, admiring the skill of the performers and the animals, laughing at the antics of the clowns and clapping as loudly as we could. But the highlight of the show was always the emergence of the orang-utan. This was what we had really come for. Knowing that our Dad had shared a drinking moment with this creature the night before made the orang-utan almost part of the family.[47]

Circuses thrived on such stories, in part because the industry was sustained through family dynasties but also because the idea of the extended "family" structured each troupe's human and animal members as a recognizable social unit. Orangutans, like other primates, lived in close proximity to their owners, often sharing small caravans and being treated very much as children. Detailed accounts of orangutans are scant in circus histories, but Mae Noell, who toured great ape acts across the southern United States with her husband from the 1940s to the 1970s, provides a window on such intimacy. Her memoir about Noell's Ark, as the family circus was called, records a nighttime incident in which her "precious little boy-orang" woke her several times with a light kiss on the cheek, as if trying to communicate something. Eventually, he placed his hand gently over her mouth, apparently to stop her snoring so he could get to sleep.[48] Noell's chronicle also suggests the violent underside of such intimacy as she describes "spanking" one juvenile because he had a habit of sprinkling her with "Eau de Orangutan," or spitting to get attention. The animal had been obtained at half price in 1959 with another infant that was also "damaged."[49] Until they retired from touring in 1971 to start a menagerie in Florida, the Noells presented orangutans as a prelude to the more renowned primate act of the circus: a boxing match in which volunteers from the audience pitted their strength against a gorilla or chimpanzee.

Noell's Ark was not an exceptional example among animal entertainment acts. In their various incarnations, circuses have always involved a commercial approach to animal (and human) Others that is only thinly veiled by the concept of the family. The most popular primate acts were often those able to suppress evidence of human coercion and control so that the material circumstances of the animals' captivity registered, if at all, as humane. The phenomenal success of Bobby Berosini's Las Vegas orangutan show derives from precisely this strategy. What was ingenious about the half-hour routine was its plausible argument, staged as a slapstick comedy, that the apes were not trained to perform but did so voluntarily and with great enjoyment. In this particular act, the figure of the knowing and willing orangutan actor of Hornaday's zoo stories "materialized" for a general audience in ways it could apprehend.

Berosini's show opened as the headliner for the Lido de Paris cabaret on the flashy stage of the Stardust Hotel Casino in the late 1970s. By this time, traditional circus had lost much of its luster in most parts of the United States, but Las Vegas was rapidly developing as "the Wild West of showbiz," where just about anything could be staged with impunity. The city was a magnet for entertainers who had built their careers working in the big-top tents, and "animal acts ruled the Strip" at this time.[50] Berosini's orangutan act, which developed from an earlier show including chimps and baby gorillas, took the popular "disobedient animal routine" (a standard trick for performing apes) to a more sophisticated level by building into it the apparently unanswerable question of how the animals had been taught to perform. "People ask me how I train them," Berosini typically pronounced as an opening line for different vignettes, each designed to parody possible answers to the question. At one point he would claim, "You have to show them who is boss," bringing forward an orangutan to demonstrate. When the animal refused to jump onto a stool, seeming not to understand orders, Berosini (as the trainer character) would be tricked into modeling the action. Once he had jumped, the orangutan would applaud him and invite the audience to do the same. In a later gag, hypnotism was the answer, so an orangutan would pretend to fall into a trance only to break into a toothy grin and nod "yes" when asked, "Are you asleep?" Bribery and trickery likewise proved to be ridiculous or ineffective against the superior intelligence of recalcitrant orangutan actors who seemed to have their own agenda (of nonperformance) for the show.[51] This comic dramatization of failed training techniques was able to broach the mostly closely guarded trade secret of the animal entertainment business—how powerful creatures such as apes are controlled[52]—without revealing "inside" information. Instead, it fostered the idea that orangutans are natural comedians who did not need coaxing to perform.

The orangutan show ran twice a night most days a week to great acclaim for more than a decade until July 1989, when a U.S. television program aired secretly

Poster for Berosini's Stardust
Hotel show

videotaped footage of Berosini hitting and punching the animals before going on stage. By then, he was earning $500,000 a year for the act, which attracted approximately $7 million worth of business to Las Vegas annually.[53] The question of the orangutans' agency—whether or not they were both knowing and willing actors—was central to the ensuing legal drama, which scarcely could have unfolded as it did if the animals had not been so convincing in their performance roles. Berosini immediately filed a defamation suit against those behind the exposé, including Ottavio Gesmundo, the dancer who had filmed the episode, and two animal rights organizations: PETA and PAWS (Performing Animals Welfare Society). He withdrew the show from the casino, claimed the animals were a much-loved part of his family, couched the beating as a simple reprimand, and even took the orangutans to the courtroom on one occasion to counter PETA's claim that they had been abused regularly and were intimidated by his presence. Anonymous death threats were sent to both parties as the public, largely sympathetic to Berosini, weighed into what would become a bitter and long-running dispute. The initial case hearing in 1990 awarded the showman $4.2 million in damages and was seen as a major victory for the animal entertainment business. In essence, the orangutan show had lent credibility to Berosini's argument that force "would never work" with the apes and was not part of his training regime, as his accusers claimed.[54] After PETA appealed the decision, it was eventually overturned by the Supreme Court of Nevada, which found in 1994 that the evidence of cruelty toward the animals was unequivocal and that the defendants had a right to express an opinion on it.

Cruelties aside, Berosini's argument that his orangutans were self-conscious comedians and co-creators of the performance may have some validity. Vicki Hearne identifies an "intelligent responsiveness" in the orangutans' performances that indicates their ability to read situations creatively and with some appreciation of humor. In her view, the routine showed the animals utilizing a "vocabulary" they shared with their trainer not only for cross-species interaction but also for self-expression and even jokes.[55] This intriguing possibility has been built into recent orangutan performances staged at San Diego's Marine World theme park in a show called "The Magic of Animals." In one sequence, drawn directly from Berosini's repertoire, an orangutan balances an orange on its upper lip and seems to make it disappear and reappear as the trainer passes a "magic" handkerchief over the animal's face. Before long, the orangutan subverts the trick by hiding the fruit in its armpit instead of its mouth.[56] Tourist shows at Universal Studios in Hollywood and Orlando also stage such acts, using apparently stern trainers and disobedient orangutans to star in more or less identical slapstick routines. Here, anthropomorphism works to highlight the disjunction between the physical animal, with its particular appearance, size and capabilities, and its choreographed human-like actions. Whether executing their roles well in the spectacle gives the orang-utans satisfaction, as Hearne's thesis would suggest, remains in question.

The more somber realities facing orangutans brought into the entertainment industry can be gleaned from 2010 biographical data for "retirees" housed at the Center for Great Apes in Florida. Among the group are hybrids born of Sumatran and Borneo subspecies, who were taken from their zoo mothers as infants and sold to trainers. Two of the sanctuary's older males, Radcliffe and Chuckie, have not developed cheek pads, beards, or throat pouches because they were castrated in cir-cuses to prevent full maturation and to make them easier to control. One female, Kiki, is recovering from severe obesity. Hollywood veterans Geri, Sammy, and Tango worked in films such as *Dunston Checks In* and *Jay and Silent Bob Strike Back* before being passed around, mostly in small cages, among roadside zoos. Other ex-actors include BamBam, who played nurse Precious in *Passions,* and Louie, once a star at Universal Studios. These precocious "child actors" exited from the limelight at about seven years old, though some were kept for a while in the industry as breeders.[57]

Although an international coalition of activists has been working for some decades to halt the use of great apes in the entertainment industry, legislation pro-tecting animals is weak in many countries and insufficient to counter the human appetite for circus-style amusements. One controversial practice, orangutan kick-boxing, has attracted particularly vocal criticism from animal welfare campaigners and wildlife protection groups but continues to draw enthusiastic crowds and In-ternet viewers. Staged daily at Safari World, an open zoo on the outskirts of Bang-kok, the kickboxing tournaments, which began in the mid-1980s, combine animal stunts with crude theater in a distinctly Westernized evocation of the Thai martial art. About a dozen juvenile orangutans appear in various roles. The show begins with one performing a break dance to music from the 1976 boxing film hit, *Rocky.* Contestants wear silk shorts and boxing gloves as they work through several rounds of kicks, punches, elbow chops, and knockouts, occasionally taking a swing at the trainer or booting him in the buttocks. A small, bikini-clad orangutan shuffles into the ring holding up a sign to announce each new round. At one point, a hairy cleaner wipes up water the contestants have spat onto the stage; at another, a raucous orangutan audience throws drink cans into the ring. After a final knockout, am-bulance officers, also played by the animal troupe, arrive to administer emergency aid to the loser. He is placed on a stretcher to be carried off, mortally wounded, but scrambles away when the canvas breaks, gesturing to the audience for applause.[58] While this odd amalgamation of cultural clichés and well-worn gags appears to be largely choreographed, as park staff have repeatedly claimed, there is some rough handling by trainers amid the fracas and no visible attempt to position the orangutans as anything but rough and ready clowns. Attempts to get the show banned on the grounds of cruelty have had no real effect, but it was temporarily closed down in 2004 under pressure of allegations by Indonesian environmentalists that many of the orangutans had been smuggled. Although DNA tests supported these claims, the investigation by Thai police was haphazard, and it took more than two years and some protracted diplomatic wrangling before 48 of the park's 110 orangutans were repatriated to a refuge at Nyaru Menteng in Kalimantan.

For the Borneo Orangutan Survival Foundation, which was centrally involved in negotiations with Thai authorities to repatriate the Safari World apes, the case was not just about illegal trade and conservation, but also about a "national symbol" and "resource."[59] Valorization of the orangutan as a national emblem is abundantly evident in advertisements for Indonesia and Malaysia as tourist destinations, despite each country's spotty record in efforts to conserve the species. Singapore has also adopted the orangutan as a national icon even though the ape is not

Orangutan kickboxing at Safari World, Thailand (Bronek Kaminski, used by permission)

part of the island's natural fauna. One singular zoo character, the legendary Ah Meng, prompted this process. She arrived as an eleven-year-old at the Singapore Zoo in 1971, having been smuggled from Sumatra and kept as a pet until confiscated from a local family. Her gentle demeanor and familiarity with humans made her an ideal zoo mascot over the thirty-seven years she spent at the park until her death in 2008 at the age of forty-eight. She became a "poster girl" for Singaporean tourism and first host of the zoo's regular Breakfast with the Orangutans event, begun in 1982 and still part of its visitor activities. As well as starring in travel films and advertisements, Ah Meng was visited by numerous international dignitaries and celebrities, including leading actors, sports stars, and musicians. As she grew older, she became a role model in the zoo's captive breeding and conservation education programs, producing five offspring and acting as a caring foster mother to two orphaned infants. The press featured hundreds of Ah Meng stories, and the Singapore government actively promoted her role as a national icon, awarding her the status of Special Tourism Ambassador in 1992, the first and only time the accolade has gone to a nonhuman. In 2001, SingPost issued a special stamp series to celebrate this "Lady of the Forest" in an initiative sponsored by multinational bank HSBC to highlight the plight of the orangutan as an endangered species. Ah Meng's death was also a high-profile affair: more than four thousand visitors attended the memorial service at the zoo; Singapore's president spoke to the media, acknowledging the orangutan's contribution to the nation; and her lifetime achievements were listed in international news reports. For ordinary Singaporeans, she had become a citizen of sorts: "Not just an ape, Ah Meng was one of us."[60]

Amid the hyperbole attached to Ah Meng's ascribed role as a national icon, another story can be discerned, one that situates her as a loved and respected friend to her keeper, Alagappasamy Chellaiyah, known also as Sam. Sam arrived at the zoo in the same year as the orangutan and looked after her for most of her tenure there. He had not been trained as a keeper but spent countless hours with the animals to develop an understanding of them and gain their trust. "The first time I saw an orangutan, I was so scared," he recalls. "It's such a big and hairy animal— truly a great ape, with the strength of four or five people. I knew it would be difficult to handle."[61] Over the years, his bond with Ah Meng developed into what appears to be a mutually experienced intimacy, however asymmetrical. When he took her out for exercise, she would greet him with a smile, touch him, or put her hand on his shoulder, unsolicited, as they ambled along. She sulked when he was absent and once ran to him to escape the attentions of a male orangutan. Sam was bitten on both arms by the male while protecting her and required numerous sutures. At another point, he slept outside Ah Meng's cage at night for three weeks

Ah Meng stamp issue, 2001 (Singapore
Philatelic Museum, used by permission)

to reassure her after she had broken her arm in a fall. Evidently, her death was an enormous personal loss for him: "I knew her for more than 36 years," he said. "That's longer than I've known my wife."[62] Sam's various accounts of his relationship with Ah Meng suggest that he came to experience her as a person, as "somebody," rather than primarily a member of a particular species.[63] In one media interview, he recounts an incident that shows how such acknowledgments of personhood might change zoo-goers' behavior. A male tourist had grabbed the breasts of a female orangutan, provoking Sam to intervene:

> I was shocked and so was the animal. People around him started scolding him, but I saw it as an opportunity to change his mind. I showed him the orang utan's eyelashes and her hands to make him understand she's like us. He got the message and immediately apologised.[64]

As if to extend the lesson to contemporary zoo visitors, this story is prominently displayed amid Ah Meng memorabilia in the area reserved for Breakfast with the Orangutans. The event demonstrates the park's "open concept" design at work as several apes come down from branches of their leafy rain forest enclosure to a feeding platform within close range of spectators.

A number of other zoos have attempted to upgrade orangutan facilities for the twenty-first century in ways that immerse the apes in apparently natural arboreal environments while offering human observers experiences of the animals that are both intimate and educational. Since 2005, for example, zoos in Melbourne, Chicago, Washington, St. Louis, and Tokyo (Tama Zoo) have invested in new, multimillion-dollar outdoor enclosures with synthetic trees or substitute vertical structures for orangutans to climb and "skywalks" encouraging them to brachiate. Some exhibits include vines, streams, or waterfalls and are presented as part of themed areas with names such as "Fragile Forest" or "Tropical Asia," conjuring nonspecific wild places in ways not so different from Hagenbeck's Tierpark a century ago. The emphasis is on the appearance of reality rather than botanical exactitude. Rothfels observes that such high-tech immersion exhibits use theatricality not for an accurate simulation of nature but "to convince people to suspend their disbelief long enough to accept what they [see] before them as an alternative but believable scene."[65] Treetop-level viewing areas where visitors are brought face-to-face with orangutans, watching them play, build nests, and solve puzzles, are a feature at the Melbourne Zoo. At the Smithsonian's animal Think Tank in Washington, apes might decide to play tug-of-war with passersby or spray them with water by pulling ropes leading into their enclosure or pressing buttons. These interactive features demonstrate orangutan intelligence in action as part of enrichment and research programs. Theme parks such as Jungle Island in Florida more overtly exploit human desires to make contact with orangutans by offering costly VIP "behind-the-scenes" tours that allow hugs and photographs with tame infants. As an adjunct attraction, the park has featured a show called "HumAnimals" in which trapeze artists, stilt walkers, and contortionists "bring alive" the park's exhibits in an "artistic fusion between human and animal worlds."[66]

Art is one enrichment activity that has recently brought attention to various zoo orangutans. The trope of the ape as artist reaches back to the seventeenth century, when the idea of ape mimicry was used to critique portrait painters whose works merely flattered their subjects.[67] Instances of actual orangutans painting date at least to the 1950s, when an art critic included works by London Zoo apes in an exhibition designed to critique abstractionism. He argued that small children and apes pursued the basic objectives of art—"esthetic composition and communicable or symbolic images"—while the nonrepresentational works of the modernist movement did not.[68] Nowadays, orangutan art is sold to raise funds for zoo facilities or conservation projects. Nonja in Vienna's Schönbrunn Zoo has painted more than 250 canvases, one of which fetched €2,000. She has also been given her own Facebook page to post photographs she takes with a special camera. Ujian, an especially creative orangutan in Heidelberg, is both a painter and a musician, having taught himself to whistle. His tunes are included as a background riff on a CD created by a local musician and sold in the zoo's shop. Many other orangutan artists have emerged in zoos around the world, including in Houston, Texas, where an exhibition called *Pongos Helping Pongos* featured the work of six red apes in 2005. According to one of the keepers, paintings by zoo residents are popular fundraisers because of "people's desire to feel closer to animals" and "to know how intelligent they are" rather than simply looking at them.[69] The Internet has become an important factor in popularizing examples of orangutan creativity by providing not only stories and footage of artists at work but also an effective way of merchandizing their products.

In many instances, this techno-cultural medium now functions as the prime means by which many humans come to "experience" orangutans in zoos and other nonnatural environments. Internet articles announcing births, deaths, and marriages in anthropomorphic terms abound, complete with blog sites and exclusive photographs, especially of newborns that are not immediately shown to actual zoo-goers. Medical disorders and surgical emergencies are also news, as are treatments for depressed or difficult orangutans such as a female at the Toledo Zoo whose anxiety has been controlled with Prozac. Unsurprisingly, orangutan "romances" are a recurrent theme, lending themselves well to representation in vernacular metaphors. Rusti, a large male in Hawai'i, for example, inspired many online posts in 2005 as he waited for an orangutan "babe" to share his "bachelor pad" in a new state-of-the-art facility. Species survival programs supply the conservation context for some of these articles. As part of an in vitro fertilization scheme at the Atlanta Zoo, we are told, a Sumatran orangutan "gave his semen voluntarily, which he learned to do after months of behavioral training."[70] Nurturing is a key part of the survival narrative in zoo terms. As well as lauding successful orangutan parents, websites chart the heroic efforts of staff members to raise youngsters rejected by their mothers after difficult births or caesarian sections, or to reintegrate them into their natural families. Keepers tell of wearing faux fur vests while caring for infants and giving orangutans stuffed "babies" with which to practice their mothering skills. Among the more unusual Internet chronicles, viewers can find details of zoo orangutans adopting pets or using computers to play games with visitors. Captivity as such is rarely visible in these modern zoo stories. Vestiges of orangutan circus acts are likewise at our fingertips, not only in archival footage but also in advertisements that show apes buying car insurance, riding watercraft, shopping for specific brands of food, and beating professionals at golf, among other unlikely feats.

This vast Internet repository suggests that the legacy of human-orangutan encounters in live exhibition contexts over the past century is mixed at best. In different but related ways, zoos and circuses have commercialized orangutans as becoming characters in populist dramas that privilege—and promise—cross-species sociality. Anthropomorphism operates very visibly in these realms as a flexible strategy for addressing human-animal relations and on occasion has led to intimacies that seem both mutual and profound. Yet this boundary-shifting process is not only contingent on direct experience of individual animals but also invested with human desires. In the circus, orangutans were (and are) commodities integrated into the human family only partially or provisionally. Owners and trainers may have felt great affection for their orangutan charges, perhaps akin to that reserved for a child, but kinship itself—in terms that accorded the red ape human status—could not be admitted to circus cultures, except as a chimera glimpsed in the ring. Helped by an education mandate, zoos have developed more enduring strategies for representing these charismatic animals, drawing their differences from, and affinities with, humans into narratives with ecological force. Yet the overall trajectory of zoo exhibits, aided by advances in display technologies, has been toward more spectacular and ever more numerous renditions of orangutan lives. In this context, the specter of the red ape's extinction in the wild gives way to an impression of plenitude.

9 ✽ On the Edge

Conservation and the Threat of Extinction

The orangutan's decline toward extinction began several thousand years ago, well before the start of human history. The red ape's dwindling in numbers is underpinned by basic features of the animal's biology. As with all apes, the natural reproduction rate of orangutans is slow.[1] Females do not normally give birth before they are fifteen years old, and they typically have no more than four offspring in a lifetime, with intervals of several years between births. For tree dwellers with few natural predators, this basic biology is unproblematic and allows for the investment of time and energy in the socialization of a small number of children. Orangutans in the wild seem to have life spans of more than fifty years.[2] The slow pace of reproduction, however, means that the loss of individuals through hunting or other catastrophe cannot quickly be made up. Both recently and in a very long historical perspective, moreover, the habitat of the orangutan has proven to be vulnerable. Tropical rain forest is a resilient ecological form, but its recovery from disruption is not swift, and local interruptions to food supply can be disastrous for a small orangutan population.

The clearest putative ancestor of the orangutan is the extinct ape genus *Sivapithecus*, found in the north of the Indian subcontinent and dated to between twelve and eight million years ago. Fossil and subfossil remains of orangutans have been found in India, China, and Southeast Asia dating from the Pleistocene era, 2.5 million to 12,000 years ago, though there is no evidence that their range encompassed the whole of this area at any one time. We can only speculate about their place in the region's ecology, but we know that there was more than one species of orangutan and more than one genus of great ape in the region and that many of the fossil and subfossil remains were of animals that appear to have been significantly larger than today's orangutan. This evidence suggests that the orangutan, like the Australian koala and the New Guinea tree kangaroo, became arboreal in fairly recent evolutionary times, since large size is not an advantage at the top of a tree. It is possible that hunting by predators such as tigers, or the greater availability of fruit in the trees, may have given evolutionary advantages to orangutans who could climb, but there is no fossil evidence to suggest the mechanism by which this shift to the trees took place.[3]

Twenty-five thousand years ago, orangutans were to be found in moist rain forest in the western Indonesian islands, over much of the Southeast Asian peninsula, and in adjacent regions in what is now eastern India and southern China.[4] The sharp contraction of the range of the orangutan to the island of Borneo and a small area in Sumatra appears to be connected with habitat loss and probably also with hunting. Eighteen to twenty thousand years ago, the earth experienced its most recent ice age, during which the climate in Southeast Asia became colder and dryer with fewer habitats suited to orangutans. It is not known when orangutans became extinct in Java or in mainland Southeast Asia. Java's dry season becomes more pronounced as one moves east, meaning that there are long periods with meager fruit supplies, so the orangutan was probably always restricted to the western half of the island. Under these circumstances local extinction was possibly caused by the eruption in AD 535 of a volcano called Kapi (very likely today's Krakatau) in the Sunda Strait region.[5] In any event, there is no recognizable trace of a folk memory of the orangutan in Javanese culture and nothing is represented in the abundant animal life depicted in carvings on Java's Hindu and Buddhist temples that can plausibly be interpreted as an orangutan.[6] Logan reported in 1849 that local people in the Malay Peninsula used the term *mawas* to denote a "race of naked savages" in the interior.[7] Since *mawas* is the common Sumatran term for the orangutan, it is not impossible that this usage reflected a folk memory of *Pongo pygmaeus* on the peninsula. The possible role of hunting in early extinctions is controversial because it potentially undermines the contemporary discourse describing indigenous peoples as traditional environmental custodians, but evidence from the Niah cave in Sarawak indicates that humans hunted and ate orangutans there 35,000 years ago. This region is now without this species.[8]

The dynamics and timing of extinction in China and mainland Southeast Asia are even less certain, though there have been persistent hints that orangutans or a related species of great ape may have survived into historical times. In the fourth century AD, the Chinese painter Jiang Guan prepared a series of illustrations of Chinese fauna that were used to illustrate the *Erya*, the oldest Chinese dictionary, which dated from the third century BC. The Erya describes two ani-

The *fei fei* and *xing xing* as depicted in the illustrated *Erya* of 1200
(Van Gulik, *The Gibbon in China*, 26–27)

mals, one called the *xing xing* (猩猩), which is the contemporary Chinese term for a great ape, the other the *fei fei* (狒狒), a term used today for the baboon.

The *xing xing* was said to have "a human face and the body of a pig, and it is able to speak. At present it is found in Chiao-chi [Jiaozhi, i.e., Vietnam] and the Fêng-hsi district [eastern Guangdong]. . . . Its call resembles the crying of a small child."[9] The same commentary describes the *fei fei* as having "a human face, with long lips; its body is black, with hair hanging down to its heels." In the fifth century a pair of *fei fei* was sent as tribute to the imperial court.[10] Early in the nineteenth century, a French missionary in southern Vietnam reported stories of a hairy, tailless animal that "almost resembles a man." It climbed trees, ate fruit, and was "sadness personified."[11]

Only in Sumatra is there evidence of the contraction of the range of the orangutan in colonial and precolonial times, though this evidence partly takes the form of persistent legends of a different kind of creature, the *orang pendek* ("short person"). The term "orang pendek" seems to have been applied to a variety of natural phenomena and cultural constructions, but it is likely that at least some identifications of the creature are based on sightings of isolated orangutans outside what is now regarded as their normal habitat.[12] The rise of human populations almost certainly played a role in the contraction of the distribution of orangutans. The clearing of lowland alluvial plains and river valleys for agriculture in recent times has significantly reduced orangutan ranges, and the same is likely to have been true in the past. Scientists generally consider a population of about five hundred to be necessary for the long-term survival of an orangutan community.[13] By fragmenting the habitat of orangutans through land clearing, early humans, like their modern-day counterparts, contributed to the slow localized extinction of the red ape.[14]

The development of a trade in live orangutans and orangutan body parts in the eighteenth century had a steadily more significant effect on wild orangutan populations, as well as on the unfortunate individuals who were captured. Foucher D'Obsonville, who lived in Sumatra in 1767, was the first to comment on the apparent contraction of their range and decline in their numbers:

> From the enquiries I made, I may venture to say that he is no longer to be seen in the peninsula on this side of the Ganges; and, likewise, that he is become very rare in the countries where he still propagates. Has this race then been confounded with others, destroyed by them, or devoured by wild beasts?[15]

Two decades later Charles Miller, also in Sumatra, wrote "The *oerang oatan*, or wild-man ... I have heard much talk of, but never seen; nor can I find any of the natives here that have seen it."[16] William Marsden, whose 1811 *History of Sumatra* is otherwise remarkable for its detail of the wildlife of the island, makes only one brief mention of the orangutan.[17] Abel reported of the juvenile orangutan that he obtained in 1817: "Captain Methuen ... brought him from Banjarmassing on the south coast of Borneo, to Java.... The natives informed Captain Methuen that he had been brought from the highlands of the interior, and that he was very rare, and difficult to take; and they evidently considered him a great curiosity, as they flocked in crowds to see him."[18] In 1841, Schlegel and Müller, who had spent twelve years as natural history collectors and observers in the archipelago, described it as one of the rarest animals in the country, even where it was most abundant.[19] Hunters in Borneo commonly commented that the ape was seldom seen and difficult to find, although this observation did not deter them from paddling up yet another river or crossing yet another ridge to find a relatively undisturbed orangutan population they could plunder. Hornaday decided to start his hunt for orangutans in Sarawak in 1878 because he heard they had "not yet" been exterminated there.[20] "This intelligent, man-like ape is probably not so common in Dutch Borneo as he is supposed to be.... It is to be hoped that these interesting animals will not soon be exterminated," wrote the Norwegian naturalist Carl Lumholtz in 1920.[21] Charles C. Miller (not to be confused with the eighteenth-century Charles Miller in Sumatra) indirectly suggests in 1942 that orangutans were becoming scarce, noting that "it had been so long since a Kanja-Dyak had killed an ape that they had no idea how it was done."[22]

The first measure to provide legal protection to the orangutan was introduced in Sarawak, which had been ruled since 1868 by Charles Brooke, nephew of James Brooke. Charles was much more deeply engaged with the culture and environment of his kingdom than James had been, and in about 1895 he issued an edict banning the hunting of orangutans, although by this time there were relatively few left in his territory.[23] Across the border in the Netherlands Indies, Brooke's action was among the factors that prompted two prominent Dutch naturalists, M. C. Piepers and P. J. van Houten, to call in 1896 for a policy to protect the wildlife of the Netherlands Indies.[24] Their plea for action by no means focused on the orangutan, which was just one of a few dozen animals and birds the two men identified as being in danger of extermination. Their case for conservation was wrapped partly in economic and geopolitical arguments. They pointed out the risk of catastrophic consequences for agriculture if insect-eating birds and animals were exterminated, and they noted that the Netherlands was falling behind other colonial powers in its policies for nature protection. The core of their argument, however, was the proposition that human existence would be poorer for the irrevocable loss of attractive and scientifically interesting species. It was not an argument reflecting direct concern for the animals and plants facing loss of habitat and individual destruction, but rather an aspect of the creeping disillusion with modernity that had given rise more than a century earlier to the idea of the noble savage.

The colonial government responded fairly promptly to the conservationist call from Piepers and Van Houten by drafting a nature protection ordinance in 1898.[25] Colonial decision making was slow, however, even at the best of times, and nature conservation did not inspire any sense of urgency. The authorities, moreover, pondered several competing considerations in trying to devise the best format for nature protection. Where commercial interests were involved, they were reluctant to implement any wholesale ban that might diminish the profitability of the colony. They hesitated to interfere with the traditional hunting activities of indigenous people or to prevent local people from defending their gardens and flocks from predators, because that might lead to political unrest. And they did not want to introduce any law that they could not enforce, because they did not want their colonial subjects to become accustomed to disregarding colonial edicts. They toyed with the registration (and thus restriction) of specific hunting weapons, with taxing the trade in endangered species, and with introducing a *pacht* ("farm") or franchise system in which hunters of particular animals would have an exclusive, long-term license to hunt in a particular area and so would have, it was hoped, an interest in sustainability. In the face of so many considerations and possibilities, the colonial authorities found it easiest to do nothing, and they let the draft sit unattended.

During the first decade of the twentieth century, however, it became clear that a nature conservation strategy should be on the agenda of any colonial power intending to present itself as a conscientious modernizing administration. In the late nineteenth century a wide range of countries and colonies had introduced nature protection measures of one kind or another for endangered species, and in 1900 conservation officially became an international concern when Britain, France, Germany, Italy, Spain, Portugal, and King Leopold's Congo signed a treaty pledging to cooperate in protecting the wildlife of their Central African territories.[26] In May 1908, the Netherlands Society for the Protection of Animals (Nederlandsche Vereeniging tot Bescherming van Dieren) wrote to the minister of colonies, pointing out that the Indies was out of step with international opinion and practice.[27]

Under this pressure, the colonial government in 1909 announced an Ordinance for the Protection of Certain Wild Mammals and Birds, to come into effect in 1910. The ordinance gave blanket protection to all wild mammals and birds in the colony, and then made a number of exceptions to that protection. The intention was to allow the continued hunting of known noxious (*schadelijk*) animals and birds—tigers, which endangered human life, along with kingfishers, ricebirds, and barbets seen as predatory on fishponds, rice fields, or orchards—as well as commercial and recreational hunting of traditional game such as wild pig, pigeons, and waterfowl. All other species, including those unknown to science, were thus automatically protected. The ordinance recognized that conditions varied from region to region, and thus partly excepted deer, elephants, birds of paradise, parrots, and a few other species from protection by giving regional administrators the right to permit the hunting of any or all of them. Completely without protection, however, were all monkeys, including orangutans, on the grounds that they were robbers of gardens.[28] Wildlife protection was a rising tide in colonial and international affairs, but it did not deliver protection to the orangutan.

It took fifteen years from the 1909 ordinance for the orangutan to be deemed preservation-worthy by the colonial government. In the 1924 hunting ordinance for colonial Indonesia, orangutans were finally included on the list of animals it was forbidden to hunt, and in the following year they were protected from killing in general.[29] There was no ban, however, on the export of orangutans, and in 1928 and 1929 sixty to seventy were sent from the colony, mainly to Europe.[30] The consequent outcry in the Netherlands pushed the colonial government into an uncharacteristic burst of energy: the Batavia administration approached its neighbors, the Brooke government in Sarawak, with a request to strengthen its protective regime and asked the British in Malaya, as well as the authorities in Brunei and North Borneo, to ban the trade in orangutans. All four authorities had complied by the end of 1930.[31] In 1931 Netherlands Indies regulations were extended with the issue of a new Animal Protection Ordinance, but the terms covering the orangutan remained unchanged.

In the international development of nature protection strategies, one of the most important transitions was from the protection of individual species by means of hunting bans to the protection of species within their habitats by means of some form of territorial protection: nature reserves, national parks, and the like.[32] The oldest impulse in territorial nature conservation was the establishment of game reserves, which protected animals for the hunting pleasure of elites. Toward the end of the nineteenth century the concept of private parks was extended to create national parks, which protected landscapes and their natural inhabitants for nations as a whole. The wildlife protection function of such parks grew out of the realization that species might also become extinct through habitat loss. The first impulse for territorial protection in the Netherlands Indies came in the spirit of the national park movement in the United States, New Zealand, and Australia. That is, it aimed to protect places of unusual scenic beauty or special scientific interest. In 1916, the colonial government issued an ordinance creating *natuurmonumenten* ("nature monuments") to protect such places.[33] The first of these reserves was created in 1919, but in the beginning they were mostly very small; they might protect a waterfall, a picturesque lake and its immediate environs, or even a single tree, but there was no attempt to protect whole ecosystems. In the 1920s, however, the *natuurmonument* idea rapidly absorbed the concept of habitat protection, and the colonial government began to create a series of larger *natuurmonumenten* with the idea of protecting the habitat of endangered animals. In 1932 the authorities recognized the possibility of allowing some human land use alongside the protection of wildlife by adding a category of wildlife reserves.[34]

Protection for the orangutan, however, was conspicuously absent. Until 1934, there was no nature reserve in any region inhabited by orangutans. In 1934, the authorities created the vast Mount Leuser reserve (then known as Löser) in Aceh in northern Sumatra, but it covered relatively little of the habitat of the orangutan and was not presented as protecting the red ape. There were two reasons for this omission. The first was constitutional: a large proportion of the orangutan's habitat lay within the domains of so-called self-governing territories (*zelfbesturende landschappen*), former Native states of the archipelago that had been incorporated into the Netherlands Indies but which had retained some degree of autonomy as long as it did not impinge on important Dutch interests. It was constitutionally within the power of the colonial government to forbid the export of orangutans, but it could not easily prohibit hunting within the Native states, and even more difficult was the abstraction of land from a Native state for the creation of a nature reserve. When the Netherlands Indies Society for Nature Protection wrote to regional governors in 1926 asking them to identify areas that might be suitable for nature reserves to protect large animals such as the orangutan, the governors in three key provinces covering orangutan habitats—Aceh, East Sumatra, and South and East Borneo—replied that there was no suitable directly ruled territory at all within their borders.[35] Establishing the Leuser reserve had involved negotiations between the colonial government and no fewer than eleven small states in southern Aceh.[36] Discussions began in 1926 about establishing a nature reserve in western Kutai, but the reserve was not officially announced until 1934, and at the end of the colonial era there was still no agreement on just where its boundaries should run, even though the Kutai sultan had a reputation as one of the most enlightened Native rulers.[37] There were similar delays in establishing a reserve at Tanjung Puting to cover a major orangutan habitat on the south coast,[38] though a reserve was indeed established at Gunung Palung in 1937.[39] Extending any Netherlands Indies regulations to Native states required them to issue what were called "concordant" regulations, but the task of negotiating these regulations lay with the regional administration (Binnenlandsch Bestuur) of the colony, an elite corps of men whose university training was largely in the laws, languages, and cultures of the archipelago, not in its natural history.[40] It was a rare official who made more than perfunctory mention of hunting-control measures in his reports, and in many cases the threat to orangutans was not the most acute risk to the natural world under his control. Rhinoceros were much more seriously endangered than orangutans; pangolins, in demand for Chinese medicine, also seemed to be seriously at risk. There were years in which collectors swept through parts of Borneo taking vast numbers of orchids; in other years gibbons were targeted.

Even less common, however, was interest in nature protection among Indonesian rulers and their indigenous subjects. Dutch colonial strategy involved the intense exploitation of lucrative natural resources and opportunities for plantation agriculture, along with a principle of avoiding intrusion into Native life as much as possible. Until the beginning of the twentieth century, Dutch policy had been to limit indigenous access to Western education, and there was thus little reason why Indonesians should have been aware of the ideas behind the Western conservation agenda. On the contrary, many Indonesians experienced the conservation program as an unwanted intrusion. When the head of the Buitenzorg Botanical Gardens in Java, who had overall responsibility for nature protection in the archipelago, wanted to create a reserve in the upper Heran and Samba river area in Borneo, he proposed removing all the local inhabitants but was prevented from doing so partly by the lack of funds needed for compensation.[41] The animals that were protected under the successive colonial regulations were mainly those with some commercial export value. Native subjects of the Dutch resented the loss of this source of income and had little compunction in continuing to hunt as far as the weak policing arrangements permitted. One government official complained that a group of illegal rhinoceros hunters turned out to have obtained their weapons from the sultan of Sambaliung in northeast Borneo.[42] The discourse of conservation was also used to justify colonial rule: the Dutch, although they were outsiders, claimed the greatest appreciation of landscape and nature in the land they had colonized and dismissed Indonesians as not to be relied upon to protect their own wildlife.[43] One of the most telling signs of the gap that had emerged between the conservationist agenda and indigenous political feeling was that the Dutch deliberately avoided using the term "national park," by then an international standard, when they launched a final reform of the nature reserve system in 1941. Instead the large areas that were given the highest degree of protection were merely called "nature parks."[44]

The conservation status of the orangutan at the end of the colonial period is perhaps most clearly reflected in the fact that they did not appear in the first *Album van Natuurmonumenten in Nederlandsch-Indië* but rather in the second, supplementary volume.[45] The orangutan's secondary status reflected more than anything else competition from species that appeared to be even more endangered, such as the rhinoceros, the Komodo dragon, rare birds of paradise. Whereas the rhinoceros in Java, Borneo, and Sumatra was reduced to a few dozen individuals, and the Komodo dragon to a few hundred, there were probably half a million orangutans in the wild in 1940. When the American conservationist C. R. Carpenter surveyed the situation of orangutans in Aceh in 1937, he concluded that even the smaller Sumatran population was in no danger of extinction, though it was under pressure from hunting and habitat loss.[46] Outside the East Indies, there was correspondingly little interest in orangutan conservation. In 1929, when the American psychologist Robert Yerkes and his wife, Ada, wrote a masterly review of what was known about orangutans, they made no mention at all of any threat to the animal in the wild.[47]

Nature conservation in general and orangutan protection in particular were poorly equipped to withstand the quarter century of military and political turmoil that beset the lands inhabited by the red ape from 1942. For twenty-five years from the Japanese occupation of Sumatra and Borneo during the Second World War until the second half of the 1960s, little was done to protect the orangutan. After the end of the Japanese occupation in 1945, the Dutch colonial government had at best weak control of most of the orangutan habitat areas in Borneo, while the smaller orangutan areas in Sumatra were largely under the control of regular and irregular forces loyal to the Indonesian Republic, founded against Dutch opposition in 1945. The British took control of northern Borneo from the Brooke family and the North Borneo Company in 1946, but the depth of local tensions over the future of the territory was reflected in the assassination of the British governor of Sarawak in 1949. Much of northern Sumatra was beyond the control of the Indonesian government in the late 1950s, while Borneo was at the front line of a political confrontation (Konfrontasi) between Indonesia and the British over the construction of a new state of Malaysia combining former British territories in Borneo with those in the Malay Peninsula.

Responsibility for nature protection in Indonesia lay with a Nature Preservation Institute (Lembaga Pengawetan Alam, LPA) under the Bogor (formerly Buitenzorg) Botanical Gardens. The head of the LPA from 1935 to 1947 was an energetic Dutch biologist, Andries Hoogerwerf, but he had only a few staff members and a small budget. His highest priority was preserving the tiny population of Javan rhinoceros on the Ujung Kulon peninsula in Java's southwest.[48] There was little he could do in the face of demand from foreign buyers for orangutans, gibbons, Komodo dragons, cassowaries, and other rare species, nor could he help the elephants that were hunted for their ivory, or the deer for their meat and skins.[49] There are few records about the experience of orangutans during these years, but what is available suggests a continuing story of hunting for foreign zoos and collectors and gradual habitat loss. In 1949 an American dealer called Rosevelt received a permit to export twenty-three baby orangutans. In 1953, twenty-four went to Blijdorp Zoo in Rotterdam.[50] Hoogerwerf noted that between 1953 and 1956, the Singapore authorities had recorded the passage of 112 orangutans through their port, none of which had legal documentation from Indonesia.[51] In the absence of resources to address the problem, he was reduced to appealing to other countries to refuse to accept smuggled orangutans. Hoogerwerf left Indonesia in 1957 and was replaced by an Indonesian conservationist, I Made Taman, but his interests, too, were focused on Java rather than on other islands. Barbara Harrisson was one of few voices expressing alarm at the decline of the orangutan in northern

Borneo, but she was unable to arouse more widespread anxiety.[52] An American researcher, Oliver Milton, published a report in 1963 in which he suggested that the population of Sumatran orangutans had been reduced to between 800 and 1,500, blaming the decline principally on the hunting for pets. He reported that 280 animals were known to have been captured in Sumatra in 1961–1962 and warned that the species could become extinct within ten years if this rate of removal continued.[53] A second investigation in Sumatra by the Swiss biologist Rudolf Schenkel and his wife, Lotte, in 1969 seemed to confirm Milton's conclusions. They noted that the preferred habitat of the orangutan in the valley floors was rapidly being cleared for agriculture and that the export of young orangutans was continuing "without any control."[54] Even the modest achievements of the colonial era came under threat in this era: in 1957, the Kutai reserve in eastern Borneo was reduced in size from two million hectares to a mere 306,000 hectares.[55] The most striking feature of these years is the dearth of conservationist attention to the orangutan. Within Indonesia itself there was no nature protection movement, and international attention was largely focused elsewhere. Indonesia's celebrity species was the Java rhino, and conservationist concern was focused on densely populated Java, where population pressure seemed to be the greatest threat to unique species. The government in Sarawak established a Maias Protection Commission in 1959, but its 1960 report rejected the idea of establishing a nature reserve for orangutans on the grounds that to do so would infringe on the rights of the indigenous Dayaks. After suggesting facetiously that the government might arm orangutans so that they could shoot back at the hunters, the report recommended tighter policing of the trade in orangutans and "intensive propaganda by means of simple, forceful and, if possible, humorous pamphlets."[56]

Until the beginning of the 1970s, thus, the story of orangutan protection was one of modest and often ineffective responses to a risk of extinction that was real but relatively remote. The range of the orangutan had been shrinking for thousands of years and orangutan numbers had been in decline. Hundreds of thousands of orangutans, however, still lived in the jungles of Borneo and thousands in Sumatra. All the trends were ominous, but the danger of extinction was not immediate. From the 1970s, however, orangutans suddenly faced a series of acute threats to their survival as a result of political and economic changes in their region. These threats stimulated the emergence of a new and versatile set of conservationist strategies.

In 1965–1966, the removal of President Sukarno from power in Indonesia put an end to the Konfrontasi with Malaysia. The military-dominated New Order of President Suharto launched an ambitious program of national development funded by abundant revenues from oil and timber exports. Vast tracts of forest land in Borneo and Sumatra were allocated to individuals associated with the regime, either for a price or in return for services rendered. Although restrictions on logging practices were formally in place to minimize adverse environmental impacts, these restrictions were virtually never enforced. The government service in charge of nature reserves and wildlife protection was a small and underfunded unit within the Forestry Department.[57] There was hardly even perfunctory recognition in official circles of the importance of nature conservation and there was no environmentalist movement at all within the country. The fate of existing reserves under these circumstances was determined more by their accessibility to loggers than by their formal legal status. The harvesting of timber within reserves was so open it could barely be called clandestine. In particular, the Kutai reserve, which had once seemed likely to be a major protection zone for orangutans in East Kalimantan, was extensively logged.[58] The authorities across the border in Malaysian Borneo were similarly rapacious. There were few orangutans left in Sarawak to be affected by the widespread logging there, but the timber industry also boomed in Sabah, and the loss of orangutan habitat gathered pace there as well.

Orangutans always fared poorly in logging operations. With habitat destroyed, their search for food was disrupted. The cutting of logging roads and the clearing of forest meant they were forced to take to the ground more frequently, where they were vulnerable to predators such as leopards (and tigers in Sumatra). Logging crews hunted them for bushmeat, occasionally keeping female orangutans for sex. Even more important, the improved lines of communication to the interior created by logging roads and the ready availability of ships carrying logs to overseas destinations meant that the smuggling of orangutans to foreign buyers was easy to carry out.[59] The intensification of habitat loss and hunting turned the threat of orangutan extinction from a serious but distant possibility to a far more immediate prospect. Just how acute the threat to orangutans has been at any specific time has always been fiercely debated. As the logging boom took off, for instance, it was sometimes argued that orangutans might be able to survive in secondary jungle after loggers had passed through. The logged-over Segama area in Sabah in fact became an important habitat for these apes after it had outlived its usefulness to the loggers.

The initially unpromising regime of the 1970s and 1980s, however, proved unexpectedly capable of supporting conservationist initiatives. Just as the first conservationist measures in colonial Indonesia had been stimulated by an unfavorable comparison with other colonies, so a new wave of environmental protection measures, especially in Indonesia, was stimulated by a sense of international obligation. When Indonesia sent a delegation to the United Nations Conference on the Human Environment in Stockholm in 1972, the Indonesian Country Report could only identify an intimidating set of problems arising from population growth, forest clearing, and pollution. There was virtually no government action on the environment that could be reported.[60] The report was embarrassing because Suharto and his supporters had defended the undemocratic character of the New Order by arguing that it was producing the best policies for the country. Indeed, Indonesian and Malaysian authorities had argued, when they paid any attention at all to conservationist protests, that the imperatives of national development and the welfare of their respective people were far more important than the interests of orangutans and other animals. As proper environmental management became one of the criteria for good governance, however, the Indonesian authorities felt pressure to conform to international norms.

Suharto's New Order was technocratically oriented and receptive in principle to coherent policy arguments. More than that, after 1974 the regime was interested in building cooperative relations with social groups that were not identified with the ideological and religious antagonisms that had plagued Indonesia under Sukarno. The enthusiasm of a generation of young environmentalists meshed well with the approach of a government whose instincts were to control and regulate in what it defined as the national interest.[61] In Malaysian Borneo, the new government was similarly oriented to addressing the complex tasks of government and similarly willing in principle to listen to the case for nature conservation, though it remained more antagonistic to civil society conservationist groups. In 1965 the International Union for the Conservation of Nature (IUCN) discussed the possible establishment of a nature reserve at Ulu Segama in Sabah, which until that time had no nature reserves at all. In the end, however, the Ulu Segama proposal failed to advance and it only became a significant orangutan reserve after it had been logged.[62] Indonesia, on the other hand, became the site of a major initiative in national park creation in the late 1970s and early 1980s. With the assistance of the World Wildlife Fund Indonesia Programme, the nation developed a national conservation strategy that included the designation of extensive national parks and many relatively undisturbed parts of the country.

The creation of national parks in Indonesia and Malaysia after 1980 followed both scientific and political imperatives. The principal scientific imperative was to protect ecosystems that provide the habitat for animals and plants that might otherwise be in danger of extinction through habitat loss. Until the 1970s, the practical strategies of conservationists focused on the prohibition of hunting and the creation of national parks from which disruptive human activities were largely excluded. As argued earlier in this chapter, in Indonesia this approach tended to produce antagonism between conservation authorities, who were determined to protect, and local people, whose capacity to earn a living by hunting or collecting in conservation areas was restricted. It was easy to argue that conservation was a burden imposed by the developed West on developing countries. In the mid-1970s, however, the international conservation movement, especially the IUCN, had begun to develop an increasingly refined strategy of identifying both the economic benefits of nature protection (notably in the form of eco-tourism) and ways in which local communities could sustain livelihoods in harmony with the existence of protected areas. Indonesia proved unexpectedly receptive to these arguments. The consequence was that in the 1980s Indonesia developed an extensive national park system characterized by complex systems of zoning, in which some areas were prohibited for all human activity while in others tourism, recreational camping, collection of firewood and plant products, or the grazing of livestock were permitted.

In the case of orangutan protection, this imperative intensified during the 1980s with the emergence of a growing scientific consensus that there were substantial differences between the Borneo and Sumatra orangutans and probably also between three subspecies in Borneo.[63] When Stiles and Orleman had pronounced in 1927 that there was only one species of orangutan, the scientific convention was that the species boundary was defined by physiological fertility: two animal communities were to be regarded as separate species only if they were physiologically unable to breed and to produce fertile offspring. Some scientists recognized two subspecies, *Pongo pygmaeus wurmbii* in Borneo and *P.p. abelii* in Sumatra, but because individuals of the two subspecies could interbreed and produce fertile offspring, the difference was considered relatively trivial. It was believed that the populations had been separated only since the last ice age, that is, for about 15,000 years. During the second half of the twentieth century, however, the sole criterion of reproductive isolation had been increasingly been replaced by an evaluation of genetic makeup in assessing whether there was a species boundary between two communities. Genetic evidence now suggests that the Sumatra and Borneo populations have been distinct for more than one million years and that there are three subspecies in Borneo, with separate lineages going back 860,000 years.[64] In 2001 the primatologist Colin Groves reclassified the Sumatra orangutan as a separate species, *P. abelii*.[65]

The most important reserve for the orangutan was the Tanjung Puting National Park on the southern coast of Central Kalimantan, an area of lowland and swamp forest with a large orangutan population. There had been a small nature reserve in the area since 1935, but in 1984 the reserve was expanded and declared to be a national park. The key planning elements for the national park were to be an expansion of employment provided by the park through the appointment of more than fifty new guards and through new construction work, the creation of buffer zones, in which local people would have the opportunity to collect forest products on a limited scale that would not harm the integrity of the park, and the promotion of eco-tourism based on the presence of orangutans.[66] In the report that led to the creation of the park, the primatologist John MacKinnon stated unambiguously that the initiative would be economically justified:

> Direct benefits to the region in the form of boosted tourism and research visitors and buffer zone utilization together with indirect benefits in the form of environmental protection of soil, waterways, climate, fisheries, protection of genetic resources and increased quality of life and increased knowledge of forest ecosystems easily justify such developments.[67]

Within this framework, however, the report made clear that the protection of the orangutan was the foremost public priority of the park, and orangutans have subsequently appeared prominently in all reports and promotions about Tanjung Puting.

The relatively positive experience of Tanjung Puting contrasted with the dismal history of Kutai. Reduced to 15 percent of its original size and then logged, the area was declared a national park in 1984, only to be trimmed again in 1991 and 1997 to less than 200,000 hectares, in the interests of expanding oil and gas exploration. Illegal settlements within the park were legalized and then excised from the park. The director of the park resigned in depair in 2001.[68] Even in Tanjung Puting, moreover, protection became weaker after the fall of Suharto in 1998. Under the political reform process that led to Indonesia's democratization, substantial powers were devolved from the center to local authorities. Although national parks remained in the end a responsibility of the central government, there was much uncertainty about their status for some years, as well as a great deal of opportunism while political attentions were distracted.[69] Illegal logging burgeoned and one report estimated that 40 percent of the park had been destroyed.[70] In vain, president Abdurrachman Wahid issued a decree in April 2001 instructing his ministers to take special measures to protect the Leuser and Tanjung Puting reserves.[71]

In the 1990s, moreover, new threats had emerged to the habitat of orangutans. First, in the mid-1990s, Suharto launched a massive project to convert one million hectares of peat swamp in Central Kalimantan to rice cultivation in order to achieve a durable national self-sufficiency in rice. When warned that the project would damage an important orangutan habitat, Suharto was reported to have exclaimed in Dutch, "Verrek de orangoetans" ("damn the orangutans").[72] In the end the project was an unmitigated disaster as the ground proved to be too acidic for rice, and the region alternated catastrophically between drought and flood.[73] Second, conditions for orangutans were made worse by the outbreak of widespread fires in Kalimantan and Sumatra in 1997. The fires did not significantly affect orangutan habitats in Sumatra, but they were calamitous in Kalimantan, sealing the ecological destruction of many areas that had already been logged over. Dense smoke, moreover, blanketed parts of the island for weeks, leading to serious respiratory disease in both humans and orangutans.[74] The fires were a consequence partly of unusually dry weather and partly of the abundance of tinder left behind after logging operations. Many of the fires were deliberately set to clear land for the planting of oil palms, *Elaeis guineensis*. The palms, the oil of which is widely used for cooking, had been cultivated in the Indies since the nineteenth century. In the late 1990s, intensive research and development in Malaysia increased yields and decreased the risk of disease, leading to a boom in production. In the 2000s, Indonesia overtook Malaysia as the world's largest palm oil producer. In only a few years, vast areas that had once been the habitat of orangutans were permanently converted to oil palm production.

The spread of oil palm cultivation raises the possibility that there will be no viable habitats for orangutans outside protected areas.[75] This risk gave rise to a novel international campaign focusing on the oil's particular markets. In 2010, Greenpeace launched a worldwide campaign against the confectionary company Nestlé over its use of palm oil obtained from the Indonesian firm Sinar Mas. The campaign focused on the popular Nestlé product Kit Kat, using the slogan "Give orangutans a break." Sensitive to its image with young people, the company announced after two months that it would cease sourcing palm oil from Sinar Mas. Greenpeace action then turned to HSBC bank, a major investor in Sinar Mas, and to Unilever, another major user of Indonesian and Malaysian palm oil.

Parallel to the threats to orangutans from habitat loss were the dangers of the continued illegal trade in live animals. The head of the Indonesian police, General Hoegeng, who was celebrated as one of the more liberal and honest figures in the early New Order regime, kept a menagerie in his back yard, including several orangutans.[76] In the 1970s and 1980s, recognizing the difficulty of adequately policing the regulations protecting orangutans within Indonesia and to a lesser extent Malaysia, the campaign to protect the orangutan was expanded to encourage the international policing of the orangutan trade, especially in Singapore and Hong Kong, which were the most important transshipment points for illegal exports. A census of the orangutan population of international zoos was also carried out so that it would be more difficult for smuggled orangutans to be lost in the substantial exile community.[77] A report by the World Wildlife Fund in 1980 painted a discouraging picture: although the formal legal framework for regulating the trade was sound, in practice there was no effective means to prevent the smuggling of protected species out of the country. Officials of the nature protection service (PPA) worked at the airport only during business hours and there was no mechanism for the inspection of ships, even if they were suspected of carrying illegal animal cargoes. Orangutans were smuggled from Indonesia on vessels carrying timber from the Kalimantan logging concessions. Destination countries (with the exceptions of the United States, Canada, and Australia) were careless in the documentation they required with animals being imported.[78]

Some progress, however, was made with controlling the illegal trade. Indonesia ratified the Convention on the International Trade in Endangered Species (CITES) in 1978.[79] Hoegeng and other elite figures surrendered their captive orangutans to zoos, and around the same time they removed once fashionable wildlife trophies such as stuffed tigers from their front rooms. By the early 1980s, respect for the environment had become fashionable in Indonesia. Orangutans, however, remained in demand in Taiwan, eastern Europe, and the United States as pets and in Thailand, where they were trained as performers, especially in Thai kickboxing (see chapter 8).[80]

In December 2007, at the time of the Bali climate change conference, President Susilo Bambang Yudhoyono announced a new orangutan conservation strategy for Indonesia, focusing on enforcement of the ban on trade in orangutans by means of additional training for conservation staff. In addition, the Indonesian government sponsored the formation of community-based patrols, called Orangutan Protection and Monitoring Units, in and around Gunung Palung National Park in West Kalimantan and at Danau Sentarum. These units were unusual in that they comprised community members, activists from the nongovernmental organization Fauna and Flora International, and forest rangers.[81] As in the case of palm oil, these official initiatives at protection have been supplemented by more radical activism focusing on the complicity of outsiders in activities that endanger orangutans. The first significant action of this kind occurred when Thai authorities intercepted a shipment of six infant orangutans at the Bangkok airport in 1990. When investigations revealed the involvement in the transaction of Matthew Block, president of a Miami-based firm specializing

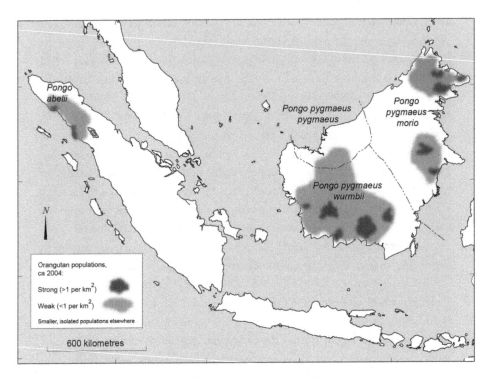

Map of orangutan populations

in the legal supply of primates for scientific research in the United States, an international campaign was launched to prevent a plea bargain that would have spared Block from jail. The orangutans were anthropomorphized as the "Bangkok Six" and a specialist was sent to nurse them back to health. In the end, Block received thirteen months in prison.[82]

The seriously threatened status of orangutans in the wild after 1990 raised the possibility that captive breeding might play a role in the long-term preservation of the species. For much of the history of human contact with orangutans, this possibility appeared remote. Only gradually were techniques developed for keeping orangutans alive for extended periods. The eccentric amateur primatologist Rosalià Abreu ran a largely successful and healthy menagerie in Havana in the 1920s by following what were then the pioneering practices of providing her animals (mainly chimpanzees alongside a few orangutans) with plenty of space, companionship, clean conditions, and protection from cold weather.[83] Since about 1980, improved standards of care in zoos have meant that captive orangutans at last have about the same life expectancy as orangutans in the wild.[84] Successful captive breeding of orangutans was slow to begin. Only in 1928 did a female orangutan successfully give birth in captivity, when Maggie, a captive orangutan at the Philadelphia Zoo, gave birth to a son. The Perth Zoo in Western Australia had early success in captive breeding, managing twenty-one births and fourteen survivals in two decades from 1968, partly because the director was a primatologist who took a special interest in orangutans and invested in well-designed housing.[85] United States zoos developed a species survival plan in 1982, focused on captive breeding. In 1985 the plan was modified to advise against the breeding of Sumatra-Borneo hybrids, and set out the aim of preserving "self-sustaining populations of . . . orang-utans that can continue to serve as ambassadors for wild conspecifics and the natural habitat, as well as a resource for conservation education and research." The plan aimed specifically to conserve 91 percent of the genetic diversity originally found in Borneo orangutans, 94 percent for Sumatra orangutans.[86] An international orangutan studbook is maintained to keep track of approximately nine hundred orangutans in legal captivity around the world.[87]

Maggie with her child, Philadelphia Zoo, 1928 (Yerkes, *Great Apes,* 142)

Captive breeding remains a contentious strategy. Carefully managed, it represents an insurance policy against the extinction of the animal in the field and has been effective in some important cases. Przewalski's horse was saved from extinction by nine captive animals (several of which had been the property of Carl Hagenbeck); their descendants have now been reintroduced into the wild in Mongolia.[88] In 1987, the IUCN endorsed the principle of captive breeding, and for many endangered species there are now complex arrangements between zoos and wildlife refuges for the exchange of breeding stock and, where it can be managed, for exchanges with wild populations, in the interests of avoiding inbreeding.[89] But the main objection to captive breeding lies in the potential affront that it represents to individual orangutans, both those who are denied the possibility of reproduction as hybrids and those who are moved from zoo to zoo to meet the needs of breeding programs.[90] Some scientists have described the segregation of orangutan species and the discrimination against hybrids as racist and reminiscent of the eugenics movement of the first half of the twentieth century.[91]

Opinions have also been divided about the value of orangutan rehabilitation. The rehabilitation idea arose out of the modest success achieved in rescuing illegally caught orangutans. These orangutans posed their rescuers with the practical problem of what was to be done with them. Whereas zoos were most often interested in adolescent or adult animals, the illegal trade in orangutans as pets focused on infants, who were more tractable and easier to conceal. In most cases the mothers of these infants had been killed in the capture of the offspring. Although saved from a probably brief life in captivity, these young animals could not simply be returned to the jungle because young orangutans learn the techniques of forest living during several years' companionship with their mothers. Without maternal training, they are helpless in their native habitat. The rescue of orangutans was reported soon after the colonial government imposed its trade ban in 1929,[92] but there is no record of how the animals were then treated. In 1962, however, Barbara Harrisson, the wife of Tom Harrisson, who was curator of the Sarawak Museum, began to experiment with the rehabilitation of young orangutans by training them in survival techniques and gradually reintroducing them to the jungle. Her initial attempts were with three young orangutans she tried to reintroduce to the Bako National Park, but her efforts were hampered by the fact that there was no local orangutan population and by the lack of scientific knowledge of orangutan behavior. Harrisson later shifted her program to Sepilok in Sabah.[93] In 1971, the Netherlands Leuser Committee set up a similar station at Ketambe, north of Kutacane in Aceh, headed by Herman Rijksen. Between October 1977 and March 1979, twenty-eight orangutans were recorded as being returned to the wild in Indonesia.[94] Just a little later, Canadian primatologist Biruté Galdikas established another rehabilitation program at Camp Leakey in the Tanjung Puting National Park. Galdikas had traveled to Kalimantan in 1971 to study orangutan behavior, and her 1978 doctoral thesis was a landmark in the study of orangutan behavior in the wild, based on thousands of hours of difficult observation in the jungle. Donna Haraway has argued that scientific research on primates is strongly gendered. Males researching great apes, she suggests, have pursued lines of investigation focusing on sexual competition, whereas female researchers have emphasized communication and family survival. The perception of women as having both a nurturing role and a more appropriately intuitive approach to great apes was reinforced by the publicity surrounding the primatological work of Jane Goodall with chimpanzees, Dian Fossey with gorillas, and to a lesser extent that of Galdikas with orangutans. All three spent long periods in the habitats of their apes, making close observations of behavior in the wild that were far more detailed than anything previously done. All three also became passionate advocates for the protection of apes from hunting and habitat loss.[95]

When Galdikas became interested in the issue of rehabilitation, the volume of her scientific work decreased. Between 1975 and 1995, she oversaw the release of approximately 180 formerly captive orangutans in the park and recorded twenty-nine births from individuals known to be rehabilitated.[96] Another rehabilitation center was set up in 1973 at Bohorok (Bukit Lawang) on the eastern edge of the Gunung Leuser National Park, with funding from the Frankfurt Zoological Society. It was run by two Swiss conservationists, Regina Frey and Monika Borner.[97] And finally, a small rehabilitation station opened at Semenggoh near Kuching in Sarawak in 1975.[98]

Rijksen, however, soon began to have misgivings about his own rehabilitation work and that of Galdikas, and he closed his program in 1979.[99] In particular, he feared that the rehabilitated animals would be vectors for the introduction of disease into the wild population and that the techniques used for rehabilitation were unable to make the released animals truly self-sufficient, so that to release them was actually to condemn them to a slow death in the jungle. These criticisms were especially sharp because the rehabilitation stations quickly became tourist attractions. Sepilok and Semenggoh were standard parts of day-trip tourist itineraries in Malaysian Borneo; Bohorok attracted overnighters. Even Tanjung Puting was regularly visited by large groups, whose presence brought the risk of disease and was likely to keep the orangutans accustomed to the company of humans. It was also argued that releasing ex-captive orangutans into an existing community would increase the burden on the local ecosystem, making life harder for the original, wild population. Others criticized the rehabilitation projects for consuming funds that might otherwise have been dedicated to true conservation. It was also suggested that the release program ignored the subspecific differences within *Pongo pygmaeus*, mixing different Bornean races and even introducing a Sumatran orangutan.[100] These concerns became more acute once the two subspecies were raised to species status. In response to these perceived problems, the Indonesian government took over all rehabilitation activities at Camp Leakey in 1991 and decided in 1994 to cease reintroduction at locations where there was a wild population.

Instead, a new rehabilitation center was established at Wanariset near Balikpapan in East Kalimantan in an area that no longer had a population of wild orangutans. Wanariset was followed by a second rehabilitation center at Nyaru Menteng near Palangka Raya in central Kalimantan, where the reintroduction of captured orangutans began in 1999. By 2008 the facility contained more than seven hundred animals. Nyaru Menteng is not a government project but is rather the initiative of two activists, Lone Drøscher Nielsen and Willie Smits, who were touched fortuitously by the plight of orphaned orangutans. To avoid the problems of government inaction and corruption, the two rehabilitation stations are run within a complex legal framework dominated by interrelated private foundations, notably the Orangutan Conservancy and the Borneo Orangutan Survival (BOS) Project.[101] These initiatives represent a conceptual challenge to the common assumption that animal protection is substantially a matter for state regulation, driven by changes in public opinion. Instead, the Orangutan Conservancy embraces the commodification of nature that took place during the twentieth century and employs the tools of exclusive private ownership and control by a self-selected, deeply committed elite to achieve public goods that the suborned political process cannot deliver.[102]

The significance of the rehabilitation endeavor lies not in its relatively meager contribution to the overall conservation needs of the orangutan but rather in the emphasis it places on individual human responsibility for the welfare of individual orangutans as distinct from the species as whole. This emphasis was highlighted in the title of the 1975 television program and associated book *Orphans of the Forest*, voiced by Peter Ustinov and featuring the work of Monica Borner and Regina Frey with infant orangutans in Sumatra. A central image of this documentary was the mothering of young orangutans by the two attractive young women. In 1992, the actress Julia Roberts appeared to similar effect in a television feature titled *Orangutans with Julia Roberts*. In 2001, Willie Smits, although less motherly than Borner, Frey, or Roberts, conjured up the same sense of needy helplessness on the part of infant orangutans in a documentary also titled *Orphans of the Forest*. The documentaries made the most of the human capacity to see something close to personhood in individual orangutans.

Spurred by the practicalities of the rehabilitation movement, and perhaps in some cases wary of the risks of hyperbole, proponents of this discourse emphasized the plight of orangutans as captives and as the victims of forest clearing and opportunistic murder.[103] In this way, the conservationist discourse has increasingly found common ground with a discourse of orangutan rights that had been prefigured by Rousseau in the eighteenth century and that began to take shape with the discovery of a close evolutionary connection between humans and great apes.

10 ❧ Faces in the Mirror
Evolution, Intelligence, and Rights

For four centuries, the West has been fascinated by the orangutan. So similar to humans, yet with so many differences, the red ape has challenged us to think about the nature of humanity and the character of human relationships with the animal world. As knowledge of the orangutan has developed, and as scientific understanding of biological processes has grown, the specific challenge presented by the orangutan has also changed. As the preoccupations of human society have changed, too, literature, theater, film, and other cultural productions have reflected on the orangutan in different ways. Yet many of the concerns the orangutan raised in the minds of Western observers in the seventeenth century are remarkably persistent. The problem of the red ape's proximity to humans has never been off the agenda.

Until the 1960s, scientific argument about the degree of affinity between humans and orangutans was based, like Tyson's seventeenth-century pioneering work, primarily on comparative anatomy. Although Darwin's encounter with Jenny is today regarded as significant in leading him to the theory of evolution by natural selection, his earliest opponents regarded the orangutan as providing empirical evidence to refute that theory. The prominent clergyman and scientist Richard Owen argued that although there were similarities of anatomical structure between humans and great apes, those similarities were analogous rather than homologous, that is to say, they had developed to resemble each other from largely different processes; the alikeness of humans and great apes did not mean they shared a common ancestor. To prove his point, Owen presented what he said was the decisive evidence that only the human brain possessed three key structures: a posterior lobe, a "posterior horn," and a hippocampus minor. Although not considered today to be of the greatest importance, these three structures had special significance in the medical understanding of the brain at the time, which was still based on the ideas of the Roman physician Galen and which identified the ventricle of the brain—where Owen's three key organs were located—as the site of the higher human qualities of sensation, cognition, and memory. On this basis, Owen announced that "[man] fulfills his destiny as the supreme master of this earth and of the lower creation."[1] In one of the most celebrated scientific feuds of the late nineteenth century, Darwin's defender, Thomas Henry Huxley, comprehensively debunked Owen's argument by showing that the structures were in fact present in the brains of great apes.

Owen had made the last serious attempt to assert a qualitative physical difference between humans and great apes. Since his time, the challenge for science has been to determine the nature of the evolutionary link between humans, apes, and their common ancestors. Toward the end of the nineteenth century, research had begun to focus on Southeast Asia as the possible cradle of humanity after German scientist Ernst Haeckel predicted that the real "missing link"—the fossil remains of an ancestor common to humans and apes—would be found in Java. He even speculatively gave the species a scientific name, *Pithecanthropus alalus,* meaning "speechless ape-man."[2] Haeckel's prediction appeared to be confirmed with paleoanthropologist Eugène Dubois' discovery of the remains of prehistoric "Java Man" in 1891, which briefly implied that the orangutan, as the great ape closest at hand, might be the closest relative of humans.[3] For several decades after Darwin and Owen, many scholars believed that the different races of humankind had descended separately from the three great apes. The German scientist Hermann Klaatsch suggested that Europeans and Asians shared a common ancestor with orangutans, Africans with gorillas. Carl Vogt believed that humankind fell into three large categories descended respectively from chimpanzees, orangutans, and gorillas.[4]

The eccentric British medical practitioner F. G. Crookshank developed this argument further in a book titled *The Mongol in Our Midst* (1924), in which he described Asians as "unfinished 'whites'" and argued that the medical condition now known as Down Syndrome, but then known as Mongolism, was a consequence of genetic recidivism or "throwback," that is, that an unfortunate and unexpected combination of recessive genes in both parents had led to the birth of a child with unusually primitive features. Crookshank attributed this condition either to the rape of European women by Mongol invaders of the thirteenth century or to still earlier couplings with orangutans, labeling the child either "Mongoloid" or "orangoid."[5] As evidence, he paid special attention to posture, asserting that orang-utans alone amongst apes naturally arrange their legs as in the Buddha or hieratic position, even when swinging in the air. . . . We then arrive at this conclusion. There is a primary posture that is naturally, habitually, and instinctively adopted by one race, and one race only, amongst the great anthropoid apes, by one division only of the human family, and by one group of imbeciles born amongst another division but conforming morphologically to the first. This is the Buddha position, natural to the orang-utan, to the racial Mongol, and to our indigenous Mongolian imbeciles and Mongoloids.[6]

The orangutan was recruited to another aberrant evolutionary view in 1912, when the British amateur paleontologist Charles Dawson announced the discovery of humanoid remains in a gravel pit in Piltdown in southeast England. The remains—fragments of a skull, part of a jawbone, and some teeth—appeared to show that part of the early evolution of humankind had taken place in Britain and that the large brain of modern humans had evolved nearly half a million years ago. Although some scholars expressed doubts about the authenticity of the remains, it was not until 1951 that they were definitively proven to be a forgery. The skull was that of a modern human and the jaw that of an orangutan. The jaw was subsequently carbon-dated to around five hundred years before present, making it likely that it was derived from a collection of specimens excavated in Bau cave in Sarawak in 1878 by an expedition under the naturalist A. Hart Everett, whose work was partly funded by Darwin himself. Everett had not discovered any ancient remains, but the recent bones he had found in the cave were presented to the British Museum, where they were under the control of Arthur Smith Woodward, Dawson's closest collaborator in publicizing the Piltdown "finds."[7]

Until the 1960s, most scientists favored the view that the three genera of great apes were more closely related to each other than to humans, and paleontologists believed that the human and ape lineages had parted around twenty million years ago. From the 1960s, however, the development of techniques for molecular comparison and, from the 1980s, for DNA sequencing revolutionized taxonomy. This technical advance coincided with, and made possible, a major change in the principles of scientific classification. Whereas Linnaean classification had arranged organisms above all according to their appearance (the literal meaning of "species"), the new approach, known as cladistics, focused exclusively on descent. The classification of an organism now depended on how recent or remote was the moment at which its ancestors had diverged from those of other species. Against the expectations of the first half of the century, it appeared that the line giving rise to orangutans parted from that leading to humans, chimpanzees, and gorillas about twelve to sixteen million years ago, whereas humans and chimpanzees had common ancestry until some six million years later.[8] Cladistics thus brought humans closer than ever to great apes, though it left orangutans at a greater distance from humankind than the others. In 1998 Watson and others formally proposed that chimpanzees and gorillas, but not orangutans, should be reclassified into the genus *Homo*.[9]

The second front on which scientists began to explore the extent of similarity between orangutans and humans was the broad area of intelligence. For much of the nineteenth century, the scientific orthodoxy was that animal behavior could be explained fully by a combination of instinct and learning, that all animals lacked the human characteristic of thought or reason, and that if they experienced emotions at all these were not complex. The behavior of "lower" animals was believed to be governed mostly by instinct, while "higher" animals also learned from their experiences by a process of trial and error. The "highest" animals, such as apes, were thought to be capable of sophisticated learning, and many animals could be induced to do remarkable things by skilled human trainers, but as in the eighteenth century the standard view was that the training of animals showed the intelligence of humans, not of the animals themselves. Only humans, it was believed, were capable of assessing a complex, new situation and devising a novel response to it. Once again, however, scientific beliefs were not always consonant with broader cultural ones. A number of writers and dramatists, as discussed in earlier chapters, had imagined orangutans, and apes generally, as possessing some of the human traits denied them by science. With the growing awareness and acknowledgment of similarities between humans and orangutans, with the abundant evidence of orangutan ingenuity from zoos and circuses, and with the development of new techniques for scientifically assessing cognitive abilities, it was only a matter of time before orangutan emotion and intelligence were restored to the research agenda.

Linnaeus, writing "know thyself" next to *Homo sapiens* in his *Systema Naturae,* had identified self-awareness as the defining character of humankind, and Darwin himself pioneered the investigation of orangutan intelligence with an early version of the so-called mirror test, in which an animal is shown itself in a mirror and observed to determine whether it recognizes the image as itself:

Many years ago, in the Zoological Gardens, I placed a looking-glass on the floor before two young orangs, who, as far as it was known, had never before seen one. At first they gazed at their own images with the most steady surprise, and often changed their point of view. They then approached close and pro-truded their lips towards the image, as if to kiss it, in exactly the same manner as they had previously done towards each other, when first placed, a few days before, in the same room. They next made all sorts of grimaces, and put themselves in various attitudes before the mirror; they pressed and rubbed the surface; they placed their hands at different distances behind it; looked behind it; and finally seemed almost frightened, started a little, became cross, and refused to look any longer.[10]

The verdict from Darwin's experiment thus appeared to be that orangutans, while capable of expressing emotion, lacked self-awareness.

In 1909, William Furness, an American medical graduate and amateur anthropologist, attempted a different approach. Like De la Mettrie in the eighteenth century, he speculated that he could apply to orangutans the techniques used to teach language to deaf and dumb children. He thus acquired an orangutan with the idea of teaching it to speak, later adding two chimpanzees and another orangutan. He spent, he reported, six hours a day with them. The first orangutan proved receptive and was eventually able say "Papa" and "cup" and to comply with complex questions such as "What is the funny sound you make when you are frightened?" The animal died, however, after a few months, and Furness was unable to repeat his success with the other apes.[11]

The systematic investigation of orangutan intelligence began in 1913, when Melvin Haggerty of the Harvard Psychological Laboratory devised a test of reasoning suited to great apes. The test was a simple one in which two orangutans, Betty and Nancy, were given the opportunity to use a tool to get access to food. Both apes were successful, and Haggerty concluded, "Had a human being exhibited this behavior for the first time, we should describe it most easily by saying that he perceived the relation between the food, himself, and the stick."[12] In other words, the orangutans showed evidence of conceptual thought. The American psychologist Robert Yerkes, who was instrumental in the development of human intelligence testing in the United States, expanded on Haggerty's research, testing the problem-solving abilities of two monkeys and an orangutan.[13] Like Betty and Nancy, the orangutan Julius demonstrated what Yerkes interpreted as ideational capacity, that is, the capacity to understand a situation abstractly and to devise appropriate actions.

Much broader in scope was the work of the German behavioral scientist Wolfgang Köhler at the Prussian Academy field research station at Tenerife in the Canary Islands.[14] Köhler observed two orangutans provided by the Dutch colonial government in 1916.[15] His comments on the pair, not published until long after his death,[16] bear a striking resemblance in tone to the descriptions of orangutan behavior offered by Vosmaer and others more than a century earlier, when living orangutans first came to the attention of scientists. In contrast with the sense of distance and distaste that marked some writing in the early nineteenth century, Köhler immediately resorted to terms usually reserved for humans. The female, soon given the name "Catalina," was "bold, not exactly cheerful, almost carefree." He described her as solicitous for the welfare of the more subdued male, comforting him with embraces and offering him food, insistent that he join in her games, agitated and distressed when he suffered an accident. The unnamed male, by contrast, was earnest and quiet, with a mournful, honest face, suffered from "baseless anxiety," and died early.[17] Köhler's work was also important for challenging the common view that chimpanzees were intellectually closer to humans than were orangutans:

> In its entire nature this creature is much nearer to Europeans than are the chimpanzees; it is less an animal than they are. This impression was not based on its intellectual achievement, in which some of the chimpanzees surely surpassed Catalina, but solely on disposition, character, and the like.... Catalina is without doubt 'finer', 'more decent,' 'more reliable' than the African species; she is often presumptuous, even impudent and astonishingly disobedient, but coarse traits, the unreliability of moods, brutal and self-indulgent emotional explosions such as are seen in chimpanzees, do not appear in this orang, and probably would not in other apes of this species. (196)

These discoveries marked the beginning of a relentless effort to determine the degree of emotional and cognitive resemblance between humans and great apes, though the primary attention was always to chimpanzees rather than orangutans. The greater evolutionary proximity of chimpanzees to humans and their greater robustness in laboratory conditions led most researchers to favor chimpanzees over orangutans as experimental subjects. Yerkes himself was the leading figure in this endeavor. In 1929 he published a detailed assessment of the psychological capacities of orangutans, trying hard to avoid projecting on the ape assessments based on the conventions of human behavior.[18] Other experimenters tested the signing orangutan, Chantek, discussed in chapter 7, with a developed version of Darwin's mirror test, marking the orangutan's forehead while he was under anesthetic and observing his response when he awoke. After seeing his image, Chantek tried to remove the mark, indicating that he was aware of himself as the being in the reflection.[19]

Observation of orangutans in the wild has also revealed impressive signs of intelligence. Until the 1970s there were no systematic field studies of orangutans because of the exceptional difficulty of observing apes who lived at the tops of trees in thick tropical jungle. Biruté Galdikas and John MacKinnon, however, pioneered a technique of dogged following, shadowing individual orangutans from the ground over periods of days and weeks, and noting in meticulous detail both their individual behavior and their social interactions.[20] Out of these and later observations came such insights as the fact that orangutans in the jungle routinely use tools to extract honey and insects from hidden nests and seeds from thick shells. They also use leaves as umbrellas, gloves, and sit pads against thorny fruit. The arboreal nests or beds that orangutans construct each night to sleep in are strikingly complex and varied pieces of construction, unlikely to be simply the product of instinct. The red apes also use a variety of tools to masturbate.[21] Tool using by orangutans appears to be significantly cultural, that is to say, it varies from community to community in a way that does not correspond to ecological opportunity. The orangutans of the Suak Blimbing region in Sumatra are significantly more adept at using a variety of tools than are orangutans in other places.[22] Orangutans in captivity have shown themselves even more adept at tool use. An orangutan called Abang was observed in 1972 knapping a flake from a piece of flint and using the resulting sharp edge to cut the cord that was binding a box of food.[23] The orangutan escape artists discussed in chapter 8 are further evidence that the imaginative invention and development of tools displayed by orangutans in captivity may be a consequence of the stresses and opportunities available in zoos.

In the decades since the 1970s, using both field observations and laboratory research, scientists and other investigators have identified a wide variety of areas in which orangutans possess forms of consciousness and mental and emotional capacities that have commonly been regarded as distinctively human. Moreover, as zoo escapees such as Ken Allen and signing orangutans such as Chantek have demonstrated, tool use, complex communication, and self-consciousness are not exclusively human traits. In 1996, Russon and Bard summarized the features of intelligence displayed by orangutans as follows: "causal and logical reasoning, counting, addition, mental maps, insight, imitation, self-awareness, ostension [conveying the meaning of a term by pointing out examples], pretense, role-reversal, teaching, planning, intentional deception, rudiments of mind-reading, and proto-language."[24] Crucially, orangutans apply cognition to cognition, thinking about the process of thinking, a phenomenon exceptionally rare in the animal world. As in discussions of human intelligence, there is disagreement over the relationship between these specific attributes: to what extent are they best understood as distinct, modular cognitive skills and to what extent as signs of broader underlying abilities or even of generalized "intelligence"?

As noted in chapter 7, an important element in the investigation of orangutan behavior is the animality of humans rather than in the humanity of apes. The core of this approach has been the controversial discipline known as sociobiology. Identified as a distinctive field by E. O. Wilson in 1975, sociobiology argues that some aspects of human behavior are to a significant degree the consequence of evolution by natural selection rather than of culture. The idea that animal behavior has a strong genetic component is generally accepted. Among animals, natural selection works at the level of communities (ants are a classic example) as well as at an individual level. Communities that are organized in ways that enhance their cooperation are more likely to survive and to pass the advantage to their offspring. There has been strong resistance, however, to suggestions that the same may be true of humans, except in trivial ways, and an enduring debate has taken place over the relative importance of nature and nurture in human behavior. Because of their evolutionary proximity to humans, great apes have been observed closely in the attempt to determine whether consistent features of human behavior might be a consequence of a shared genetic origin.[25]

One of the first topics to be considered was the difficult question of orangutan rape. Early travelers had been intrigued by the stories of the rape of human women by orangutans, but in 1974 John MacKinnon reported that he had observed male orangutans forcing sexual intercourse on females.[26] Subsequent research has indicated that in stable orangutan communities, the dominant figure is a large male with a prominent facial flange. Females appear to prefer to mate with these males, and when the females are fertile they will come to the males in response to a long male call. Sex between females and flanged males is normally consensual. Younger males in the vicinity of a dominant male generally lack the facial flange. They are not sought after by females and adopt a sexual strategy that resembles rape, creeping up on an isolated female and forcing sex on her. Not all matings by flanged males are consensual and not all matings by unflanged younger males are forced, but a pattern is clear. A significant number of the rapes lead to pregnancy, too, meaning that this sexual strategy is an effective one in terms of reproduction. This pattern of dual sexual strategy is unknown among other great apes (apart from humans), and the key issue is whether such similar behavior might have been generated by the same or comparable evolutionary forces.[27] The orangutan case is controversial because it could imply that the rape of women by men has its origins in an instinctive reproductive strategy rather than simply an exercise of dominance and power.[28]

More recent investigations, however, have suggested that humans and great apes have inherited not so much biological imperatives as social flexibility from their shared ancestors. It appears that the ancestors of today's humans and great apes lived in sociable multi-adult communities of males and females. These communities were multilevel: individuals were likely to be closer to some members of the community than to others. They were also volatile: they repeatedly broke up and reformed in different configurations. The consequence was that the ancestral hominids developed unusually strong capacities for complex social relationships of amity (including probably polyamory), negotiation, and enmity. Social interactions also involved conscious manipulation, sometimes called "Machiavellian intelligence," which depends upon recognizing and making use of the presumed mental processes of other animals. Examples include deliberately distracting others from food sources.[29] This complexity has been apparent, too, in the performances of orangutans in zoos and circuses. A consequence of such social alliances and interactions appears also to be the holding of long-term grudges against those who betray such alliances.[30] In practical terms, this complex sociality meant that some of these ancestral hominids liked each other, others did not, and still others would intervene in disputes to help achieve a peaceful settlement. Social outcomes were thus a product both of individual emotion and choice and of the operation of networks of alliance and association within which sophisticated techniques for reconciliation were developed. In particular, female hominids like some males and not others, in a way that is rare outside the hominid family. This evolutionary observation is consistent with the field observation that social groups among orangutans are largely constructed by female choice: even though females are normally half the size of males, it is female preference that sustains a sexual relationship, not male control. It appears possible that a female orangutan's choice of partner is related also to the need for protection from the unwanted attention of other males.[31]

The early hominids did not gather instinctively in communities, but rather developed the ability to adapt to a wide range of social conditions.[32] This circumstance helps reconcile the generally solitary character of orangutan life in the wild with the highly social behavior that occurs on the rare occasions when abundant food allows them to gather, and with the known examples of cooperative problem solving.[33] It also helps explain the affection captive orangutans have shown for their carers, as discussed earlier in this book. Complex social engagement led to selection in favor of larger brains, sending the hominids, and later especially the genus *Homo,* careering down a path of rapidly growing intelligence. For all hominids, this pressure led to a shift in diet from leaves to high-energy foods, such as fruit and meat, needed to sustain larger brains.[34] It also led to extended periods of active care for offspring, to allow for training and maturation after birth. An orangutan mother often suckles her child for eight years while instructing him or her in the complexities of food gathering and arboreal life.

This analysis suggests that the common ancestors of humans and great apes did not succeed because they developed a particular social form that made them more likely to survive, but rather that their source of success was social flexibility. In the case of humans, the capacity to gather a group of people together into an orchestra, a political party, a tutorial, a family, a mob, or any of the countless other social forms we recognize is the huge advantage that our species enjoys over all others. The companionability that we still feel with orangutans may not be just a projection of human values on a nonhuman creature but rather a surviving trace of this ancient bond. The scientific discoveries of the past forty years have confounded the once-common view that orangutans have only rudimentary cognitive abilities, but nonetheless the sum total of the intelligences attributed to them still falls far short of those that humans value in themselves. And yet a by-product of the investigation of orangutan intelligence has been an unavoidable recognition of their complex emotional lives. The issues at stake in these investigations are critically important to the understanding of human and great ape evolution and to the understanding of cognition in general, but they are invariably also invested with a question of meaning: just what moral and philosophical consequences flow from such a degree of resemblance between orangutans and humans?

In 1861, the British magazine *Punch* confronted its readers with an ape who apparently claimed human status using almost the same words as had appeared on Josiah Wedgwood's antislavery slogan "Am I Not a Man and a Brother?"

Punch was a satirical magazine, and the treatment of orangutans was not then a social issue, but the cartoon prefigured the twentieth century's growing unease with the separation of humans and animals, especially great apes, into separate moral universes.

When seventeenth- and eighteenth-century scientists and philosophers considered the resemblance of orangutans to humans, they were primarily interested in the moral status of humans and not in that of orangutans. The actual treatment of orangutans by humans was a matter of indifference to most commentators, with Rousseau the only person to suggest that recognizing humanity in orangutans might demand that humans treat them differently. Once anatomical science had categorized the red apes as nonhuman, the only protection for them lay in a minority discourse about animal rights in general. Not until late in the nineteenth century did anyone publicly ask a reflective question about the rights of orangutans. In 1880, the French journalist Aurélien Scholl wrote an essay titled "L'orangoutang" in which he commented on the anomaly that slavery was forbidden in France, yet orangutans could be kept in cages and allowed to die in misery, despite what he described as their greater intelligence than the Fuegans.[35]

"Am I a Man and a Brother?"
(1861)

MONKEYANA.

Although the moral status of animals had been an issue in ancient Greece, ancient India, and other civilizations, modern discourse on the topic commenced with René Descartes.[36] In his *Meditations* (1641), Descartes proposed that animals were no more than automata with neither soul nor mind nor reason. In contrast, human rights, for Descartes, rested on what he saw as the unique human capacity to think and feel. Although purely secular, Descartes' argument also accorded with the standard Christian position that God had given humans dominion over animals, which therefore had no inherent moral standing. A succession of other philosophers, including Rousseau, challenged the Cartesian position, arguing that the sentience of animals gave them rights, though, as Corbey points out, it is by no means clear why ethical regard for great apes should be based on their resemblance to us, rather than on respect for difference.[37] In the late eighteenth century, Jeremy Bentham proposed that animals should be protected from suffering:

> The day may come when the rest of the animal creation may acquire those rights which never could have been withholden from them but by the hand of tyranny. The French have already discovered that the blackness of the skin is no reason a human being should be abandoned without redress to the caprice of a tormentor.

It may one day come to be recognized that the number of the legs, the villosity of the skin, or the termination of the os sacrum are reasons equally insufficient for abandoning a sensitive being to the same fate. What else is it that should trace the insuperable line? Is it the faculty of reason or perhaps the faculty of discourse? But a full-grown horse or dog, is beyond comparison a more rational, as well as a more conversable animal, than an infant of a day or a week or even a month, old. But suppose the case were otherwise, what would it avail? The question is not, Can they reason?, nor Can they talk? but, Can they suffer?[38]

Since the seventeenth century, laws and regulation to protect animals from cruelty had occasionally been introduced in various European jurisdictions, but the case for animal protection was always most effective when it related to what was considered wanton cruelty that served no human interest apart from the perpetrator's self-gratification, as, for example, in public entertainments such as dog fighting and bear baiting. The same argument that cruelty ought not be tolerated where it was unnecessary underpinned a late nineteenth-century campaign against the vivisection of animals for scientific research. In 1894, Henry Salt advanced the discussion considerably by arguing that animal rights did not exist only where human interests were absent. Rather, "animals, as well as men, though, of course, to a far less extent than men, are possessed of a distinctive individuality, and, therefore, are in justice entitled to live their lives with a due measure of . . . 'restricted freedom'."[39] During the twentieth century, however, Salt's proposition was overwhelmed by the rapidly growing medical interest in experimentation on animals and by the growing industrialization of animal production for food.

Despite the small but persistent demand for orangutans as bushmeat, the commercial animal production industry has, of course, had no direct interest in orangutans. In medical research, orangutans have been much less important than chimpanzees, macaques, and rhesus monkeys because they are scarce and relatively fragile in laboratory conditions.[40] In 1905, a German scientist, Albert Neisser, based in Batavia, announced that he needed "a large number" of orangutans and gibbons for research aimed at finding a cure for syphilis.[41] His experiments showed that it was possible to transmit the spirochete that causes syphilis from animal to animal, but his research was hampered by the outbreak of other illnesses among his experimental subjects.[42] More recently, the use of orangutans in invasive medical experiments has been virtually nonexistent, though the possibility remains that they will receive attention in niche areas in which they have a resemblance to humans not shared with other apes. Orangutans have a propensity for diabetes and cardiovascular disease, for instance, and might be used as test subjects for therapeutic experiments. The possibility has also arisen that they may be raised to provide organs for transplant into humans.[43]

In the 1960s, the sheer scale of the issues of factory farming and medical experimentation sparked a new wave of concern with animal welfare. In 1965, the novelist Brigid Brophy published an article in the London Sunday *Times* titled "The Rights of Animals":

> The relationship of homo sapiens to the other animals is one of unremitting exploitation. We employ their work; we eat and wear them. We exploit them to serve our superstitions: whereas we used to sacrifice them to our gods and tear out their entrails in order to foresee the future, we now sacrifice them to science, and experiment on their entrails in the hope—or on the mere off-chance—that we might thereby see a little more clearly into the present.[44]

Brophy's article catalyzed a spate of writings, including *Animal Liberation* (1975) by Australian philosopher Peter Singer, that grappled afresh with the philosophical bases on which the exploitation and harming of animals could be rejected. Singer argued for what he called an "equal consideration of interest" for animals, which would make it unethical for humans to exploit them or to cause them to suffer. The philosophical character of the debate over animal rights is complex. Singer holds the position that the central principle is not to cause suffering; this lays an obligation upon humans but does not confer rights on animals and is independent of whether we consider them intelligent or evolutionarily close to us. Tom Regan argues by contrast that animals have rights because they possess cognition; that is, consciousness of themselves as living beings. They possess these rights even though they are incapable of comprehending them or acting on the basis of a moral consciousness.

Two general social conditions assisted in the spread of this new sensitivity to the importance of ethical treatment of animals. First, the scope of the rights discourse in general has broadened immensely during the period since the Second World War. Modern societies have undergone what Ignatieff calls a "rights revolution,"[45] in which a richer and more generous array of rights has been discovered and accorded to, or claimed by, the disadvantaged. It is hardly surprising that animal rights should be among them. Second, a series of profound shocks to modern Western self-confidence—the carnage of the First World War, the Great Depression, the Holocaust, and the vast, multifaceted environmental crisis—have combined to undermine the assurance with which the global project of modernization entered the twentieth century. These events seriously challenged the claims to civilized human exceptionalism by those who believed that Western ways marked the highest point yet in human achievement and that the rest of the world aspired, and was destined, to follow the path of the West. As it became easier for the dominant intellectual groups in society to see things from the point of view of the less powerful, discourses that endeavored to take account of animal perspectives became more prominent.

The defenders of animals may have managed a series of philosophical and ethical breakthroughs, but there appear to be at least four important hindrances to their future success. One is the enduring power of Thomas Hobbes' characterization of the life of "primitive" humankind as "solitary, poor, nasty, brutish, and short."[46] The extent to which orangutans can win respect for their resemblance to humans is limited by the deep-seated contempt for the primitive that is expressed in the almost universally negative image of "Neanderthal Man," commonly classified as another species, *neanderthalensis,* within the genus *Homo.* Recognizing the humanity of great apes is unlikely by itself to bring an ethical change in their treatment.[47]

Another hindrance is the difficulty of uniting to present a single powerful message. The defenders of animals often have deeply divergent messages, ranging from modest concern with acts of cruelty to revolutionary plans for a new paradigm in animal-human relations. Particularly important is the divided counsel between proponents of animal rights (in the broad sense) and proponents of nature conservation. Even though both offer critiques of the dominant developmentalist ideology in modern society, supporters of ethical behavior emphasize that animals have rights (or something resembling rights), whereas conservationists consider, as the chief character in J. M. Coetzee's *The Lives of Animals* (2001) puts it, the interests of species as a whole:

> In the ecological vision, the salmon and the river-weeds and the water-insects interact in a great, complex dance with the earth and the weather. The whole is greater than the sum of the parts. In the dance, each organism has a role: it is these multiple roles, rather than the particular beings who play them, that participate in the dance. As for the actual role-players, as long as they are self-renewing, as long as they keep coming forward, we need pay them no heed. (53–54)

Recently some activists have sought to bridge the gap between species protection and the welfare of individual animals by adopting the rhetoric of genocide prevention to call attention to the plight of the orangutan. Craig Stanford, in particular, draws a strong connection between the treatment of great apes and the treatment of indigenous peoples in the expansion of the modern world.[48]

A third hindrance to the campaign for animal rights is the relative absence of a discourse that can identify material advantages to humans in a more ethical treatment of animals. Whereas the campaigns to liberate slaves, colonized peoples, and women could all point to social and even commercial advantages in admitting oppressed people to the fellowship of humankind, the campaign for ethical behavior toward animals cannot easily offer material benefit to humans. On the contrary, the current widespread exploitation and killing of animals for commercial profit would necessarily be threatened. Thus any recognition of animal rights would involve widespread costs to current global economics. Related to this hindrance is the fact that animals themselves appear unlikely ever to take a role in the struggle; except in salutary works of fiction, animals rely upon well-disposed humans to claim and defend their rights. A major rationale for the "rights revolution" has been that it empowers the weak to take legal action in their own defense[49]; orangutans, however, will always have to rely on human champions.

Fourth, it is difficult to assess the scale of the cognitive reorientation needed to achieve a broad recognition of animal rights. The issue of the moral status of animals has been on the philosophical agenda for more than two millennia, not much less than the time we have been grappling with issues relating to the moral status of humans. The relationship between humans and pets ("companion animals") demonstrates the human capacity to recognize personhood in individual animals. Writing in 1965, Brophy commented that we find it remarkable that Greek philosophers could be blind to the immorality of slavery, and Singer has made explicit comparison between the liberation of women and the liberation of animals. The animal rights movement emerged in an era earmarked by the recognition of new rights and the establishment of new ethical standards in the treatment of humans.[50] The vocabulary that categorizes human mistreatment of other humans—genocide, concentration camps, torture, cruel and unusual punishment—can easily be employed to describe human treatment of animals. Nevertheless resistance remains to the idea of widening the net of rights to include any animals at all.

Because of these formidable hindrances, some advocates of animal rights have argued for a step-by-step approach, with the rights of great apes being the first target for recognition, setting a precedent and establishing a principle that will later help other species. Great apes are an especially plausible bridgehead because of their evolutionary proximity to humans and their closeness in terms of intelligence and emotional sophistication. In 1993, in order to create a clear programmatic goal, Paola Cavalieri and Peter Singer proposed that the United Nations should adopt a Declaration on the Rights of Great Apes, containing just three rights:

- the right to life, meaning that any killing of a great ape would constitute murder under international law;

- the protection of individual liberty, meaning that no great ape could be detained in a zoo, for entertainment, or for research; and

- the prohibition of torture, meaning that cruelty toward apes would likewise be criminalized.[51]

In practice, however, legislative and regulatory advances have been limited to restrictions on using great apes in medical research and, to a lesser extent, on their use for other purposes. In 1997, the British home secretary, Jack Straw, instructed the government's advisory committee on animal experimentation that further use of great apes should not be permitted.[52] Two years later, the Animal Welfare Act in New Zealand formally prohibited the use of nonhuman hominids (thus including gibbons as well as great apes) in research and teaching, unless such use was in the animals' own interests.[53] In 2007, the Spanish province of the Balearic Islands also banned research on great apes, and Spanish legislation in 2008 banned both medical research and the use of great apes in film and circuses, but not in zoos.[54] The European Union followed suit in September 2010 by banning the use of great apes in experiments in all twenty-seven member countries, though the regulation allowed a two-year period of grace for compliance and provided exemptions for conditions such as Alzheimer's disease, cancer, and Parkinson's disease, where it could be argued that research required the use of great apes.[55] A Great Ape Protection and Cost Savings Act was introduced into the U.S. Congress in April 2011, with the aim of limiting invasive research on great apes while also reducing the cost of keeping them in custody.[56]

There is a paradox in this process and the issues at stake are profound. If we accept that great apes have rights, then the processes currently visited upon them are criminal and are not remedied by such qualified and piecemeal measures. Yet there remains a philosophical difficulty in according to orangutans rights they cannot themselves claim and which they cannot comprehend, even if they would clearly benefit from their application. (Nevertheless, infants and mentally disabled persons, for instance, do have rights that they do not understand and in many instances cannot claim.) The fluidity of the criteria by which orangutans could be recognized as sharing rights that humans have won for themselves remains a profound obstacle to the systematic extension of those rights.

☙ Afterword

In many ways the great apes are very like us, yet we simultaneously hold the conviction that they are not like us. It may be true that they lack the capacity to feel remorse, a sense of duty, and historical consciousness, but they can feel sadness, curiosity, and love, and they can demonstrate mischievousness. It may be true that they do not speak in the sense in which humans understand language, but they communicate, listen, and comprehend, and they teach. It may be easy to construct a list of traits that humans alone possess in serious measure; it is hard nonetheless to escape the feeling that the list is tendentious, a grab bag of attributes assembled to assert human distinctiveness rather than incontrovertible evidence of separate moral standing. Should apes be ineligible because they cannot read? Are *Homo sapiens'* twenty-three pairs of chromosomes so fundamentally superior to the twenty-four pairs that *Pongo, Pan,* and *Gorilla* possess that we exist in separate moral universes? Although philosophers and scientists are divided on the issue, maintaining a fundamental distinction between human and animal actions and feelings has become increasingly difficult to sustain. In 1996 Pope John Paul II recognized evolution as a valid scientific doctrine and ruled that it was not in conflict with the special status of humans because God had conferred personhood on humans in an "ontological leap" that had been denied to animals.[1] Human thought had come close to full circle, returning to an approximation of Buffon's undefined "superior principle."

What Jeremy Bentham in 1780 ironically described as "the insuperable line"[2] is weakening. This waning has occurred not only through the revelations of complex animal emotions (and the increasing scholarly valorization of our own) but because, as the divide weakens, our arbitrary and contradictory uses of the logics of similarity and difference are exposed. We experiment on animals (especially other animals and primates) because they are not like us, while the experiments are only of value because the animals are like us.

This book, a cultural history of the orangutan, has examined the changing pattern of ways human beings—scientists, novelists, entertainers, and others—have grappled with the specific and unsettling similarities and differences between orangutans and humans. The usual name of the animal itself was appropriated from Malay as a means of expressing without endorsing the red ape's status as an animal that looked and behaved like a human. For four centuries, scientists, philosophers, writers, and dramatists have endeavored to pin down the nature of the difference between orangutans and humans. They focused initially on physical form and later turned to behavior, to ancestry, to biochemistry, and to emotional responses and cognitive ability. Consistently, they found grounds to separate orangutans conceptually from humans while always acknowledging an astounding and intriguing resemblance.

What they attributed to orangutans varied widely (and a few withheld any aspects of humanity from the red ape altogether), but it included a sense of justice, reverence for the dead, sexual modesty, erotic impulses, historical consciousness, nostalgia, and altruism, as well as the capacity to write and mechanical skills such as the ability to set a fire and broil a fish. In some contexts, these attributions would be called anthropomorphism—the (implicitly inappropriate) ascription of human characteristics to nonhumans—but we have been parsimonious with that term in this book because the scientific verdict on just what characteristics orangutans share with humans has been so fluid and because the philosophical definition of humanity often ran counter to scientific orthodoxy.

Of all the great apes, so science tells us, orangutans are evolutionarily furthest from *Homo sapiens,* having diverged along a different evolutionary path before the human-chimpanzee divide. Yet many observers, scientists included, have seen in the eyes, expressions, and behavior of the red apes a greater similarity to ourselves than we see in the chimpanzees. In tool use, contemplative behavior, and modes of problem solving they seem to resemble us more. It is possible that a convergent evolutionary path, rather than direct descent, has produced these perceived similarities and brought us in the end closer to *Pongo pygmaeus* than to our other near relatives. Yet in the contemporary world, our major difference is a potentially catastrophic one for the orangutan. As the character Anne noted in *What the Orangutan Told Alice,* human overpopulation and commercial greed threaten the orangutans' continuing existence.

This book has explored orangutan encounters with human culture and what humans have made of this large, gentle animal of the Borneo and Sumatra forests. Collected, observed, hunted, imagined, represented, abused, and, more rarely, cherished and protected, the orangutan is under the gravest threat from something neither Tulp nor Monboddo would ever have imagined: the loss of natural habitat through irreversible deforestation. We can only hope that its future is not that of a slow disappearance of orangutans from the wild until the last are to be found only in zoos, a shameful condemnation of our attitudes and practices and a tragedy not only for the "Wild Man from Borneo" but for us as well.

✺ Notes

INTRODUCTION

1. Tyson, "A Philological Essay" (1699), 1.
2. Gimma, *Dissertationum Academicarum* (1714), 145.
3. Le Comte, *Nouveaux mémoires* (1696), 502.
4. CITES, "Enforcement Efforts." See also Ancrenaz et al., "Aerial Surveys."
5. "Orangutan Population Down 50 Percent," March 9, 2010, http://naturealert.blog spot.com/2010/03/orangutan-population-down-50-percent.html, accessed April 16, 2011.
6. Janson, *Apes and Ape Lore,* 199–326. Monkeys had a similar range of cultural meanings in Japan and China; see Ohnuki-Tierney, *The Monkey as Mirror,* and Van Gulik, *The Gibbon in China.*
7. Variants of the phrase include "Wild Man from Borneo."
8. DeGenaro, "The Little Men," 34.
9. Australian National Dictionary Centre, "Word of the Month Nov 2012: Ranga," http://andc.anu.edu.au/sites/default/files/WOTM%20-%20Nov.%202012.pdf, accessed Dec. 22, 2012.
10. Lévi-Strauss, *Totemism,* 89.

1. FROM SATYR TO PONGO

1. Tulp, *Geneeskundige Waarnemingen* (1740 [1641]), 371–372.
2. Scientific opinion in modern times has been sharply divided over the identification of Tulp's ape. The often authoritative Yerkes and Yerkes, for instance, describe the ape unambiguously as a chimpanzee (*The Great Apes,* 103). Montagu, *Edward Tyson,* 253, suggests that the picture is of a chimpanzee modified by artistic license. Reynolds, "On the Identity of the Ape Described by Tulp 1641," suggests that it was a bonobo.

Rijksen and Meijaard, by contrast, argue forcefully that Tulp's picture agrees in most respects with the orangutan. Much less plausibly, they suggest that Tulp's mention of the creature's place of origin as Angola actually referred to the Sumatran region of Angkola. Whereas Angola was a well-known geographical location in Tulp's time, we have been unable to find a contemporary map of Sumatra that shows Angkola as a location. See Rijksen and Meijaard, *Our Vanishing Relative*, 424–425. Groves, *Extended Family*, 68–69, regards the issue as unresolved.

3. In this book, we follow the convention of referring to the island as a whole as Borneo, using the Indonesian term Kalimantan to denote the Indonesian portion of the island.

4. Tulp, *Geneeskundige Waarnemingen*, 373–374. Bloemaart's own account does not include any hint of this story. See Bloemaart, "Discourse ende ghelegentheyt van het Eylandt Borneo" (1646), 98–107.

5. Malay Concordance Project, http://www.anu.edu.au/asianstudies/ahcen/proud foot/MCP_/tapis.pl.

6. Bowrey, *A Dictionary English and Malayo, Malayo and English*, n.p.

7. Abdullah, *The Hikayat Abdullah*, 77.

8. See Yule and Burnell, *Hobson-Jobson*, 643–644; Scott, "Malayan Words in English, Part II," 86–89.

9. The word *wurangutan* appears, evidently as a loan word from Malay, in the ninth century Javanese poem, *Ramayana Kakawin*, where it seems to refer to some kind of ape; Zoetmulder and Robson, *Old Javanese-English Dictionary*, part 2, 2333. The section of the text in which the word appears, however, is considered by specialists to be a later interpolation. We are grateful to Helen Creese and Ben Arps for their assistance on this point.

10. For detailed accounts of what is known of Bontius' life, see Von Römer, "Dr. Jacobus Bontius," and Van Andel, "Introduction," ix–xlii.

11. Bontius, *On Tropical Medicine* (1642), 285.

12. Maple, *Orang-utan Behavior*, 35.

13. Groeneveldt, *Notes on the Malay Archipelago*.

14. Genesis 2:16–25, 3:1–7.

15. Janson, *Apes and Ape Lore*, 261–276. For a single, ambiguous exception, see p. 75.

16. Bontius also fails to note that the hair of his "ourang outang" was red, but he lived in an age when red-green color-blindness, which affects 5–10 percent of the male population, was not recognized as a condition. There is no other reason to suppose that Bontius was color-blind, but we cannot assume that any early male observer of an orangutan had sufficient color vision to be struck by the redness of its hair.

17. Fissell, "Hairy Women and Naked Truths," 43–74.

18. Janson, *Apes and Ape Lore*, 332–334. Indeed, as early as the late seventeenth century, one observer suggested that Bontius' illustration had little basis in direct observation and that it might be derived from Reuwich's drawing. See Tyson, *Orang-Outang sive Homo Sylvestris* (1699), 19.

19. See Spencer, "From Pithekos to Pithecanthropus," in Corbey and Theunissen, *Ape, Man, Ape-Man,* 25.

20. This conclusion has been suggested in Lach, *Asia in the Eyes of Europe,* though the catalogue does not suggest sources for the illustration. It was also suggested by Rijksen and Meijaard, *Our Vanishing Relative,* 422.

21. We are grateful to Dr. David Mitchell for first drawing our attention to this possibility.

22. Querido, "Epidemiology of Cretinism," 9–18; Merke, *Geschichte und Ikonographie des endemischen Kropfes und Kretinismus,* 214–215; Reeve, "Some Account of Cretinism," 111.

23. Radermacher, "Proeve nopens verschillende Gedaante en Coleur der Menschen" (1784), 225.

24. Albrecht von Haller, *Elementa physiologiæ corporis humani* (1763), quoted in Miles, "Goitre, Cretinism and Iodine in South Asia," 50.

25. See Husband, *The Wild Man,* 1–18.

26. We follow here the modern convention that scientific names are italicized, the generic name being capitalized and the specific epithet being in lowercase, although this convention was not in place in Linnaeus' time.

27. Under the conventions of scientific nomenclature, when a generic name has already been mentioned, it can be abbreviated in subsequent references when accompanied by a species name, known as the specific epithet.

28. Linnaeus, *Systema naturae* (1758), 24–25.

29. Linnaeus, *Systema naturae* (1758), 14–19.

30. *Wonderen der Natuur.*

31. Valentijn, 1726, quoted in Groves and Holthuis, "The Nomenclature of the Orang Utan," 415.

32. I.e., *arak,* or distilled palm sugar liquor.

33. Dobson, "John Hunter and the Early Knowledge of the Apes," 4.

34. On Linnaeus' lack of knowledge of Tyson's work, see Meijer, *Race and Aesthetics,* 46, and Broberg, "*Homo sapiens,*" 182.

35. Tyson, *Orang-Outang sive Homo Sylvestris* (1699), 2. Presumably these "Sea-Captains and Merchants" had indeed encountered orangutans, but it is most unlikely that any would have encountered "a great many of them," given the elusive character of the ape in the wild. Subsequent references to this source appear in parentheses in the main text.

36. Allamand, "Eerste byvoegzel" (1783), 43.

37. Edwards, *Gleanings of Natural History* (1758), 6.

38. See, for instance, Houttuyn, *Natuurlyke Historie* (1761), 330–350.

39. As authority for *S. lucifer,* Hoppius cited both Bontius' chapter on the "ourang outang," which had separately mentioned that there were tailed men in Borneo, and the writings of a Swedish traveler, Mats Nilson Kiöping (sometimes spelled Kjöping). Kiöping's account, however, is brief and confused. He gives the term "OranghGutans"

as a local name for baboons, locates them in Java (or perhaps Ceylon), and describes them as commonly damaging fruit gardens. He repeats the stories that they refuse to speak to avoid being put to work, that they are strong enough to attack men, and that they capture and impregnate women, but offers no evidence that he actually saw them. See Kiöping, *Een kort beskriffning* (1667), 95.

40. On this chimpanzee, see chapter 3 and Rousseau, "Madame Chimpanzee," 198–209.

41. For an extensive list of names, see Stiles and Orleman, "The Nomenclature for Man," 11–43. For *Ourangus outangus,* see Zimmerman, *Specimen Zoologiae* (1777), 398.

42. Stiles and Orleman, "The Nomenclature for Man," 7.

43. Groves and Holthuis, "The Nomenclature of the Orang Utan"; Röhrer-Ertl, "Zur Erforschungsgeschichte und Namengebung."

44. The modern scientific discovery of the gorilla is attributed to Savage and Wymans, writing in 1847 (Savage, "Notice of the External Characteristics," and Wyman, "Osteology of the Same"), but many scholars accept that Battell's description was of gorillas.

45. Buffon, *Histoire naturelle,* vol. 14 (1766), 43–83.

46. Camper, "Account of the Organs of Speech" (1779), 144–145.

47. [Camper], *Natuurkundige Verhandelingen* (1782), 34.

48. [Camper], *Natuurkundige Verhandelingen* (1782), 12.

2. "A MORE THAN ANIMAL INTELLIGENCE"

1. Plato, "Protagoras," 189; Carlino, *Paper Bodies.*

2. Broberg, *"Homo sapiens,"* 176.

3. Tulp, *Geneeskundige Waarnemingen,* 373–374.

4. Hesse, *Ost-Indische Reise-Beschreibung* (1690), 185–186.

5. Le Comte, *Memoirs and Observations* (1698), 500–501.

6. [Leguat], *The Voyage of François Leguat* (1708), vol. 2, 234.

7. Adams, *Travelers and Travel Liars,* 100–102; Atkinson, *The Extraordinary Voyage,* 35–65.

8. [Hamilton], *A New Account of the East Indies* (1728), 132.

9. [Noble], *A Voyage to the East Indies* (1762), 61–62.

10. Beeckman, *A Voyage to and from the Island of Borneo* (1718), 37–38.

11. [Battell], *The Strange Adventures of Andrew Battell,* 54–55.

12. Quoted in Smellie, *The Philosophy of Natural History,* vol. 1, 440.

13. Buffon, *Histoire naturelle,* vol. 14 (1766), 55–56.

14. De la Mettrie, *Man a Machine,* 38–39.

15. Diderot, "Suite de l'Entretien," 190. If not entirely imaginary, the creature Diderot has Polignac encounter was almost certainly a chimpanzee rather than an orangutan.

16. Rousseau, *Discourse on the Origin of Inequality,* 104.

17. *Histoire générale des voyages* (1747), between 411 and 412.

18. Rousseau, *Discourse on the Origin of Inequality*, 106.

19. Burnet, *Of the Origin and Progress of Language* (1773), 289–290.

20. Cloyd, *James Burnett Lord Monboddo*, 43; Lovejoy, "Monboddo and Rousseau," 281–289; McKay, "Peacock, Monboddo," 422–424; Barnard, "*Orang Outang* and the Definition of Man," 95–112.

21. Niekerk, "Man and Orangutan," 489.

22. Buffon, *Histoire naturelle*, vol. 14, 43–83.

23. [Buffon], *Natural history*, vol. 8, 96.

24. Buffon, *Histoire naturelle*, vol. 14, 43.

25. Vosmaer, *Beschryving* (1778), 9–13.

26. Vosmaer, *Beschryving* (1778), 13.

27. Mazel, "Van een Aap in 1777," 363–364.

28. Vosmaer, *Beschryving* (1778), 7–9.

29. Vosmaer referred to the orangutan as "she" (*zij*), but since he was writing in Dutch, which uses grammatical gender, we cannot be sure that he intended the anthropomorphism that this pronoun might imply in English.

30. Vosmaer, *Beschryving* (1778), 13.

31. [Camper], *Natuurkundige Verhandelingen* (1782), 33. Subsequent references to this source appear in parentheses in the main text.

32. Mazel, "Van een Aap in 1777."

33. Meijer, "The Century of the Orangutan," 64.

34. Forster, *A Voyage round the World*, vol. 2 (1777), 553–554. The dispute over the orangutan's remains took another twist when its skeleton was removed in 1795 by French revolutionaries who confiscated the prince's cabinet and held it until the restoration of Dutch rule in 1813; Pieters, "Notes on the menagerie," 540–542.

35. [Camper], *Natuurkundige Verhandelingen*, 34.

36. Oskamp, *Naauwkeurige Beschryving* (1803), 1.

37. Jouffroy, "Primate Hands and the Human Hand," 7.

38. Cuvier, "Description of an Ourang Outang" (1811), 188–199.

39. Domeny de Rienzi, *Océanie* (1836), vol. 1, 29.

40. Cuvier, "Description of an Ourang Outang" (1811), 189. Subsequent references to this source appear in parentheses in the main text.

41. Ballard, "Strange Alliance," 134.

42. Corbey, *The Metaphysics of Apes*, 172–173.

43. Donovan, *Naturalist's Repository*, vol. 2 (1824), unpaginated.

44. Meijer, *Race and Aesthetics*, 105–109.

45. [Camper], *The Works of the Late Professor Camper*, 32.

46. See Bindman, *Ape to Apollo*, 201–209.

47. Oskamp, *Naauwkeurige Beschryving*, 3.

48. Tiedemann, "On the Brain of the Negro" (1836), 497.

49. White, *An Account of the Regular Gradation in Man* (1799), 34.

50. Gibbons of the genus *Hoolock* are native to Bengal and northeastern India. In some early accounts, they were given nearly human attributes.

51. Jeffries, "Some Account of the Dissection of a Simia Satyrus" (1825), 570–580; Grant, "Account of the Habits and Structure" (1831), 27–46.

52. Savage, "Notice of the External Characteristics," and Wyman, "Osteology of the Same" (1847), 441.

53. Quoted in Granier, *Conferences on Homœopathy* (1859), 97–98.

54. Roger, *Buffon: A Life in Natural History*, 316.

55. Radford, "The Value of Embryonic and Fœtal Life," (1848), 89.

56. Dekkers, *Dearest Pet*.

57. Rousseau, *Discourse on the Origin of Inequality*, 105.

58. Greene, "The American Debate," 387.

59. [Long], *History of Jamaica* (1774), 364.

60. See for instance, "Observations on Equivocal Generation" (1844).

61. See Etkind, "Beyond Eugenics"; McGrady, *The Youth Doctors*, 55; and Voronoff's own *Rejuvenation by Grafting* (1925).

62. Raphael, "America's Disastrous Invasion of Canada," 37.

63. [Benkowitz], *Der Orang-Outang in Europa* (1780).

64. [Goldsmith], "The Natural History of Animals" (1782), 591–592.

65. Donovan, *Naturalist's Repository*, vol. 2, unpaginated, with plate 57.

66. See Engels, *The Part Played by Labour*.

67. [Sibly], *An Universal System of History* (1795), 24–25.

68. [Buck], "Jungle Laundress," in *Bring 'Em Back Alive*, 31–39.

69. Carter, *Don't Tell Mum*, 201.

70. [Long], *The History of Jamaica*, vol. 2, 355.

71. Perry, *Apes and Angels*, 29–67.

72. Rosenthal, Tauber, and Uhlir, *The Ark in the Park*, 120.

73. Brooker, "The Great War."

74. See Corbey, *The Metaphysics of Apes*, 23–35; 153–160; Wilson, *Sociobiology* (1975); Pinker, *The Blank Slate* (2003).

75. Corbey, *The Metaphysics of Apes*, 5.

76. Sens, "Dutch Debates on Overseas Man," 83.

3. WANTED DEAD OR ALIVE

1. Pieters and Rookmaaker, "Arnout Vosmaer," 11.

2. Pieters, "De Menagerie van Stadhouder Willem V," 40.

3. Niekerk, "Man and Orangutan," 492.

4. Haynes, "Natural Exhibitioner," 12.

5. "A Description of the Holophusicon, or, Sir Ashton Lever's Museum," *European Magazine, and London Review*, January 1782, 17–21.

6. These clippings are compiled in Lysons, *Collectanea*, vol. 1, n.p.

7. Mizelle, "'Man Cannot Behold'" 159–161.

8. The chief source for this account of the mermaid is Bondeson, *Feejee Mermaid*, 36–63.

9. Wonders, *Habitat Dioramas*, 23.

10. Baratay and Hardouin-Fugier, *Zoo*, 65.

11. Ward, *Charles Willson Peale*, 104.

12. *Claypoole's American Daily Advertiser*, April 13, 1799, n.p. See also Brigham, *Public Culture*, 130.

13. Yanni, *Nature's Museums*, 28–29.

14. Sellers, *Charles Willson Peale*, 241.

15. Richardson, Hindle, and Miller, *Charles Willson Peale and His World*, 123.

16. From "Lectures in Natural History," Lecture 2, quoted in Miller, Hart, and Appel, *Selected Papers*, 248.

17. "Singular Species of Monkey," *Observer*, August 28, 1803, 2.

18. "Most Wonderful Wonder of Wonders!!" handbill, 1803, shelfmark 806.k.1.(61), Rare Books Collection, British Library, London. The document, written in the style of John Bull cartoons, is attributed to "Napolean I, Emperor of the French."

19. "Bartholomew Fair," *Observer*, September 7, 1817, 3.

20. *London Times*, March 16, 1818, 3.

21. Donovan, *Naturalist's Repository*, n.p.

22. Donovan, *Naturalist's Repository*, n.p.

23. Costeloe, *William Bullock*, 39–40.

24. Yanni, *Nature's Museums*, 27.

25. See Costeloe, *William Bullock*, 40–41, for reviews of the museum; also Yanni, *Nature's Museums*, 25.

26. Thornbury and Walford, *Old and New London*, 257.

27. "Descriptions and Anecdotes of the Orang Outangs, now exhibiting at Egyptian Hall, Picadilly. With portraits of the animals, drawn and engraved by Thomas Landseer, esq." (London: Henry Baylis, 1831), 3–4.

28. "Descriptions and Anecdotes," 6.

29. "Descriptions and Anecdotes," 7–8.

30. "Descriptions and Anecdotes," 6–7.

31. "Descriptions and Anecdotes," 8.

32. *London Times*, April 24, 1818, 3.

33. Jesse, *Gleanings in Natural History* (1834), 42.

34. Blunt, *Ark in the Park*, 38.

35. Osborne, "Zoos in the Family," 35.

36. Osborne, "Zoos in the Family," 36.

37. Bernard and Couailhac, *Jardin des Plantes*, 84.

38. Lamarck, *Zoological Philosophy* (1809), 161.

39. Gilman, *Difference and Pathology*, 85–88.
40. See Jobling, "Daumier's *Orang-Outaniana*," 231–246.
41. The article is reproduced in Jobling, "Daumier's *Orang-Outaniana*," 243.
42. Baratay and Hardouin-Fugier, *Zoo*, 149.
43. Bernard and Couailhac, *Jardin des Plantes*, 37–38.
44. Barthélemy, *Jardiniers du Roy*, 209.
45. Bernard and Couailhac, *Jardin des Plantes*, 84–85.
46. Brauer, "Wild Beasts," 201.
47. Ritvo, *Animal Estate*, 209–214.
48. *Mirror of Literature, Amusement and Instruction*, January 13, 1838, 17.
49. Broderip, "Recreations in Natural History," 98.
50. *The Life of Richard Owen*, quoted in Vevers, comp., *London's Zoo*, 81.
51. Rousseau, *Enlightenment Crossings*, 202–203.
52. Letter to the Duchess of Somerset, April 18, 1820, quoted in Sophia Raffles, *Memoir of the Life* (1835), 447.
53. "The Orang-outan," *Penny Magazine of the Society for the Diffusion of Useful Knowledge*, Feb. 3, 1838, 44.
54. The *Complete Works of Charles Darwin* online, CUL-DAR122. (page sequence 69) Note: Darwin Charles Robert 1838.02.00--1838.07.00 Notebook C: [Transmutation of species], http://darwin-online.org.uk, accessed May 22, 2011.
55. Darwin to Susan Darwin, April 1, 1838, in Burkhardt and Smith, *Correspondence of Charles Darwin*, 80.
56. In Kohn, ed., *Charles Darwin's Notebooks*, 300.
57. Bennett, "Exhibitionary Complex," 73.
58. Robinson, Foreword to *New Worlds, New Animals*, x.
59. Veltre, "Menageries, Metaphors and Meanings," 27.
60. The innovation of the "Midway Plaisance" (meaning a pleasure ground) was introduced into nineteenth-century world fairs at the Paris Universal Exhibition and became as much an amusement alley as an educative exhibit.
61. "The Orang-Outang," *New York Times*, September 5, 1880, 2.
62. *Manchester Times*, November 24, 1893, n.p.
63. *New York Times*, November 5, 1897, 6.
64. See Betts, "P. T. Barnum," 353–368.
65. Betts, "P. T. Barnum," 353.
66. Advertisement for American Museum, 1845, "Boxed Playbills," Harvard Theatre Collection, New York.
67. Goodall, *Performance and Evolution*, 195.
68. Selected advertisements for American Museum, 1845, "Boxed Playbills," Harvard Theatre Collection.
69. Advertisement for American Museum, 1845, "Boxed Playbills," Harvard Theatre Collection.

70. *New York Times*, June 14, 1852, 3.
71. Goodall, *Performance and Evolution*, 36.
72. Quoted details are derived from "A Wild Girl of Sumatra," *New York Times*, August 6, 1881, 2.
73. Dolph, "Wildlife to the Millions," 251.
74. "An Orang-Otang Group," *Washington Post*, August 17, 1883, 1.
75. Hornaday, *Taxidermy and Zoological Collecting*, 230–31.
76. Dolph, "Bringing Wildlife to the Millions," 289.
77. Hornaday, "On the Species of Bornean Orangs," 455.
78. Hornaday, "Passing of the Buffalo," 85.
79. "97 Primates Back in Museum Cases," *New York Times*, August 27, 1965, 31.
80. "Kenneth E. Behring Family Hall of Mammals," Smithsonian National Museum of Natural History Report, http://www.mnh.si.edu/press_office/annual_reports/annualreport2003/pdfs/Mammals.pdf, accessed December 5, 2010.

4. DARKEST BORNEO, SAVAGE SUMATRA

1. On this trade, see Wolters, *Early Indonesian Commerce*; on the history of Borneo, see King, *The Peoples of Borneo*; Wadley, ed. *Histories of the Borneo Environment*.
2. Kruk, "Traditional Islamic Views of Apes and Monkeys."
3. Eder, *Batak Resource Management*, 14–18; Geddes, *The Land Dayaks of Sarawak*, 84–90.
4. Schlegel and Müller, "Bijdragen tot de natuurlijke Historie van den Orang-oetan" (1839–1844), 24; Rijksen and Meijaard, *Our Vanishing Relative*, 112–113.
5. Perry, W. J., *The Megalithic Culture of Indonesia*, 159; Boomgaard, *Frontiers of Fear*.
6. Evans, *Among Primitive People*, 174.
7. See, for example, Bernstein, *Spirits Captured in Stone*, 56. Sutlive and Sutlive give meager attention to orangutans in *The Encyclopaedia of Iban Studies*, 1312–1313.
8. Gudgeon, *British North Borneo* (1913), 69; Rijksen and Meijaard, *Our Vanishing Relative*, 113. Miller, *Black Borneo* (1942), 209, asserted, "A fresh orangutan head is every bit as powerful spiritually as a human head."
9. See Hoskins, "Headhunting as Practice"; Obeyesekere, *Cannibal Talk*, 264.
10. Dammerman, "De Orang Oetan" (1937), 28. For a modern equivalent, see Caldecott and Miles, *World Atlas of Great Apes*, 226.
11. Hose, *Fifty Years of Romance & Research* (1927), 89–90; MacKinnon, "The Orang-utan in Sabah Today," 147.
12. Clifford, "The Strange Elopement of Châling the Dyak" (1895); Blakeney, *The Journal of an Oriental Voyage* (1841), 110; Lumholtz, *Through Central Borneo* (1920), 414; "Sergeant ontvoerd door Orang oetan," *De Sumatra Post*, December 27, 1930, 4; "Door een Orang oetan geschaakt," *Tilburgsche Courant*, December 14, 1925, 6.
13. Walker's translation of De Torquemada, quoted in George T. Dodds, "Burrough's

Sailor Among Apes," *ERBzine* 1474, www.erbzine.com/mag14/1474.html, consulted May 9, 2010.

14. See Jobson, *The Golden Trade* (1623), 153.

15. On the pervasiveness of the head-hunting trope in analyses of Dayak society, see Peluso and Harwell, "Territory, Custom, and the Cultural Politics of Ethnic War," 83–118.

16. Harrisson, "Maias vs. Man," 621–627.

17. This difference is the consequence of the fusion into a single human gene of genetic material that is organized in two separate genes in apes. There has been no plausible attempt yet to date this fusion, but it must have taken place since the human and chimpanzee lineages parted, that is, within the last five million years or so. There is no evidence at this stage that significant genetic material was lost in this fusion and thus no evidence that the fusion is connected with the evolutionary success of humans. See Fan et al., "Genomic Structure and Evolution." Wild horses have thirty-three chromosomes whereas domestic horses have thirty-two, but the two are able to interbreed. See Kavara and Dovčb, "Domestication of the Horse," 1.

18. Aelian, *On the Characteristics of Animals*, 235.

19. Tulp, *Geneeskundige Waarnemingen*, 373–374.

20. "Ein von Affen geraubtes Mädchen," *Der Jäger: allgemeine Jagdzeitung für Deutschland* 3 (1841): 156. See also Sureng, "Six Orang Stories," 455.

21. Galdikas, *Reflections of Eden*, 293–294.

22. Donovan, *Naturalist's Repository* (1824), vol. 2, unpaginated near plate 59.

23. Heidhues, "The First Two Sultans of Pontianak," 283–284.

24. Radermacher, "Beschrijving van het Eiland Borneo" (1784), 142–143.

25. Van Wurmb, "Beschryving van de groote Borneoosche Orang outang" (1784).

26. On the interpretation of the skeleton, see Barsanti, "L'orang-outan."

27. Abel, "Some Account of an Orang Outang" (1825), 489–498. Donovan's 1824 account may well have been based on a private report of the same killing. At least one published report of the killing predates that of Abel. See "Enormous Orangutan Found in Sumatra," *The Edinburgh Journal of Science* 3 (1825): 144. No such celebrity surrounds the subsequent report of the killing of what was presumed to be the orangutan's mate a few months later. See *The Comparative Coincidence of Reason and Scripture* (1832), vol. 3: 217–218.

28. Hornaday, *Two Years in the Jungle* (1926), 361, 371.

29. "Een muitende lading," *Het nieuws van den dag voor Nederlandsch-Indië*, June 25, 1927, zesde blad B.

30. [Buck], *Bring 'Em Back Alive*, 14.

31. Corbey, *The Metaphysics of Apes*, 10.

32. Vosmaer, *Beschryving* (1778), 9.

33. Vosmaer, *Beschryving* (1778), 10.

34. Radermacher, "Beschrijving van het eiland Borneo," 141.

35. Hose, *Fifty Years of Romance & Research* (1927), 90–91.

36. Gudgeon, *British North Borneo*, 68–69; Yerkes and Yerkes. *The Great Apes*, 130–132. Schlegel and Müller, "Bijdragen tot de natuurlijke Historie van den Orang-oetan," 23–24. Frank Buck reports Malays as telling him that they captured a large orangutan by substituting alcohol for water that it had become accustomed to drink. "Giant Jungle Man," in [Buck], *Bring 'Em Back Alive*, 19–21.

37. In addition to the five orangutans discussed in chapters 1 and 2, the Swede Carl Tersmeden claimed to have seen an orangutan in Amsterdam in about 1735, the German scientist Johann Friedrich Blumenbach saw one in Jena in 1770, and an otherwise unknown orangutan from "Java" was brought to Uppsala in Sweden in 1785 by one of Linnaeus' students, Carl Peter Thunberg. See Broberg, "*Homo sapiens*," 181; Blumenbach, *De generis humani varietate nativa liber* (1781), 31; and Rookmaaker, "A Living Orang Utan in Uppsala in 1785," 275–276. Another Swede, Carl Gethe, records seeing an orangutan at Canton in about 1745 (Broberg, "*Homo sapiens*," 181), and the public display at the site of the Dutch East India Company post of Dejima in Japan mentions that orangutans arrived there from the Indies in 1792 and 1800, but we have been unable to find further information on these animals.

38. Buffon, *Histoire naturelle* (1766), vol. 14, 72.

39. Allamand, "Eerste Byvoegzel" (1783), 46.

40. [Polo], *The Travels of Marco Polo*, 148.

41. Schlegel and Müller, "Bijdragen tot de natuurlijke Historie van den Orang-oetan," 2.

42. Schlegel and Müller, "Bijdragen tot de natuurlijke Historie van den Orang-oetan," 24. One hundred forty guilders was about half a year's wages for a working man in the Netherlands at the time.

43. Jones, "The Orang Utan in Captivity," 18.

44. Belcher, *Narrative of the Voyage of H.M.S. Samarang* (1848), 15.

45. "Wambo," *De Indische Courant*, March 11, 1930, 2.

46. Introduction to *On Jungle Trails*, cited in Steven Lehrer, "Introduction," in [Buck], *Bring 'Em Back Alive*, xix.

47. *Proceedings of the Zoological Society of London* 102 (1841): 56 [55–60].

48. Wallace, *The Malay Archipelago* (1869), 46–57.

49. Beccari, *Wanderings in the Great Forests of Borneo* (1904), 137–152, 162–165, 167–168.

50. Hornaday, *Two Years in the Jungle* (1926), 363.

51. Hornaday, *The Man Who Became a Savage* (1896), 16.

52. Allamand, "Eerste Byvoegzel," 43.

53. Abel, *Narrative of a Journey in the Interior of China* (1819), 319.

54. Grant, "Account of the Habits and Structure" (1831), 41.

55. Abel, *Narrative of a Journey in the Interior of China* (1819), 366.

56. Schlegel and Müller, "Bijdragen tot de natuurlijke Historie van den Orang-oetan," 2–8, 13–22, 27–28.

57. Wallace, *My Life* (1905), 179–181.

58. Wallace, *The Malay Archipelago*, 46. See also Benton, "Where to Draw the Line."

59. Hornaday, *Two Years in the Jungle*, 368, 370.

60. Hornaday, *Two Years in the Jungle*, 91–93.

5. IMAGINING ORANGUTANS

1. Wasserman, "Re-inventing the New World," 132.

2. In Matton, ed., *Tintinnabulum Naturae*, 164–174.

3. The letter appeared as an appendix to Bretonne's novel, *La Découverte australe par un Homme-volant ou le Dédale français; Nouvelle très-philosophique*. Subsequent references to this source appear in parentheses in the main text.

4. "The Orang Outang," 497. Subsequent references to this source appear in parentheses in the main text.

5. *Haut ton* (French), "high tone."

6. Peacock, *Melincourt*, 37. Subsequent references to this source appear in parentheses in the main text.

7. Even though Sir Oran Haut-ton's origins are philosophic rather than geographical and biological, it is worth noting that *Melincourt* is one of the earliest novels in which environmental questions are aired, in conversations between Forester and Fax. Though their discussions and conclusion have nothing to do with orangutans or forests outside England, they prefigure those fictions of the twentieth and twenty-first centuries wherein orangutan habitat destruction is directly linked to human greed and overpopulation.

8. We have thus been conditioned to interpret thinking, talking animals in fiction as representing humans, and only recently have some fiction writers begun to portray animals as speaking and thinking subjects in their own right, deliberately including literary devices to forestall or preclude their being read as allegorical subjects. Novels in this category include Barbara Gowdy, *The White Bone* (1998); Peter Goldsworthy, *Wish* (1995); Timothy Findley, *Not Wanted on the Voyage* (1985); and Yann Martel, *Life of Pi* (2001). For a critical discussion of this issue, see Huggan and Tiffin, *Postcolonial Ecocriticism*.

9. Now known as *Presbytis rubicunda*, the red leaf monkey or maroon langur is one of the large family of Old World monkeys and is not considered especially close to humans. Within the context of the novel, the Beast's identity is disputed by "a noted scientist" and it is unlikely that he could overcome Power, as he does later in the story, if he were indeed *Semnopithecus*. His "dresser" is said to be the only one who knows whether or not he has a tail.

10. Constable, *Curse of Intellect*, 37. Subsequent references to this source appear in parentheses in the main text.

11. See Achebe's critique of Conrad's deployment of African figures in *Heart of Darkness*, in "An Image of Africa."

12. *The Curse of Intellect* is said to have influenced H. G. Wells' novel *The Island of Dr Moreau* (1896), which portrays an island on which Dr. Moreau has vivisected several animals to make them more like humans. See Suvin, *Victorian Science Fiction,* 60. The idea of grafting an orangutan head on to a human body appears in a comic in the Mandrake the Magician series entitled "L'homme orang-outang," *Mandrake hebdomadaire* 176, August 8, 1968.

13. Roland, *Le Presqu'homme,* 24–37.

14. Summary of Google translation from *La Conquete d'Anthar,* Anticipations Anciennes, http://jlbrodu.free.fr/aaa3.html, accessed May 24, 2011.

15. Greene, *Planet of the Apes,* 1.

16. Greene, *Planet of the Apes,* 1. Explicit racial themes were absent from the 2011 film *Rise of the Planet of the Apes.*

17. Diski, *Monkey's Uncle,* 9. Subsequent references to this source appear in parentheses in the main text.

18. In fact the Malay word *suka* has the relatively prosaic meaning "to like."

19. The dates and details in Diski's novel do not precisely match historical details of Jenny or the first orangutans acquired by the London Zoo, which are discussed in chapter 3.

20. The Librarian's simian transformation occurs in the second book of the series, *The Light Fantastic* (1986). Subsequently, he appears periodically in his orangutan form as a minor character.

21. Smith, *What the Orangutan Told Alice,* 68. Subsequent references to this source appear in parentheses in the main text.

6. CLOSE ENCOUNTERS AND DANGEROUS LIAISONS

1. See, for instance, Carroll, "Anglo-American Aesthetics," 249–251.

2. Stump, "Translator's preface," xxvii.

3. Verne, *Mysterious Island,* 319. Page numbers for further citations of this source are given in parentheses.

4. Sheridan, *Young Marooner,* 25. Page numbers for further citations of this source are given in parentheses.

5. Cited in Pougens, *Jocko,* 142–143. Page numbers for further citations of this source are given in parentheses.

6. Wallace, *My Life,* 179–180.

7. Pogorel'skii, "Journey," 112.

8. In "A Russian Tarzan," Leblanc reads the story differently, suggesting that it might be a deliberate parody of the sentimentalism of Pougens' story.

9. Berthet, *Wild Man,* 50. Page numbers for further citations of this source are given in parentheses.

10. Robida, *Adventures of Saturnin Farandoul.* Page numbers for further references to Robida's story are given in parentheses.

11. These islands are now known as the Tuamoto archipelago.

12. An engaging Italian silent film version of the novel was released in 1913 under the title *Le Avventure straordinarissime di Saturnino Farandola.*

13. Jefferson, *Notes on the State of Virginia,* 230.

14. See McClintock, *Imperial Leather.*

15. Summarized in Harrisson, "Maias vs. Man," 612–617.

16. Flaubert, "Whatever You Want," 81–82.

17. Roland, *Le Presqu'homme,* 67.

18. Champsaur, *Ouha,* 34.

19. Champsaur, *Ouha,* 49.

20. Berliner, "Mephistopheles and Monkeys," 322.

21. Jensen, *Waarom Vrouwen van Apen houden,* 212–222.

22. Pogorel'skii, *The Double,* 111.

23. Hone, *Table Book,* 378–379.

24. On Scott's influence, see Moore, "Poe, Scott." For an account of the possibilities of Poe's having seen an orangutan, see Mitchell, "Natural History."

25. Creed has argued, contrary to this position, that the orangutan's interest in shaving implies self-consciousness and would have been understood by Poe's nineteenth-century readers in such terms. See *Darwin's Screens,* 138.

26. Lee, *Slavery,* 14–51.

27. See Frank, "'The Murders in the Rue Morgue': Edgar Allan Poe's Evolutionary Reverie," 181.

28. Frank, "'The Murders in the Rue Morgue': Edgar Allan Poe's Evolutionary Reverie," 169.

29. Greenwood, *Reuben Davidger,* 144.

30. Greenwood, *Reuben Davidger,* 144.

31. Reid, *Castaways,* 223.

32. Gilmour, *Long Recessional,* 92.

33. Kipling, "Bertran and Bimi," 339.

34. Hall, *Gone Wild,* 6. Subsequent references to this source appear in parentheses in the main text.

35. See Benton, "Where to Draw the Line."

36. The *nouvelle manga* is a form of adult printed cartoon developed in France and Belgium and drawing inspiration from the Japanese manga.

7. MONKEY BUSINESS

1. After the premiere production, a text was published under the authorship of Messrs. Gabriel and Rochefort, though commentators sometimes attributed the work to Mazurier, who is assumed to have played a signal role in its development.

2. The common French name for the capuchin monkey.

3. Winter, *Theatre of Marvels,* 154.

4. Cohen, "Monkey Show," n.p.

5. See Taylor, *British Pantomime Performance,* on the development of harlequinades in the nineteenth century.

6. *New Monthly Magazine,* December 1, 1825, 535.

7. *The Literary Gazette,* November 12, 1825, 733.

8. *The Literary Gazette,* December 17, 1825, 813.

9. Goodall, *Performance and Evolution,* 52.

10. Goodall, *Performance and Evolution,* 50.

11. Playbill for *The Brazilian Ape, or The Man Monkey,* Highland Council Archive, Inverness. Online at Ambaile Highland History & Culture, http://www.ambaile .org.uk/en/item/item_page.jsp?item_id=16002, accessed January 2, 2011.

12. Snigurowicz, "Sex, Simians and Spectacle," 53.

13. Thompson, *Jack Robinson and His Monkey,* 21.

14. Quotations are from an adapted 1842 script: Somerset, *Bibboo, or the Shipwreck,* 46.

15. Somerset, *Bibboo, or the Shipwreck,* 88.

16. Teasdale, *Life and Adventures,* 70.

17. Lindfors, "Ethnological Show Business," 213.

18. Rodwell, *The Ourang Outang,* 11. All subsequent references to this script are given in parentheses.

19. Snigurowicz, "Sex, Simians and Spectacle," 55. Versions of this (Lamarckian) idea animated debates among French anatomists in the early nineteenth century, notably in a confrontation between Cuvier and Geoffroy St Hilaire in 1830.

20. Teasdale, *Life and Adventures,* 67–68.

21. Barnum denied his involvement in this ruse, possibly because it was rumored to have contributed to Leech's death some months later, but his letters strongly suggest he was the exhibit's instigator. See Cook, *Arts of Deception,* 126–127.

22. *Illustrated London News,* August 29, 1846, 143.

23. Altick, *Shows of London,* 266.

24. Bondeson and Miles, "Julia Pastrana, the Nondescript," 199.

25. The pamphlet is quoted at length in Lindfors, "P. T. Barnum and Africa," 21–22.

26. Quoted in Lindfors, "P. T. Barnum and Africa," 20.

27. Cook, *Arts of Deception,* 122–125.

28. These plot details are derived from Snigurowicz, "Sex, Simians and Spectacle," 68–71.

29. "Vari-whiskered Celt under a Judicial Ban," *New York Times,* December 30, 1904, 9. As the racist characterization of blacks as orangutans receded, it was supplemented by a similar derogation of the Irish. See the popular song by Grace Carleton "Too Thin, or Darwin's Little Joke," with music attributed to O'rangoutang (New York: Wm A. Pond [1874]).

30. "A Wild Animal Film," *Bioscope,* August 12, 1915, 655.

31. See Mitman, *Reel Nature,* 6–8.

32. Peary, "Missing Links," n.p.

33. *Variety,* June 4, 1930, p. 21.

34. See Lehrer, "Introduction," xvi, and New York World's Fair, 1939–1940, Medicus collection, http://www.archive.org/details/Medicus1939_18, accessed February 22, 2010.

35. The trope of the orangutan as apparently knowing witness to human actions also figures briefly in the original 1942 film of Kipling's *The Jungle Book,* albeit in a different light. As one of the villagers murders another, the camera cuts to close-up shots of an orangutan, teeth bared, screaming. In Disney's cartoon version (1967), the orangutan morphs into an intimidating character, King Louie, who pressures Mowgli to give him the gift of fire. His song, "I Want to Be Like You," is sung as a black American jazz tune reminiscent of Louis Armstrong's work.

36. See Torgovnik, *Gone Primitive,* 42–72.

37. Rooy, "In Search of Perfection," 204–205.

38. Tudor, *Monsters and Mad Scientists,* 139–140.

39. Woolf, "The Movies in the Rue Morgue," 47–59.

40. Examples include *King Kong* (1933), *Mighty Joe Young* (1949), and *Gorilla at Large* (1954). On the ape monster trope in film, see Creed, *Phallic Panic.*

41. Chris Wood, *Link,* British Horror Films, last modified February 24, 2010, http://www.britishhorrorfilms.co.uk/link.shtml, accessed July 23, 2011.

42. See Vint, "Simians, Subjectivity and Sociality," 226–227.

43. Smith, *Clint Eastwood,* 176–177.

44. Miles, "Language and the Orangutan," 46–50.

45. Hillix and Rumbaugh, *Animal Bodies, Human Minds,* 189–191.

46. Miles, "Language and the Orangutan," 51, drawing terms from psychological definitions of personhood.

47. See, for example, "Mr. Smith: CJ the Ape," 1980s Sitcoms, last modified Aug. 15, 2006, http://www.sitcomsonline.com/photopost/showphoto.php/photo/43323, accessed March 31, 2010.

48. "Passions and the Daytime Emmys: The Precious Problem," June 16, 2008, http://soaps.sheknows.com/passions/news/id/2029/Passions_and_the_Daytime_Emmys_The_Precious_Proble//, emphasis added.

49. "Nurses Find Orangutan 'Nurse' on NBC's *Passions* Less than Precious," The Truth about Nursing: Changing How the World Thinks about Nursing, last modified December 16, 2003, http://www.truthaboutnursing.org/press/releases/2003/2003dec16_passions.html, accessed May 21, 2004.

50. Burt, *Animals in Film,* 41.

51. The film includes two scenes focusing on Darwin's visit to the first Jenny at the London Zoo, presenting the ailing orangutan as a tragic double for the scientist's beloved daughter, Annie.

8. ZOO STORIES

1. Zuckerman, "The Rise of Zoos and Zoological Societies," 15.
2. Robinson, "Foreword," vii.
3. Rothfels, *Representing Animals*, 214.
4. Reichenbach, "Tale of Two Zoos," 55–57.
5. Reichenbach, "Tale of Two Zoos," 59.
6. Alexander, *Museum Masters*, 330–332.
7. Hagenbeck, *Beasts and Men*, 277.
8. Hagenbeck, *Beasts and Men*, 283–284.
9. Hagenbeck, *Beasts and Men*, 278–279.
10. Hagenbeck, *Beasts and Men*, 292.
11. "Orang-Outangs at Dinner," *New York Times*, July 26, 1901, 5.
12. "Star Role of 'Rajah,'" *New York Times*, August 18, 1901, 3.
13. "Rajah Gets a Letter," *New York Times*, August 29, 1901, 8.
14. "Rajah the Attraction," *New York Times*, September 2, 1901, 2.
15. "Trained Orang-Outangs Dead," *New York Times*, October 22, 1901, 5.
16. Hornaday, *Popular Official Guide*, p. 16.
17. See Ditmars, "Training Orangs and Chimpanzees," 93–96, and his "Collection of Great Apes," 756–758.
18. Such sentiments are scattered through the 1901 *New York Times* articles cited above.
19. "Orang-Outangs at Dinner," *New York Times*, July 26, 1901, 5.
20. See Sorenson, *Ape*, 85; also Rothfels, *Savages and Beasts*, 191.
21. See Allen, Park, and Watt, "Chimpanzee Tea Party," 45–54.
22. Ditmars, "Collection of Great Apes," 756.
23. Hornaday, *Minds and Manners*, 10–11.
24. Hornaday, *Minds and Manners*, 72.
25. Hornaday, *Minds and Manners*, 74–76, 209–210.
26. Hornaday, *Minds and Manners*, 205.
27. Hornaday, *Wild Animals*, 62.
28. Details about Mollie are drawn from Courcy, *Zoo Story*, 108–109.
29. "A Piece of History," "Primate House," St. Louis Zoo, http://www.stlzoo.org/yourvisit/thingstoseeanddo/historichill/primatehouse.htm, accessed December 21, 2010.
30. Benchley, *My Friends the Apes*, 80.
31. Benchley, *My Life*, 63.
32. Benchley, *My Life*, 60–61.
33. Benchley, *My Life*, 63.
34. See Mitchell, "Humans, Nonhumans and Personhood," 240–241.
35. Rothfels, *Savages and Beasts*, 191.

36. Linden, *Octopus and the Orangutan*, 8–13.

37. On Ken Allen's escapes, see Hribal, "Orangutans, Resistance and the Zoo," n.p.

38. Linden, *Parrot's Lament*, 18.

39. Linden, 'Can Animals Think?' n.p.

40. Burt, "Illumination of the Animal Kingdom," 213.

41. "Red-Headed Ape Opens $386,000 Zoo House," *New York Times*, October 13, 1950, 41.

42. "Orang-utan on the Wagon for Royal Visit to Zoo," *London Times*, February 24, 1972, 18.

43. Peterson and Goodall, *Visions of Caliban*, 136.

44. Bouissac, *Circus and Culture*, 118.

45. Beers, *For the Prevention of Cruelty*, 105.

46. Morris and Morris, *Men and Apes*, 92; figures are converted to U.S. dollars.

47. Val Mills, "The Beer Drinking Orangutan," *Issues*, September 26, 2010, http://socyberty.com/issues/the-beer-drinking-orang-utan/#ixzz1CyZUlZMT, accessed February 7, 2011.

48. Noell, *Noell's Ark Gorilla Show*, 240–241.

49. Noell, *Noell's Ark Gorilla Show*, 237.

50. Reed, "Hear No Evil," n.p.

51. Performance details are drawn largely from Hearne, *Animal Happiness*, 176–178.

52. Peterson and Goodall, *Visions of Caliban*, 143.

53. Peterson and Goodall, *Visions of Caliban*, 176.

54. Peterson and Goodall, *Visions of Caliban*, 167.

55. Hearne, *Animal Happiness*, 173–196.

56. See "Orangutan Funny Show," YouTube, http://www.youtube.com/watch?v=oySbzIHuxgg, accessed February 16, 2011; and "Bobby Berosini's Orangutans," YouTube, http://www.youtube.com/watch?v=gI9EdBt5bUo, accessed February 16, 2011.

57. See Centre for Great Apes, http://www.centerforgreatapes.org/residents.aspx, accessed February 16, 2011.

58. This description is drawn from YouTube video excerpts. See for, instance, "Orangutan Kickboxing," http://www.youtube.com/watch?v=xNIM453_dgg&feature=related, accessed February 18, 2011.

59. "Thai Zoo to Be Charged over Illegal Orangutan," *Environmental News Network*, December 16, 2004, http://www.enn.com/wildlife/article/599/print, accessed February 18, 2011.

60. "Love Bite Needed 22 Stitches," *New Paper*, August 11, 2004, 12.

61. "Love Bite Needed 22 Stitches," *New Paper*, August 11, 2004, 12.

62. "Goodbye, Ah Meng," *The Straits Times*, February 9, 2008, n.p.

63. See Knight, "Introduction," 2.

64. "Love Bite Needed 22 Stitches," *New Paper*, August 11, 2004, 12.

65. Rothfels, *Representing Animals*, 218.

66. "Jungle Island Presents HumAnimals," *Ad Hoc News,* January 3, 2011, 2011, http://www.ad-hoc-news.de/jungle-island-presents-humanimals--/de/News/21821365, accessed February 21, 2011.

67. Morris and Morris, *Men and Apes,* 43.

68. "Artists and Apes Exhibit in London," *New York Times,* September 18, 1958, 40.

69. "Art Goes Wild at the Zoo," *Houston Chronicle,* August 10, 2005, http://www.chron.com/cs/CDA/printstory.mpl/ae/art/3303484, accessed August 11, 2005.

70. "Test Tube Apes," *Atlanta Journal-Constitution,* July 13, 2005, E1.

9. ON THE EDGE

1. Indeed, with the single, spectacular, and recent exception of *Homo sapiens*, the apes as a group appear to have been in decline since the Miocene era, fifteen million years ago. See Gould, *Eight Little Piggies,* 289.

2. Atmoko and Van Schaik, "The Natural History of Sumatran Orangutan," 42.

3. Schwartz, *The Red Ape,* 78–81; Andrews and Cronin, "The Relationships of *Sivapithecus* and *Ramapithecus*."

4. Brown and Ward, "Basicranial and Facial Topography," 247; Kelley, "The Hominoid Radiation."

5. Keys, *Catastrophe.*

6. See Steinmann, "De Dieren op de Basreliefs van de Boroboedoer."

7. Logan, "Five Days in Naning" (1849), 490. See a cryptic reference that might also refer to an orangutan from Johor in Frederickson, *Ad Orientem* (1890), 276–277.

8. Hooijer, "The Orang-utan in Niah Cave Pre-history," 408.

9. Van Gulik, *The Gibbon in China,* 26.

10. Van Gulik, *The Gibbon in China,* 28.

11. Grillet to Mgr de Vanes, 18 avril 1806, Archives de la Société des Missions-Étrangères de Paris, Cochinchine, Vol. 747, pp. 421–422. We are grateful to Nola Cooke for providing and translating this reference. Bao Ninh's novel *The Sorrow of War* describes a soldier who killed an orangutan in what was then South Vietnam: "when it was killed and shaved the animal looked like a fat woman with ulcerous skin, the eyes, half-white, half-grey, still rolling" (p. 5). Although it might seem improbable that orangutans could have survived in that chronically war-ravaged country, the possibility is made plausible by the fact that a small community of Javan rhinoceros was discovered in 1988; see Schaller et al., "Javan Rhinoceros in Vietnam."

12. See Forth, *Images of the Wildman*; and Rijksen and Meijaard, *Our Vanishing Relative,* 59–64.

13. Franklin, "Evolutionary Changes." See, however, Harcourt, "Empirical Estimates," who estimates that the smallest viable population is at least twenty thousand.

14. Atmoko and Van Schaik, "The Natural History of Sumatran Orangutan," 42;

Delgado and Van Schaik, "The Behavioural Ecology and Conservation of the Orangutan," 203–204.

15. D'Obsonville, *Philosophic Essays* (1784), 354.

16. Miller, "An Account of the Island of Sumatra &c." (1778), 170.

17. Marsden, *The History of Sumatra* (1811), 117.

18. Abel, *Narrative of a Journey in the Interior of China*, 319.

19. Schlegel and Müller, "Bijdragen tot de natuurlijke Historie van den Orang-oetan," 12.

20. Hornaday, *Two Years in the Jungle*, 335.

21. Lumholtz, *Through Central Borneo*, 99–100.

22. Miller, *Black Borneo*, 218.

23. Curiously, this ban, although reported from several sources, does not appear in the official *Sarawak Gazette*.

24. Piepers, "Door welke maatregelen?" (1896); Van Houten, *Nota aangeboden aan het Bestuur der Maatschappij ter Bevordering van het Natuurkundig Onderzoek* (1896), 26.

25. *Koloniaal Verslag* 1898, 83.

26. *Correspondence relating to the Preservation of Wild Animals in Africa* (1906); Mackenzie, *Empire of Nature*, 207–209; Swadling, *Plumes from Paradise*, 93–96.

27. Archief Ministerie van Koloniën (AMK) verbaal 1.9.1908 no. 62, Nederlandsche Vereeniging tot Bescherming van Dieren to Minister of Colonies, May 30, 1908.

28. Ordonnantie tot Bescherming van Sommige in het Wild Levende Zoogdieren en Vogels (Ordinance for the Protection of Certain Wild Mammals and Birds), *Indisch Staatsblad* [hereafter *Ind.Stbl.*] 1909 nos. 497 and 594 and 1910 no. 337. The regulations were issued on October 14, 1909, but came into effect only on July 1, 1910.

29. *Ind.Stbl.* 1924 no. 234 Jachtordonnantie; *Ind.Stbl.* 1925 no. 566 Zoogdieren. Vogels.

30. Dammerman, "De Orang Oetan," 32. A Dutch newspaper report from 1928 suggests that the figure may have been higher, referring to sixty animals sent to the Riviera and forty-six to London. See "Gevangen Orang-oetans," *Het Vaderland: Staat- en Letterkundig Nieuwsblad*, July 12, 1928, avondblad C 2.

31. Coomans de Ruiter, "Natuurbeschermingsmaatregelen noodig geacht," 28–30; and documents in the National Archives, Kew, CO273/560/9, CO273/565/23, and CO717/72/11.

32. Adams, *Against Extinction*, 4.

33. *Ind.Stbl.* 1916 no. 278, cited in Coomans de Ruiter, "Natuurbescherming in Nederlandsch-Indië," 141.

34. *Ind.Stbl.* 1932 no. 17 Natuurmonumenten- en wildreservatenordonnantie 1932.

35. Nederlandsch-Indische Veneeniging tot Natuurbescherming (hereafter NIVN), *Verslag over den Jaren 1924–1928*, 13–14.

36. NIVN, *Verslag over den Jaren 1933–1934*, 24–25.

37. Jhr B.C.C.M.M. van Suchtelen, Nota van overgave, 1931–1933, Residentie Zuider- en

Oosterafdeeling van Borneo, 196, Memories van overgave. Collecties van het Ministerie van Koloniën (MMK), Lisse, The Netherlands: MMF Publications; NIVN, *Verslag over den Jaren 1933±1934,* 12–13, 15; Arnscheidt, *"Debating" Nature Conservation,* 346.

38. H. J. Koerts, "Aanvullende memorie van overgave van de onderafdeling Sampit" n.d. [1937 or after], 35, Memories van overgave. Collecties van het Ministerie van Koloniën (MMK), Lisse, The Netherlands: MMF Publications.

39. J. Oberman, Algemeene Memorie Westerafdeeling van Borneo, n.d. after 1937, 62, Memories van overgave. Collecties van het Ministerie van Koloniën (MMK), Lisse, The Netherlands: MMF Publications.

40. Jhr B.C.C.M.M. van Suchtelen, Nota van overgave, 1931–1933, Residentie Zuider- en Oosterafdeeling van Borneo, 192; "Memorie van overgave van den sedert overleden Resident der Zuider- en Oosterafdeeling van Borneo, J. de Haan," 1929, 79; Coomans de Ruiter, "Natuurbescherming in Nederlandsch-Indië," 146–147.

41. Koerts, "Aanvullende memorie van overgave van de onderafdeling Sampit," 35.

42. C. Ensing, "Memorie van overgave van de onderafdeling Beraoe," 1937, 95, Memories van overgave. Collecties van het Ministerie van Koloniën (MMK), Lisse, The Netherlands: MMF Publications.

43. Cribb, "Conservation in Colonial Indonesia."

44. *Ind. Stbl.* 1941 no. 187 Natuurbeschermingsordonnantie 1941.

45. Dammerman, "De Orang Oetan," 32.

46. Carpenter, *A Survey of Wild Life Conditions* (1938), 19.

47. Yerkes and Yerkes, *The Great Apes,* 135.

48. The naturalist F. J. Appelman commented in a private letter in 1953 on Hoogerwerf's lack of interest in orangutans. Appelman to Eshuis, July 23, 1953, *Archief Nederlands Commissie voor Internationale Naruurbescherming* [hereafter *Archief NCIN*], inv. nr. 30, Gemeentearchief, Amsterdam.

49. "Overzicht activiteit op gebied v/d Natuurbescherming v/d Kebun Raya Indonesia over de afgelopen jaren" [1951?]; A. Hoogerwerf, "Annual report 1955 of the Department for Natureprotection and wildlife management of the Botanical Gardens of Indonesia," 1956, Nationaal Archief, Den Haag, Collectie 579 A. Hoogerwerf, 1899–1974, nummer toegang 2.21.281.27 [hereafter *Archief Hoogerwerf*], inv. nr. 29.

50. Appelman to van Tienhoven, 7 Dec. 1949; Appelman to van Tienhoven, 25 Jan. 1953, *Archief NCIN* inv. nr. 30

51. A. Hoogerwerf, "Jaarverslag over 1956 van de Afdeling Jacht en Natuurbescherming van Kebun Raya Indonesia, Bogor," 12, *Archief Hoogerwerf* inv. nr. 29.

52. Harrisson, "Orang-utan—What Chances of Survival?" (1961); Harrisson and Harrisson, "Has the Orang a Future?"

53. Milton "The Orang-utan and Rhinoceros in North Sumatra."

54. Prof. Dr. Rudolf Schenkel and Dr. Lotte Schenkel, "Report on a survey trip to Riau

area and the Mt. Leuser reserve to check the situation of the Sumatran rhino and the Orang Utan (21 August to 13 September 1969)," typescript generously provided by Professor Schenkel, 4.

55. Arnscheidt, *"Debating" Nature Conservation,* 346.

56. Bruen and Haile, *Report of the Maias Protection Commission* (1960), 5–7.

57. The Direktorat Perlindungan dan Pengawetan Alam (Directorate of Nature Protection and Preservation) was created in about 1966 as a merger of the forest protection section of the Forestry Department and the Nature Conservation section of the Bogor Botanical Gardens. I Made Taman, interview (RC), Bogor, June 4, 1997.

58. Arnscheidt, *"Debating" Nature Conservation,* 346.

59. "The Animal Trade in Indonesia."

60. "Environmental Problems of Indonesia: Country Report," prepared within the framework of the United Nations Conference on the Human Environment, Stockholm, June 1972.

61. Cribb, *The Politics of Environmental Protection in Indonesia.*

62. MacKinnon, "The Orang-utan in Sabah Today," 142, 148, 190.

63. Atmoko and Van Schaik, "The Natural History of Sumatran Orangutan."

64. Warren, "Speciation and Intrasubspecific Variation."

65. Groves, *Primate Taxonomy,* 298–300.

66. MacKinnon, *Tanjung Puting National Park,* 1–2.

67. MacKinnon, *Tanjung Puting National Park,* 2.

68. Arnscheidt, *"Debating" Nature Conservation,* 346–351.

69. Casson, "Decentralisation of Policies," 20.

70. *Penebangan Liar*; "Tanjung Puting Timur Kian Marak Dijarah," *KoranTempo,* October 16, 2004; "40 Persen Taman Nasional Tanjung Puting Rusak," *Tempointeraktif,* April 21, 2001, http://www.tempo.co.id/hg/nasional/2001/04/21/brk,20010421-06 ,id.html, accessed October 21, 2010.

71. Instruksi Presiden Republik Indonesia Nomor 5 Tahun 2001 Tentang Pemberantasan Penebangan Kayu Illegal (Illegal Logging) Dan Peredaran Hasil Hutan Illegal Di Kawasan Ekosistem Leuser Dan Taman Nasional Tanjung Puting, April 19, 2001, http://legislasi.mahkamahagung.go.id/docs/Inpres/Inpres_2001_5_Pemberantasan %20Penebangan%20Kayu%20Illegal%20di%20Kawasan%20Ekosistem%20 Leuser%20dan%20Taman%20Nasional%20Tanju.pdf, accessed October 20, 2010.

72. Confidential interview (RC).

73. "The World's Biggest Ecological Disaster," *Tempo* 22/IV, February 3–9, 2004, http:// www.sea-user.org/news-detail.php?news_id=1092, accessed October 22, 2010.

74. On the fires, see Dauvergne, "The Political Economy of Indonesia's 1997 Forest Fires"; Gellert, "A Brief History and Analysis of Indonesia's Fire Crisis."

75. [Brown and Jacobson], *Cruel Oil.*

76. Hoegeng, interview (RC), Jakarta 1982.

77. MacKinnon, "The Orang-utan in Sabah Today," 189.

78. "The Animal Trade in Indonesia," 4–5.

79. "The Animal Trade in Indonesia," v.

80. Smits, Heriyanto, and Ramono, "A New Method," 70.

81. Sugardjito and Adhikerana, "Measuring Performance."

82. Cantor, "Items of Property."

83. Wynne, "Rosalià Abreu and the Apes of Havana," 300.

84. Wich et al., "Captive and Wild Orangutan (Pongo sp.) Survivorship."

85. Markham, "Breeding Orangutans at Perth Zoo."

86. Perkins and Maple, "North American Orangutan Species Survival Plan"; Perkins, "Conservation and Management of Orang-utans." The enculturated orangutan Chantek, discussed in chapter 7, was one of the hybrids affected by this ban. See Raloff, "Caste-off Orangs."

87. The *2002 International Studbook of the Orangutan (Pongo pygmaeus, Pongo abelii)*, online at http://library.sandiegozoo.org/studbooks/primates/orangutan2002.pdf, records 869 orangutans in captivity (379 Borneo, 298 Sumatra, 174 hybrids, and 18 unknown taxon), housed in 215 facilities.

88. "An Extraordinary Return from the Brink of Extinction for Worlds [*sic*] Last Wild Horse," http://www.zsl.org/info/media/press-releases/null,1790,PR.html, accessed May 18, 2011.

89. Kennedy, *Australasian Marsupials and Monotremes*, 40–41.

90. "Great Ape Project (Proyecto Gran Simio) Pleads to Keep Orangutan 'Silvestre' in Spain," September 1, 2010, http://www.greatapeproject.org/en-US/noticias/Show/3247,great-ape-project-proyecto-gran-simio-pleads-to-keep-orangutan-silvestre-in-spain, accessed May 18, 2011.

91. "Orangutan Hybrid, Bred to Save Species, Now Seen as Pollutant," *New York Times*, Feb. 28, 1995, http://www.nytimes.com/1995/02/28/science/orangutan-hybrid-bred-to-save-species-now-seen-as-pollutant.html?src=pm, accessed May 18, 2011.

92. "Faunabescherming in Nederlandsch-Indië," *Het Vaderland: staat- en letterkundig nieuwsblad*, Feb. 4, 1931.

93. Harrisson, *Orang-utan,* ix–xvii, 91–135; MacKinnon, "The Orang-utan in Sabah Today," 149, 190; Schaller, "The Orang-utan in Sarawak"; Harrisson, "The Immediate Problem of the Orangutan"; De Silva, "The East-Coast Experiment"; Harrisson, "Orang-utans in Sabah."

94. MacKinnon, "Orang-utans in Sumatra," 241; "The Animal Trade in Indonesia," 2.

95. Haraway, *Primate Visions*, 279–303; Montgomery, *Walking with the Great Apes*.

96. On Galdikas' career, see her memoirs, *Reflections of Eden*, and Montgomery, *Walking with the Great Apes*, 142–163. For a hostile assessment of Galdikas, see Spalding, *A Dark Place in the Jungle*.

97. See Borner, *Orang Utan*.

98. Kaplan and Rogers, *The Orang-utans*, 140.

99. Rijksen and Meijaard, *Our Vanishing Relative*, 156–157, 163–170; Aveling, "Orang Utan Conservation in Sumatra," 300, 305; Aveling and Mitchell, "Is Rehabilitating Orangutans Worthwhile?"

100. Yeager, "Orangutan Rehabilitation in Tanjung Puting"; MacKinnon, "The Future of Orang-utans."

101. Van Dyke, *Conservation Biology*, 52; Bertelsen, *From Forest Kindergarten to Freedom*, 18–20.

102. See Braun and Castree, *Remaking Reality*.

103. Schuster, Smits, and Ullal, *Thinkers of the Jungle*.

10. FACES IN THE MIRROR

1. Owen, "On the Characters, Principles of Division, and Primary Groups of the Class Mammalia" (1857), quoted in Gross, "Hippocampus Minor and Man's Place in Nature," 404.

2. Richards, *The Tragic Sense of Life*, 254–245.

3. Rooy, "In Search of Perfection," 195–198.

4. Delisle, "Welcome to the Twilight Zone," 60–61.

5. Crookshank, *The Mongol in Our Midst* (1925), 29–30.

6. Crookshank, *The Mongol in Our Midst* (1925), 44.

7. Weiner, *The Piltdown Forgery*; Sherratt, "Darwin among the Archaeologists."

8. Begun, "Hominid Family Values," 13–14. On the significance of cladistics for the classification of humans, see Corbey, *The Metaphysics of Apes*, 103–105, 146–147, 169–170. Recent evidence suggests that hybridization between the human and chimpanzee lines continued well after the initial separation, possibly as recently as four million years ago. See Patterson et al., "Genetic Evidence for Complex Speciation of Humans and Chimpanzees." Research on the history of human parasites indicates that humans acquired the pubic louse from gorillas three to four million years ago, possibly as a result of sexual contact but more likely through predation or the sharing of nest sites. See Reed et al., "Pair of Lice Lost."

9. Watson, Eastael, and Penny, "Homo Genus." The American physical anthropologist Jeffrey H. Schwartz has argued persistently against this view and in favor of a human-orangutan line of descent that diverged from that of chimpanzees and gorillas. This interpretation, however, is not widely accepted, and the completion of the sequencing of the orangutan genome in January 2011 appears to have made it untenable. See Schwartz, *The Red Ape*; Grehan and Schwartz, "Evolution of the Second Orangutan."

10. Darwin, *The Expression of the Emotions in Man and Animals* (1872), 142.

11. Furness, "Observations on the Mentality of Chimpanzees and Orang-Utans" (1916), 282–285. See also Hillix and Rumbaugh, *Animal Bodies, Human Minds*, 53–55.

12. M. E. Haggerty, "Plumbing the Minds of *Apes*," *McClure's Magazine* 41 (1913): 151–154, cited in Yerkes and Yerkes, *The Great Apes*, 181–183.

13. Yerkes, "Ideational Behavior of Monkeys and Apes." On the work of Yerkes, see Dewsbury, *Monkey Farm*.

14. On this research station, which mainly investigated chimpanzees, see Knauer, *Menschenaffen*, 91–92.

15. Köhler, "The Mentality of Orangs."

16. Köhler's better-known work, *The Mentality of Apes* (1926), makes only four brief mentions of orangutans.

17. Köhler, "The Mentality of Orangs," 190–193.

18. Yerkes and Yerkes, *The Great Apes*, 150–194.

19. Hillix and Rumbaugh, *Animal Bodies, Human Minds*, 194–200.

20. Galdikas, "Adult Male Sociality and Reproductive Tactics"; MacKinnon, "The Orang-utan in Sabah Today."

21. Atmoko and Van Schaik, "The Natural History of Sumatran Orangutan," 46–47; Bard, "'Social Tool Use' by Free-Ranging Orangutans."

22. Van Schaik and Knott, "Geographic Variation in Tool Use."

23. Wright, "Imitative Learning of a Flaked Tool Technology."

24. Russon, "The Nature and Evolution of Intelligence in Orangutans," 485.

25. While cognitive and emotional similarities between orangutans and humans might suggest the possibility of these traits in a common ancestor, Anne Russon, for example, has argued persuasively for a convergent rather than directly lineal evolution, with orangutan intelligence developing independently in response to its tropical environment and not from a common ancestor. Russon, "The Nature and Evolution of Intelligence in Orangutans."

26. MacKinnon, *In Search of the Red Ape*, 176.

27. Knott and Kahlenberg, "Orangutans in Perspective," 296–301; Harrison and Chivers, "The Orang-utan Mating System."

28. For further discussion of this issue, see Wrangham and Peterson, *Demonic Males*, 132–143; Thornhill and Palmer, *A Natural History of Rape*. For powerful criticism of this position, see Travis, *Evolution, Gender, and Rape*.

29. Machiavellian intelligence is better documented among chimpanzees than among orangutans, because it is significantly easier to observe subtle, manipulative behavior at ground level than in the trees. Miles noted that Chantek repeatedly used deception, but this was in the context of captivity. See Miles, "Language and the Orangutan," 51; Miles, "How Can I Tell a Lie?" 253–258; Byrne and Whiten, *Machiavellian Intelligence*; Whiten and Byrne, *Machiavellian Intelligence II*; Russon, "Evolutionary Reconstructions," 4–6.

30. The holding of grudges by chimpanzees is well recorded in scientific literature, but instances of grudges held by orangutans are so far only anecdotal. See "Saving the

Great Apes," http://www.centerforgreatapes.org/news-detail.aspx?id=9, accessed December 22, 2012; Shawn Thompson, "How to Apologize to an Orangutan," http://www.psychologytoday.com/blog/the-intimate-ape/201004/how-apologize-orangutan, accessed December 22, 2012.

31. Van Schaik and van Hooff, "Toward an Understanding of the Orangutan's Social System," 3–15.

32. Malone, Fuentes, and White, "Variation in the Social Systems of Extant Hominoids;" Grueter, Chapais, and Zinner, "Evolution of Multilevel Social Systems in Nonhuman Primates and Humans"; Krützen, Willems, and van Schaik, "Culture and Geographic Variation in Orangutan Behavior."

33. Chalmeau et al., "Cooperative Problem Solving by Orangutans (*Pongo pygmaeus*)."

34. Russon has argued that orangutan intelligence is strongly linked to the challenges of obtaining food in their tropic arboreal environment, an environment they did not share with humans in evolutionary terms. If this analysis is correct, then the evolution of orangutan intelligence ran parallel to, rather than being an aspect of, the evolution of human intelligence. See Russon, "The Nature and Evolution of Intelligence in Orangutans (*Pongo pygmaeus*)."

35. Scholl, *Fleurs d'adultère*, 54–62. "Fuegans" were the indigenous people of Tierra del Fuego and were considered by some to be the most primitive of all human beings.

36. On the history of human attitudes to animal rights, see Ryder, *Animal Revolution*.

37. Corbey, *The Metaphysics of Apes*, 175.

38. Bentham, *Introduction to the Principles of Morals and Legislation* (1789), 283.

39. Salt, *Animals' Rights* (1894).

40. Kaplan and Rogers, "Of Human Fear and Indifference," 10.

41. "In het belang der wetenschap," *De Sumatra Post*, Sept. 14, 1905.

42. Weindling, *Health, Race and German Politics*, 169.

43. Nick Taylor, "Heart to Heart."

44. Cited in Ryder, *Animal Revolution*, 5.

45. Ignatieff, *The Rights Revolution*.

46. Hobbes, *Leviathan*, 62.

47. Bowler, "The Geography of Extinction," argues that our perception of Neanderthals as an evolutionary dead-end exterminated by humans is a largely unsubstantiated transfer of social Darwinist interpretations of the era of Western imperialism.

48. Stanford, *Planet without Apes*, 8–11.

49. Banik, *Rights and Legal Empowerment*, 32–33.

50. These include the Universal Declaration of Human Rights and the Convention on the Prevention and Punishment of the Crime of Genocide (both 1948), the Declaration on the Granting of Independence to Colonial Countries and Peoples (1961), Convention on the Elimination of all Forms of Discrimination against Women (1979), the Declaration of the Rights of the Child (1989), and the Declaration on the Rights of Indigenous Peoples (2007).

51. Cavalieri et al., "A Declaration on Great Apes," 4.

52. "Medical Tests on Great Apes Should Not Be Banned, Says Research Chief," *The Telegraph* (London), June 3, 2006, http://www.telegraph.co.uk/news/uknews/ 1520152/Medical-tests-on-great-apes-should-not-be-banned-says-research-chief.html, accessed May 15, 2011.

53. Taylor, "A Step at a Time." Although plans were announced for a Non-human Hominid Protection Bill to be proposed as a private member's bill, the proposal remains in draft.

54. "June 2008 Update: Spain's Latest Ape Decision Welcomed—With Caution," Great Ape Standing and Personhood, http://www.personhood.org/, accessed May 15, 2011.

55. "New EU Rules on Animal Testing Ban Use of Apes," *The Independent on Sunday*, Sept. 12, 2010, http://www.independent.co.uk/life-style/health-and-families/new-eu -rules-on-animal-testing-ban-use-of-apes-2077443.html, accessed May 15, 2011.

56. H.R. 1513 (S. 810) Great Ape Protection and Cost Savings Act, http://www.govtrack .us/congress/bills/112/hr1513#, accessed Dec. 22, 2012.

AFTERWORD

1. Pope John Paul II, "Evolution and the Conception of Man," Message to the Pontifical Academy of Sciences, October 22, 1996, http://conservation.catholic.org/magiste rium_is_concerned_with_qu.htm, accessed August 14, 2011.

2. Bentham, *Introduction to the Principles of Morals and Legislation,* 143.

✺ Bibliography

Abdullah bin Abdul Kadir. *The Hikayat Abdullah*. Kuala Lumpur: Oxford University Press, 1990.

Abel, Clarke. *Narrative of a Journey in the Interior of China and of a Voyage to and from that Country in the Years 1816 and 1817*. London: Longman, Hurst, Rees, Orme, and Browne, 1819.

————. "Some Account of an Orang Outang of Remarkable Height Found on the Island of Sumatra; together with a Description of Certain Remains of this Animal, Presented to the Asiatic Society by Capt. Cornfoot, and at Present Contained in its Museum." *Asiatic Researches* 15 (1825): 489–498.

Achebe, Chinua. "An Image of Africa: Racism in Conrad's *Heart of Darkness*." In *Heart of Darkness, An Authoritative Text, Background and Sources, Essays in Criticism*, ed. Robert Kimbrough, 251–261. 3rd ed. New York: Norton, 1988.

Adams, Percy G. *Travelers and Travel Liars 1660–1800*. Berkeley: University of California Press, 1962.

Adams, William M. *Against Extinction: The Story of Conservation*. London: Earthscan, 2004.

Aelian. *On the Characteristics of Animals*. London: Heinemann, 1959.

Africa Screams. Directed by Charles Barton. Huntington Hartford Productions, 1949. DVD, 79 min.

Alexander, Edward P. *Museum Masters: Their Museums and Their Influence*. Nashville, Tenn.: American Association for State and Local History, 1983.

Alexander, Richard D. *How Did Humans Evolve? Reflections on the Uniquely Unique Species*. Museum of Zoology (Special Publication No. 1). Ann Arbor: University of Michigan, 1990.

Allamand, Jean. "Eerste Byvoegzel tot de natuurlyke Historie van de Orangs-outangs, door den Heer Professor Allamand." In *De Algemeene en byzondere natuurlyke*

Historie met de Beschryving van des Konings Kabinet door de Heeren De Buffon en Daubenton, vol. 14. Amsterdam: J. H. Schneider, 1783 (separately paginated).

Allen, John S., Julie Park, and Sharon S. Watt. "The Chimpanzee Tea Party: Anthropomorphism, Orientalism and Colonialism." *Visual Anthropology* 10, no. 2 (1994): 45–54.

Altick, Richard. *The Shows of London*. Cambridge, Mass.: Harvard University Press, 1978.

Ancrenaz, Marc, et al. "Aerial Surveys Give New Estimates for Orangutans in Sabah, Malaysia." *PLoS Biology* 3, no. 1 (January 2005): e3, 30–37.

Andel, M. van. "Introduction." In Jacobus Bontius, *On Tropical Medicine*. Amsterdam: Nederlandsch Tijdschrift voor Geneeskunde, 1931, ix–xli.

Andrews, Peter, and J. E. Cronin. "The Relationships of *Sivapithecus* and *Ramapithecus* and the Evolution of the Orang-utan." *Nature* 297 (June 17, 1982): 541–546.

"The Animal Trade in Indonesia." Unpublished report, World Wildlife Fund Indonesia Programme, Bogor, 1980.

Any Which Way You Can. Directed by Buddy Van Horn. Malpaso Company/Warner Bros. Pictures, 1980. DVD, 115 min.

Arnscheidt, Julia. *"Debating" Nature Conservation: Policy, Law and Practice in Indonesia*. Leiden: Leiden University Press, 2009.

Atkinson, Geoffroy. *The Extraordinary Voyage in French Literature from 1700 to 1720*. New York: Columbia University Press, 1920.

Atmoko, Sri Suci Utami, and Carel P. van Schaik. "The Natural History of Sumatran Orangutan (*Pongo abelii*)." In *Indonesian Primates*, ed. Sharon Gursky-Doyen and Jana Prijatna, 41–55. New York: Springer, 2010.

Audebert, J. B. *Histoire naturelle des Singes et des Makis*. Paris: Desray, year 8 [1801].

Aveling, Rosalind, and Arthur Mitchell. "Is Rehabilitating Orangutans Worthwhile?" *Oryx* 16 (1982): 263–271.

Aveling, R. J. "Orang Utan Conservation in Sumatra, by Habitat Protection and Conservation Education." In *The Orang Utan: Its Biology and Conservation*, ed. Leobert E. M. de Boer, 299–315. The Hague: W. Junk, 1982.

Le Avventure straordinarissime di Saturnino Farandola. Directed by Marcel Fabre. Società Anonima Ambrosio, 1913. http://www.europafilmtreasures.eu/research/extra ordinary_adventures_of_saturnino_farandoul-0.htm. Accessed May 25, 2011.

Babe: Pig in the City. Directed by George Miller. Kennedy Miller Productions, 1998. DVD, 97 min.

Balaoo: The Demon Baboon. Directed by Victorin-Hippolyte Jasset. Société Fraçaise des Films Éclair, 1913. VHS, 24 min.

Bali, The Unknown: or Ape Man Island. Directed by Harold H. Horton. Prima, 1921, 50 min.

Ballard, Chris. "Strange Alliance: Pygmies in the Colonial Imaginary." *World Archaeology* 38, no. 1 (2006): 133–151.

Banik, Dan. *Rights and Legal Empowerment in Eradicating Poverty*. Franham: Ashgate, 2008.

Baratay, Eric, and Elisabeth Hardouin-Fugier. *Zoo: A History of Zoological Gardens in the West*. London: Reaktion, 2000.

Bard, Kim A. "'Social Tool Use' by Free-Ranging Orangutans: A Piagetian and Developmental Perspective on the Manipulation of an Inanimate Object." In *"Language" and Intelligence in Monkeys and Apes: Comparative Developmental Perspectives,* ed. S. T. Parker and K. R. Gibson, 356–378. Cambridge: Cambridge University Press, 1980.

Barnard, Alan "*Orang Outang* and the Definition of Man: The Legacy of Lord Monboddo." In *Fieldwork and Footnotes: Studies in the History of European Anthropology,* ed. Han F. Vermeulen and Arturo Alvarez Roldán, 95–112. London: Routledge, 1995.

Barsanti, Giulio. "L'Orang-outan déclassé (Pongo wurmbi Tied.): Histoire du premier Singe à Hauteur d'Homme (1780–1801) et ébauche d'un Théorie de la Circularité des Sources." *Bulletins et Mémoires de la Société d'Anthropologie de Paris* n.s 1, nos 3–4 (1989): 67–104.

Barthélemy, Guy. *Les Jardiniers du Roy: Petite Histoire du Jardin des Plantes de Paris*. Paris: Le Pélican, 1979.

[Battell, Andrew]. *The Strange Adventures of Andrew Battell of Leigh, in Angola and the Adjoining Regions*. London: Hakluyt Society, 1901.

The Beast of Borneo. Directed by Harry Garson. Far East Productions, 1934. VHS, 63 min.

Beccari, Odoardo. *Wanderings in the Great Forests of Borneo*. Singapore: Oxford University Press, 1989, first published London: Archibald Constable, 1904.

Beeckman, Daniel. *A Voyage to and from the Island of Borneo in the East-Indies*. London: T. Warner and J. Batley, 1718.

Beers, Diane L. *For the Prevention of Cruelty: The History and Legacy of Animal Rights Activism in the United States*. Athens, Ohio: Swallow Press, 2008.

Begun, David R. "Hominid Family Values: Morphological and Molecular Data on Relations among the Great Apes and Humans." In *The Mentalities of Gorillas and Orangutans: Comparative Perspectives,* ed. Sue Taylor Parker, Robert W. Mitchell, and H. Lyn Miles, 3–42. Cambridge: Cambridge University Press, 1999.

Belcher, Sir Edward. *Narrative of the Voyage of H.M.S. Samarang during the Years 1843–46,* vol. 1. London: Reeve, Benham, and Reeve, 1848.

Benchley, Belle. *My Friends the Apes*. Boston: Little Brown, 1942.

——. *My Life in a Man-Made Jungle*. Boston: Little, Brown, 1940.

Beneath the Planet of the Apes. Directed by Ted Post. APJAC Productions/Twentieth Century Fox, 1970. DVD, 95 min.

[Benkowitz, Karl Friedrich]. *Der Orang-Outang in Europa oder die Pohle, nach seiner wahren Beschaffenheit: eine methodische Schrift, welche im Jahre 1779 einen Preis in der Naturgeschichte davon getragen hat* (Kalifornien [i.e., Berlin]: n.p., 1780).

Bennett, Tony. "The Exhibitionary Complex." *New Formations* 4 (Spring 1988): 73–102.

Bentham, Jeremy. *Introduction to the Principles of Morals and Legislation* (1789), chapter 17. In *The Collected Works of Jeremy Bentham,* ed. J. H. Burns and H. L. A. Hart. Oxford: Oxford University Press, 1996.

Benton, Ted. "Where to Draw the Line: Alfred Russel Wallace in Borneo." *Studies in Travel Writing* 1 (1997): 96–116.

Berliner, Brett A. "Mephistopheles and Monkeys: Rejuvenation, Race, and Sexuality in Popular Culture in Interwar France." *Journal of the History of Sexuality* 13, no. 3 (July 2004): 306–326.

Bernard, Pierre, and Jean Couailhac. *Le Jardin des Plantes: Description complète, historique et pittoresque du Muséum d'Histoire naturelle.* Paris: L. Curmer, 1842.

Bernstein, Jay H. *Spirits Captured in Stone: Shamanism and Traditional Medicine among the Taman of Borneo.* Boulder, Colo.: Lynne Rienner Publishers, 1997.

Bertelsen, Pia Lykke. *From Forest Kindergarten to Freedom.* [Ålborg, Denmark]: PLB Network, 2006.

Berthet, Élie. *The Wild Man of the Woods: A Story of the Island of Sumatra.* London: Seeley, Jackson & Halliday, 1868.

Betts, John Rickards. "P. T. Barnum and the Popularization of Natural History." *Journal of the History of Ideas* 20, no. 3 (1959): 353–368.

Bignon, Édouard. *L'orang-outang, ou les Amans du Désert.* Paris and Brussels: Gambier, 1806.

Bindman, David. *Ape to Apollo: Aesthetics and the Idea of Race in the 18th Century.* London: Reaktion, 2002.

Blakeney, Richard. *The Journal of an Oriental Voyage in His Majesty's Ship Africaine.* London: Simkin, Marshall, 1841.

Bloemaart, Samuel. "Discourse ende Ghelegentheyt van het Eylandt Borneo, ende 't gene daervoor ghevallen is int Iaer 1609." In *Begin ende Voortgangh van de Vereenighde Neederlandtsche Geoctroyeerde Oost-Indische Compagnie, tweede deel.* N.p.: n.p., 1646.

Blumenbach, Johann Friedrich. *De generis humani varietate nativa liber.* Göttingen: Abr. Vandenhoek, 2nd ed. 1781.

Blunt, Wilfrid. *Ark in the Park.* London: Tryon Gallery, 1976.

Bondeson, Jan. *The Feejee Mermaid and Other Essays in Natural and Unnatural History.* Ithaca, N.Y.: Cornell University Press, 1999.

Bondeson, Jan, and A. E. W. Miles. "Julia Pastrana, the Nondescript: An Example of Congenital, Generalized Hypertrichosis Terminalis with Gingival Hyperplasia." *American Journal of Medical Genetics* 47 (1993): 198–212.

Bontius, Jacobus. *On Tropical Medicine.* Amsterdam: Nederlandsch Tijdschrift voor Geneeskunde, 1931.

Boomgaard, Peter. *Frontiers of Fear: Tigers and People in the Malay World, 1600–1950.* New Haven, Conn.: Yale University Press, 2001.

Boon, James A. *Affinities and Extremes: Crisscrossing the Bittersweet Ethnology of East Indies History, Hindu-Balinese Culture, and Indo-European Allure.* Chicago: University of Chicago Press, 1990.

Borner, Monica (with Bernard Stonehouse). *Orang Utan: Orphans of the Forest.* London: W. H. Allen, 1979.

Bouissac, Paul. *Circus and Culture: A Semiotic Approach.* Bloomington: Indiana University Press, 1976.

Boulle, Pierre. *Monkey Planet.* Harmondsworth: Penguin, 1975. Originally published as *La Planète des Singes.* Paris: Julliard, 1963.

Bowler, Peter J. "The Geography of Extinction: Biography and the Expulsion of Ape Men from Human Ancestry in the Early Twentieth Century." In *Ape, Man, Ape-Man: Changing Views since 1600,* ed. Raymond Corbey and Bert Theunissen, 185–193. Leiden: Department of Prehistory, Leiden University, 1995.

Bowrey, Thomas. *A Dictionary English and Malayo, Malayo and English.* London: Sam. Bridge, 1701.

Brauer, Fae. "Wild Beasts and Tame Primates: 'Le Douanier' Rousseau's Dream of Darwin's Evolution." In *The Art of Evolution: Darwin, Darwinisms, and Visual Culture,* ed. Barbara Larson and Fae Brauer, 194–225. Hanover, N.H.: University Press of New England, 2009.

Braun, Bruce, and Noel Castree, eds. *Remaking Reality: Nature at the Millenium* [*sic*]. Abingdon: Routledge, 1998.

Brigham, David R. *Public Culture in the Early Republic: Peale's Museum and Its Audience.* Washington, D.C.: Smithsonian Institution Press, 1995.

Bring 'Em Back Alive. Directed by Clive E. Elliot. Van Beuren Studios, 1932.

Broberg, Gunnar. "*Homo sapiens:* Linnaeus's Classification of Man." In *Linnaeus: The Man and His Work,* ed. Tore Frängsmyr, 156–194. Canton, Mass.: Science History Publications, 1994.

Broderick, Colin. *Orangutan: A Memoir.* New York: Three Rivers Press, 2009.

Broderip, William. "Recreations in Natural History—No. VI. Monkeys of the Old Continent, &c." *New Monthly Magazine and Humourist,* January–April 1838, 88–99.

Brooker, Jewel Spears. "The Great War at Home and Abroad: Violence and Sexuality in Eliot's 'Sweeney Erect.'" *Modernism/modernity* 9, no. 3 (September 2002): 423–438.

Brown, Barbara, and Stephen C. Ward. "Basicranial and Facial Topography in *Pongo* and *Sivapithecus.*" In Jeffrey H. Schwartz, *Orang-utan Biology.* New York: Oxford University Press, 1988, 247–260.

[Brown, Ellie, and Michael F. Jacobson]. *Cruel Oil: How Palm Oil Harms Health, Rainforest & Wildlife.* Washington, D.C.: Center for Science in the Public Interest, 2005.

Bruen, Desmond Lingard, and Neville Seymour Haile. *Report of the Maias Protection Commission.* [Kuching: Maias Protection Commission], 1960.

[Buck, Frank]. *Bring 'Em Back Alive: The Best of Frank Buck.* Lubbock: Texas Tech University Press, 2000.

Buffon, Georges-Louis Leclerc. *Histoire naturelle générale et particulière avec la Description du Cabinet du Roi,* vol. 14. Paris: Imprimerie Royale, 1766.

[Buffon, Georges-Louis Leclerc]. *Natural History, General and Particular, by the Count de Buffon,* vol. 8. Edinburgh: William Creech, 1780.

Burnet, James [Lord Monboddo]. *Of the Origin and Progress of Language,* vol. 1. Edinburgh: J. Balfour, 2nd ed. 1774.

Burkhardt, Frederick, and Sydney Smith, eds. *The Correspondence of Charles Darwin,* vol. 2, *1837–1843.* Cambridge: Cambridge University Press, 1985.

Burroughs, Edgar Rice. *Tarzan and "The Foreign Legion."* Tarzana, Calif.: Edgar Rice Burroughs Inc., 1947.

Burt, Jonathan. *Animals in Film.* London: Reaktion, 2002.

———. "The Illumination of the Animal Kingdom: The Role of Light and Electricity in Animal Representation." *Society and Animals* 9, no. 3 (2001): 203–228.

Butler, Elizabeth M. *The Fortunes of Faust.* University Park: Pennsylvania State University Press, 1998.

Byrne, Richard W., and Andrew Whiten, eds. *Machiavellian Intelligence: Social Expertise and the Evolution of Intellect in Monkeys, Apes and Humans.* Oxford: Clarendon Press, 1988.

Caldecott, Julian Oliver, and Lera Miles. *World Atlas of Great Apes and Their Conservation.* Berkeley: University of California Press, 2005.

Camper, Peter. "Account of the Organs of Speech of the Orang Outang." *Philosophical Transactions of the Royal Society of London,* 69 (1779): 139–159.

[Camper, Petrus]. *Natuurkundige Verhandelingen van Petrus Camper over den Orang Outang en eenige ander Aap-soorten.* Amsterdam: Erven P. Meijer en G. Warnars, 1782.

[Camper, Petrus]. *The Works of the late Professor Camper on the Connexion between the Science of Anatomy and the Arts of Drawing, Painting, Statuary &c. &c.* London: C. Dilly, 1794.

Cantor, David. "Items of Property." In *The Great Ape Project: Equality beyond Humanity,* ed. Paola Cavalieri and Peter Singer, 280–290. New York: St. Martin's Griffin, 1993.

Carlino, Andrea. *Paper Bodies: A Catalogue of Anatomical Fugitive Sheets.* London: Wellcome Institute for the History of Medicine, 1999.

Carpenter, C. R. "A Survey of Wild Life Conditions in Atjeh North Sumatra, with Special Reference to the Orang-utan." Report prepared in cooperation with H. J. Coolidge Jr., former secretary of the American Committee for International Wild Life Protection, February 1938.

Carroll, Noel. "Anglo-American Aesthetics and Contemporary Criticism: Intention and the Hermeneutics of Suspicion." *Journal of Aesthetics and Art Criticism* 51, no. 2 (1993): 245–252.

Carter, Paul. *Don't Tell Mum I Work on the Rigs: She Thinks I'm a Piano Player in a Whorehouse.* London: Nicholas Brealey, 2005.

Casson, Anne. "Decentralisation of Policies Affecting Forests and Estate Crops in Kotawaringin Timur District, Central Kalimantan." CIFOR Reports on Decentralisation and Forests in Indonesia, Case Study no 4. Jakarta: Center for International Forestry Research, 2001.

Cavalieri, Paola, et al. "A Declaration on Great Apes." In *The Great Ape Project: Equality beyond Humanity,* ed. Paola Cavalieri and Peter Singer, 4–7. London: Fourth Estate, 1993.

Chalmeau, Raphaël, et al., "Cooperative Problem Solving by Orangutans (*Pongo pygmaeus*)." *International Journal of Primatology* 18, no. 1 (1997): 23–32.

Champsaur, Felicien. *Ouha, Roi des Singes.* Paris: Charpentier et Fasquelle, 1923.

Cheney, Dorothy L., and Robert M. Seyfarth, *Baboon Metaphysics: The Evolution of a Social Mind.* Chicago: University of Chicago Press, 2007.

CITES MA Indonesia. "Enforcement Efforts on the Conservation of Orangutan in Indonesia." SC57 Doc. 30 Annex 2 [2008], http://www.cites.org/common/com/ SC/57/E57-30A2.pdf. Accessed October 21, 2010.

Clifford, Hugh. "The Strange Elopement of Châling the Dyak." In Hugh Clifford, *Studies in Brown Humanity: Being Scrawls and Smudged in Sepia, White, and Yellow.* London: Grant Richards, 1898, 234–264.

Cloyd, E. L. *James Burnett Lord Monboddo.* Oxford: Clarendon Press, 1972.

Coetzee, J. M. *The Lives of Animals.* Princeton, N.J.: Princeton University Press, 1999.

Cohen, Matthew. "The Monkey Show." *Latitudes* 48 (2005). http://216.67.229.254/main/ article/vol48-6.html. Accessed February 3, 2011.

The Comparative Coincidence of Reason and Scripture, vol. 3. London: J. Hatchard and Son, 1832.

Comte, Louis le. *Memoirs and Observations Made in a Late Journey through the Empire of China.* London: Benj. Tooke, 2nd ed., 1698.

———. *Nouveaux Mémoires sur l'État Present de la Chine, tome second.* Paris: Jean Anisson, 1696.

[Constable, Frank Challice]. *The Curse of Intellect.* Edinburgh: William Blackwood, 1895.

Cook, James W. *The Arts of Deception: Playing with Fraud in the Age of Barnum.* Cambridge, Mass.: Harvard University Press, 2001.

Coomans de Ruiter, L. "Natuurbescherming in Nederlandsch-Indië." *Indonesië* 2 (1948): 140–162.

———. "Natuurbeschermingsmaatregelen noodig geacht voor de Westerafdeeling van Borneo." *Nederlandsch-Indische Veneeniging tot Natuurbescherming, Verslag over het jaar 1932,* 23–30.

Corbey, Raymond. *The Metaphysics of Apes: Negotiating the Animal–Human Boundary.* Cambridge: Cambridge University Press, 2005.

Corbey, Raymond, and Bert Theunissen, eds. *Ape, Man, Ape-Man: Changing Views since 1600.* Leiden: Department of Prehistory, Leiden University, 1995.

Correspondence Relating to the Preservation of Wild Animals in Africa. London: HMSO, 1906.

Cortesão, Armando, ed. *The Suma Oriental of Tomé Pires.* London: Hakluyt Society, 1944.

Costeloe, Michael. *William Bullock: Connoisseur and Virtuoso of the Egyptian Hall.* Bristol: HiPLAM, 2008.

Courcy, Catherine de. *The Zoo Story: The Animals, the History, the People.* Melbourne: Penguin, 1995.

Creation. Directed by Jon Amiel. Recorded Picture Company/BBC Films, 2009. DVD, 108 min.

Creed, Barbara. *Darwin's Screens: Evolutionary Aesthetics, Time and Sexual Display in the Cinema.* Melbourne: Melbourne University Press, 2009.

———. *Phallic Panic: Film, Horror and the Primal Uncanny.* Melbourne: Melbourne University Press, 2005.

Cribb, Robert. "Conservation in Colonial Indonesia." *Interventions* 9, no. 11 (2007): 49–61.

———. *The Politics of Environmental Protection in Indonesia.* Clayton, Vic.: Monash University Centre of Southeast Asian Studies, 1988.

Crookshank, F. G. *The Mongol in our Midst.* London: Kegan Paul, Trench, Trubner, 2nd ed. 1925.

Cuvier, Frederick [*sic*]. "Description of an Ourang Outang: With Observations on Its Intellectual Faculties." *The Philosophical Magazine* 38 (1811): 188–199.

Dammerman, K. W. "De Orang Oetan." In *Album van Natuurmonumenten in Nederlandsch-Indië: Album Serie II,* ed. C. G. G. J. van Steenis, 27–32. Batavia: Nederlandsch-Indische Vereniging tot Natuurbescherming, 1937.

Darwin, Charles. *The Descent of Man, and Selection in Relation to Sex,* vol. I. London: John Murray, 1871.

———. *The Expression of the Emotions in Man and Animals.* London: John Murray, 1872.

Dauvergne, Peter. "The Political Economy of Indonesia's 1997 Forest Fires." *Australian Journal of International Affairs* 52, no. 1 (1998): 13–17.

DeGenaro, Stephan A. "The Little Men." *Timeline,* November/December 2002, 32–35.

Dekkers, Midas. *Dearest Pet: On Bestiality.* London: Verso, 1994.

Delgado, R. A., and C. P. van Schaik. "The Behavioural Ecology and Conservation of the Orangutan (*Pongo pygmaeus*): A Tale of Two Islands." *Evolutionary Anthropology* 9 (2000): 201–218.

Delisle, Richard G. "Welcome to the Twilight Zone: A Forgotten Early Phase of Human Evolutionary Studies." *Endeavour* 36, no. 2 (2012): 55–64.

Desmond, Jane C. *Staging Tourism: Bodies on Display from Waikiki to Sea World.* Chicago: University of Chicago Press, 1999.

Dewsbury, Donald A. *Monkey Farm: A History of the Yerkes Laboratories of Primate Biology, Orange Park, Florida, 1930–1965.* Lewisburg, Pa.: Bucknell University Press, 2006.

Diderot, Denis, "Suite de l'Entretien." In *Pensées sur l'Interprétation de la Nature 1754* (Oeuvres complètes de Diderot: Revues sur les éditions originales, Tome 2) Paris: Garnier frères, 1875, 182–191.

Diski, Jenny. *The Monkey's Uncle.* London: Weidenfeld and Nicolson History, 1994.

Ditmars, Raymond L. "The Collection of Great Apes." *Zoological Society Bulletin* 45 (1911): 756–758.

———. "Training Orangs and Chimpanzees." *Zoological Society Bulletin* 10 (1903): 93–96.

Dobson, Jessie. "John Hunter and the Early Knowledge of the Apes." *Proceedings of the Zoological Society of London* 123 (1953): 1–12.

D'Obsonville, Foucher. *Philosophic Essays on the Manners of Various Foreign Animals.* London: John Johnson, 1784. Translation of *Essais philosophiques sur les Moeurs de divers Animaux étrangers.* Paris: Couturier fils, 1783.

Dr. Renault's Secret. Directed by Harry Lachman. Twentieth-Century Fox, 1942. VHS, 58 min.

Dodillon, Émile. *Hémo.* Paris: Alphonse Lemerre, 1886.

Dolph, James Andrew. "Bringing Wildlife to the Millions: William Temple Hornaday, the Early Years: 1854–96." PhD diss., University of Massachusetts, 1975.

Domeny de Rienzi, M. G. L. *Océanie, ou cinquième partie du monde,* vol. 1. Paris: Firmin Didot frères, 1836.

Donovan, Edward. *Naturalist's Repository, or Monthly Miscellany of Exotic Natural History, Exhibiting Rare and Beautiful Specimens of Foreign Birds, Insects, Shells, Quadrupeds, Fishes, and Marine Productions,* vol. 2. London: W. Simpkin and R. Marshall, 1824.

Ducrotay de Blainville, Henri-Marie. *Ostéographie ou Description Iconographique compare du Squelette et du Système dentaire des Mammifères récents et fossiles,* vol. 1. Paris: J. B. Baillière [1863].

Dunston Checks In. Directed by Ken Kwapis. Twentieth-Century Fox, 1996. DVD 88 min.

Dwyer, James Francis. "The Orang-Outang Fight on the Papuan Queen: The Story of a Fourteen Day and Night Vigil." In *Breath of the Jungle.* Chicago: A. C. McClurg, 1915, 241–254.

Dyke, Fred van. *Conservation Biology: Foundations, Concepts, Applications.* Dordrecht: Springer, 2008.

Eder, James F. *Batak Resource Management: Belief, Knowledge and Practice.* Gland: IUCN, the World Conservation Union and World Wide Fund for Nature, 1997.

Edgar Allan Poe's Murders in the Rue Morgue. Directed by Robert Florey. Universal Pictures, 1932.

Edwards, George. *Gleanings of Natural History, Exhibiting Figures of Quadrupeds, Birds, Insects, Plants &c., Most of which have not, Till Now, been either Figured or Described.* London: published by the author, 1758.

Engel-Ledeboer, M. S. J., and H. Engel, eds. *Carolus Linnaeus, Systema Naturae 1735: Facsimile of the First Edition.* Nieukoop: B. de Graaf, 1964.

Engels, Frederick. *The Part Played by Labour in the Transition from Ape to Man.* Peking: Foreign Languages Press, 1975. Originally published 1876.

Etkind, Alexander. "Beyond Eugenics: The Forgotten Scandal of Hybridizing Humans and Apes." *Studies in History and Philosophy of Biological and Biomedical Sciences* 39 (2008): 205–210.

Evans, Ivor H. N. *Among Primitive People in Borneo.* Singapore: Oxford University Press, 1990. First published London: Seeley Service, 1922.

Every Which Way but Loose. Directed by James Fargo. Malpaso Company, 1978. DVD, 110 min.

Fan, Yuxin, Elena Linardopoulou, Cynthia Friedman, Eleanor Williams, and Barbara J. Trask. "Genomic Structure and Evolution of the Ancestral Chromosome Fusion Site in 2q13–2q14.1 and Paralogous Regions on Other Human Chromosomes." *Genome Research* 12 (2002): 1651–1662.

Fissell, Mary E. "Hairy Women and Naked Truths: Gender and the Politics of Knowledge in *Aristotle's Masterpiece.*" *William and Mary Quarterly* 60, no. 1 (2003): 43–74.

Flaubert, Gustave. "Whatever You Want." In *Early Writings.* Translated by Robert Griffin. Lincoln: University of Nebraska Press, 1991, 76–102.

Forster, George. *A Voyage round the World in His Britannic Majesty's Sloop, Resolution, Commanded by Capt. James Cook, during the Years 1772, 3, 4, and 5,* vol. 2. London: B. White et al., 1777.

Forth, Gregory. *Images of the Wildman in Southeast Asia: An Anthropological Perspective* London: Routledge, 2008.

Frank, Lawrence. " 'The Murders in the Rue Morgue': Edgar Allan Poe's Evolutionary Reverie." *Nineteenth-century Literature* 50, no. 2 (September 1995): 168–188.

Franklin, Ian Robert. "Evolutionary Changes in Small Populations." In *Conservation Biology: An Evolutionary-Ecological Perspective,* ed. Michael E. Soulé and Bruce A. Wilcox, 135–149. Sunderland: Sinauer, 1980.

Frederickson, A. D. *Ad Orientem.* London: W. H. Allen, 1890.

Furness, William H. "Observations on the Mentality of Chimpanzees and Orang-Utans." *Proceedings of the American Philosophical Society* 55, no. 3 (1916): 281–290.

Gabriel et Rochefort. *Jocko, ou le Singe du Brésil.* Paris: Chez Quoi, 1825.

Galdikas, Biruté M. F. "Adult Male Sociality and Reproductive Tactics among Orangutans at Tanjung Puting." *Folia Primatologica* 45 (1985): 9–24.

———. *Reflections of Eden: My Years with the Orang-utans of Borneo.* Boston: Little, Brown, 1995.

Geddes, W. R. *The Land Dayaks of Sarawak: A Report on a Social Economic Survey of the Land Dayaks of Sarawak.* [London]: H.M.S.O. for the Colonial Office, 1954.

Gellert, Paul K. "A Brief History and Analysis of Indonesia's Fire Crisis." *Indonesia* 65 (April 1998): 63–85.

Gilman, Sander L. *Difference and Pathology: Stereotypes of Sexuality, Race and Madness.* Ithaca, N.Y.: Cornell University Press, 1985.

Gilmour, David. *The Long Recessional: The Imperial Life of Rudyard Kipling.* London: John Murray, 2002.

Gimma, Hyacinthi. *Dissertationum academicarum tomus primus qui duas exhibit dissertationes, nempe I. De hominibus fabulosis. II De fabulosis animalibus, in qua legitur de fabulosa generatione viventium, et fabulae in philosophia-experimentali . . . ; tomus secundus, qui duas exhibit dissertations, nempe I. De brutorum anima et vita II. Miscellanea de hominibus et animalibus fabulosis.* Naples: Michaelis Aloysii Mutio, [1714].

Going Ape. Directed by Jeremy Joe Kronsberg. City Films/Paramount Pictures, 1981.

[Goldsmith, Oliver]. "The Natural History of Animals That Most Nearly Approach Humanity." *The Hibernian Magazine, or, Compendium of Entertaining Knowledge,* January 1782, 587–592.

Goodall, Jane. *Performance and Evolution in the Age of Darwin: Out of the Natural Order.* London: Routledge, 2002.

Gould, Stephen Jay. *Eight Little Piggies: Reflections in Natural History.* London: Vintage, 2007.

Granier, Michel. *Conferences on Homœopathy.* London: Leath and Ross, 1859.

Grant, J. "Account of the Habits and Structure of a Male and Female Orang-outang, that Belonged to George Swinton, Esq. Secretary to the Government, Calcutta." *Edinburgh Journal of Science* 4 (new series) (1831): 27–46.

Green, Eda. *Borneo: Land of River and Palm.* Westminster: Society for the Propagation of the Gospel in Foreign Parts, 1912.

Greene, Eric. *Planet of the Apes as American Myth: Race, Politics, and Popular Culture.* Middletown, Conn.: Wesleyan University Press, 1998.

Greene, John C. "The American Debate on the Negro's Place in Nature, 1780–1815." *Journal of the History of Ideas* 15, no. 3 (1954): 384–396.

Greenwood, James. *The Adventures of Reuben Davidger; Seventeen Years and Four Months Captive among the Dyaks of Borneo.* London: S. O. Beeton, 1865.

Grehan, John R., and Jeffrey H. Schwartz. "Evolution of the Second Orangutan: Phylogeny and Biogeography of Hominid Origins." *Journal of Biogeography,* 2001, doi: 10.1111/j.1365-2699.2009.02141.x.

Grijzenhout, Frans. " 'Na 't Leven Geteekent in de Diergaarde van Zijn Hoogheid': Kunstenaars rond de Menagerie van Prins Willem V." In *Een Vorstelijk Dierentuin: de Menagerie van Willem V,* ed. B. C. Sliggers and A. A. Wertheim, 61–86. [Zutphen]: Walburg Pers, 1994 [bilingual text, Du and Fr].

Groeneveldt, W. P. *Notes on the Malay Archipelago and Malacca Compiled from Chinese Sources.* Batavia: n.p., 1876.

Gross, Charles G. "Hippocampus Minor and Man's Place in Nature: A Case Study in the Social Construction of Neuroanatomy." *Hippocampus* 3, no. 4 (1993): 403–416.

Groves, Colin. *Extended Family: Long Lost Cousins, a Personal Look at the History of Primatology.* Arlington, Va.: Conservation International, 2008.

———. *Primate Taxonomy.* Washington, D.C.: Smithsonian Institution Press, 2001.

Groves, Colin P., and L. B. Holthuis. "The Nomenclature of the Orang Utan." *Zoologische Mededelingen* 59, no. 31 (1985): 411–417.

Grueter, Cyril C., Bernard Chapais, and Dietmar Zinner, "Evolution of Multilevel Social

Systems in Nonhuman Primates and Humans." *International Journal of Primatology* 33 (2012): 1002–1037.

Gudgeon, L. W. W. *British North Borneo.* London: A. C. Black, 1913.

Gulik, R. H. van. *The Gibbon in China: An Essay in Chinese Animal Lore.* Leiden: E. J. Brill, 1967.

Hagenbeck, Carl. *Beasts and Men: Being Carl Hagenbeck's Experiences for Half a Century among Wild Animals.* London and New York: Longmans, Green, 1912.

Hall, James W. *Gone Wild.* New York: Random House, 1996.

[Hamilton, Alexander]. *A New Account of the East Indies: Being the Observations and Remarks of Capt. Alexander Hamilton.* Edinburgh: John Mosman, 1728.

Haraway, Donna. *Primate Visions: Gender, Race, and Nature in the World of Modern Science.* New York: Routledge, 1989.

Harcourt, A. H. "Empirical Estimates of Minimum Viable Population Sizes for Primates: Tens to Tens of Thousands?" *Animal Conservation* 5 (2002): 237–244.

Harrison, Mark E., and David J. Chivers. "The Orang-utan Mating System and the Unflanged Male: A Product of Increased Food Stress during the Late Miocene and Pliocene?" *Journal of Human Evolution* 52 (2007): 275–293.

Harrisson, Barbara. "The Immediate Problem of the Orangutan." *Malayan Nature Journal* 16 (1962): 4–5.

———. *Orang-utan.* Singapore: Oxford University Press, 1987. Originally published London: Collins, 1962.

———. "Orang-utan—What Chances of Survival?" *Sarawak Museum Journal* 10 (1961): 238–261.

———. "Orang-utans in Sabah." *Oryx* 9 (1968): 256.

Harrisson, Barbara, with Tom Harrisson. "Has the Oranga Future?" In Barbara Harrisson, *Orang-utan,* 199–208. Singapore: Oxford University Press, 1987. Originally published London: Collins, 1962.

Harrisson, Tom. "Maias v. Man." *Sarawak Museum Journal* 6, no. 6 (1955): 598–632.

Hauff, Wilhelm. "The Young Englishman." "Der Affe als Mensch (Der junge Engländer)." *Märchen-Almanach auf das Jahr 1827.* http://www.gutenberg.org/ebooks/6639. Accessed May 25, 2011.

Haynes, Clare. "A 'Natural Exhibitioner': Sir Ashton Lever and His Holophusikon." *British Journal for Eighteenth-Century Studies* 24 (2001): 1–13.

Hearne, Vicki. *Animal Happiness: A Variety of Irresistible Creatures.* New York: Harper, Collins, 1994.

Heidhues, Mary Somers. "The First Two Sultans of Pontianak." *Archipel* 56 (1998): 273–294.

Hensler, Karl Friedrich. *Der Orang Utang oder der Tigerfiest.* Vienna: Goldhannschen, 1792.

Hesse, Elias. *Ost-Indische Reise-Beschreibung oder Diarium.* Leipzig: Michael Günther, 1690.

Hillix, W. A., and Duane Rumbaugh. *Animal Bodies, Human Minds: Ape, Dolphin, and Parrot Language Skills.* New York: Kluwer, 2004.

Histoire générale des Voyages: Ou, Nouvelle Collection de toutes les Relations de Voyages par Mer et par Terre, vol. 6. The Hague: Pierre de Hondt, 1747.

Hobbes, Thomas. *Leviathan Or The Matter, Forme, and Power of a Common-wealth, Ecclesiasticall and Civil.* London: Andrew Crooke, 1651.

Hone, William. *The Table Book: or, Daily Recreation and Information.* London: William Tegg, 1827.

Hooijer, D. A. "The Orang-utan in Niah Cave Pre-history." *Sarawak Museum Journal* 9, no. 15 (1960): 408–421.

Hoppius, Christianus Emmanuel. *Anthropomorpha.* Stockholm: Amoenitates Academiensis, 1760.

Hornaday, William T. *The Man who Became a Savage: A Story of our Own Times.* Buffalo, N.Y.: Peter Paul, 1896.

———. *The Minds and Manners of Wild Animals: A Book of Personal Observations.* New York: Charles Scribner's Sons, 1922.

———. "On the Species of Bornean Orangs, with Notes on Their Habits." *Proceedings of the American Association for the Advancement of Science* 28 (1879): 438–455.

———. "The Passing of the Buffalo." *Cosmopolitan* 4 (October 1887): 51–89.

———. *Popular Official Guide to the New York Zoological Park,* June 1. New York: New York Zoological Society, 1906.

———. *Two Years in the Jungle: The Experiences of a Hunter and Naturalist in India, Ceylon, the Malay Peninsula and Borneo.* New York: Charles Scribner's Sons, 1926.

———. *Wild Animals: Interviews and Wild Animal Opinions of Us.* New York: Charles Scribner's Sons, 1929.

Hose, Charles. *Fifty Years of Romance & Research or a Jungle-wallah at Large.* London: Hutchinson, 1927.

Hoskins, Janet. "Headhunting as Practice and Trope." In *Headhunting and the Social Imagination in Southeast Asia,* ed. Janet Hoskins, 1–49. Stanford, Calif.: Stanford University Press, 1996.

Houten, P. J. van. *Nota aangeboden aan het Bestuur der Maatschappij ter Bevordering van het Natuurkundig Onderzoek der Nederlandsche Koloniën: Staatsbescherming van Nuttige of Merkwaardige Dieren en Planten in de Nederlandsche Koloniën.* The Hague, n.p., 1896.

Houttuyn, Martinus. *Natuurlyke Historie of Uitvoerige Beschryving der Dieren, Planten en Mineraalen, volgens het Samenstel van de Heer Linnaeus,* vol. 1. Amsterdam: F. Houttuyn, 1761.

Hribal, Jason. "Orangutans, Resistance and the Zoo: The Story of Ken Allen and Kumang." *CounterPunch,* December 16, 2008. http://www.counterpunch.org/hribal12162008.html. Accessed December 21, 2010.

Huggan, Graham, and Helen Tiffin. *Postcolonial Ecocriticism: Literature, Animals, Environment.* London: Routledge, 2010.

Husband, Timothy. *The Wild Man: Medieval Myth and Symbolism.* New York: Metropolitan Museum of Art, 1980.

Huxley, Thomas Henry. "On the Natural History of the Man-like Apes" (1863). In Thomas Henry Huxley, *Man's Place in Nature and Other Essays.* London: Dent, 1906, 1–56.

Ignatieff, Michael. *The Rights Revolution.* Toronto: House of Anansi Press, 2nd ed. 2007.

Janson, H. W. *Apes and Ape Lore in the Middle Ages and the Renaissance.* London: Warburg Institute, University of London, 1952.

Jefferson, Thomas. *Notes on the State of Virginia.* London: John Stockdale, 1787.

Jeffries, John. "Some Account of the Dissection of a Simia Satyrus, Ourang Outang, or Man of the Woods." *Boston Journal of Philosophy and Arts* 2 (1825): 570–580.

Jensen, Stine. *Waarom Vrouwen van Apen houden: De vergeefse Mensenliefde voor Bokito en andere Apen.* Amsterdam: Prometheus, 2007.

Jesse, Edward. *Gleanings in Natural History.* 2nd ed., vol. 1. London: John Murray, 1834.

Jobling, Paul. "Daumier's *Orang-Outaniana.*" *Print Quarterly* 10, no. 3 (1993): 231–246.

Jobson, Richard. *The Golden Trade: or, a Discovery of the River Gambia, and the Golden Trade of the Aethiopians.* London: Nicholas Okes, 1623.

Jones, M. L. "The Orang Utan in Captivity." In *The Orang Utan: Its Biology and Conservation,* ed. Leobert E. M. de Boer, 17–38. The Hague: W. Junk, 1982.

Jouffroy, Françoise K. "Primate Hands and the Human Hand: The Tool of Tools." In *The Use of Tools by Human and Non-human Primates,* ed. Arlette Berthelet and Jean Chavaillon, 6–33. New York: Oxford University Press. 1993.

Jungle Cavalcade. Directed by William C. Ament, Armand Denis, and Clyde E. Elliot. RKO Radio Pictures, 1941. 76 min.

Kalof, Linda. *Looking at Animals in Human History.* London: Reaktion, 2007.

Kaplan, G., and L. Rogers. "Of Human Fear and Indifference: The Plight of the Orangutan." In *The Neglected Ape,* ed. Ronald D. Nadler, Birute F. M. Galdikas, Lori K. Sheeran, and Norm Rosen, 3–12. New York: Plenum Press, 1995.

Kaplan, Gisela, and Lesley J. Rogers. *The Orang-utans.* St Leonards, N.S.W.: Allen & Unwin, 1999.

Kavara, Tatjana, and Peter Dovčb, "Domestication of the Horse: Genetic Relationships between Domestic and Wild Horses." *Livestock Science* 116, nos. 1–3 (2008): 1–14.

Kelley, Jay. "The Hominoid Radiation in Asia." In *The Primate Fossil Record,* ed. Walter C. Hartwig, 369–384. Cambridge: Cambridge University Press, 2002.

Kennedy, Michael. *Australasian Marsupials and Monotremes: An Action Plan for their Conservation.* Gland: IUCN, 1992.

Keys, David. *Catastrophe: An Investigation into the Origins of the Modern World.* New York: Ballantine, 2000.

King, Victor T. *The Peoples of Borneo.* Cambridge, Mass.: Blackwell, 1993.

Kiöping, Nils Matson. *Een Kort Beskriffning vppå trenne Resor och Peregrinationer, sampt Konungarijket Japan.* Stockholm: Johann Kankel, 1667.

Kipling, Rudyard. "Bertran and Bimi." In *Life's Handicap: Being Stories of Mine Own People*. London: Macmillan, 1891, 336–342.

Knapman, Gareth. "Orang-utans, Tribes, and Nations: Degeneracy, Primordialism, and the Chain of Being." *History and Anthropology* 19, no. 2 (2008): 143–159.

Knauer, Friedrich. *Menschenaffen, ihr Frei- under Gefangenleben*. Leipzig: Deutsche Naturwissenschaftliche Gesellschaft, [1915].

Knight, John. "Introduction." In *Animals in Person: Cultural Perspectives on Human–Animal Intimacies*, ed. John Knight, 1–13. Oxford: Berg, 2005.

Knott, Cheryl D., and Sonya M. Kahlenberg. "Orangutans in Perspective: Forced Copulations and Female Mating Resistance." In *Primates in Perspective*, ed. Christina J. Campbell et al., 290–305. New York: Oxford University Press, 2007.

Köhler, Wolfgang. *The Mentality of Apes*. New York: Harcourt, Brace, 1926.

———. "The Mentality of Orangs." *International Journal of Comparative Psychology* 6, no. 4 (1993): 189–229.

Kohn, David, ed. *Charles Darwin's Notebooks, 1836–1844: Geology, Transmutation of Species, Metaphysical Enquiries*. Transcribed by David Kohn. Notebook C. Cambridge: Cambridge University Press, 1987.

Kruk, Remke. "Traditional Islamic Views of Apes and Monkeys." In *Ape, Man, Ape-Man: Changing Views since 1600*, ed. Raymond Corbey and Bert Theunissen, 29–41. Leiden: Department of Prehistory, Leiden University, 1995.

Krützen, Michael, Erik P. Willems, and Carel P. van Schaik. "Culture and Geographic Variation in Orangutan Behavior." *Current Biology* 21, no. 21 (2011): 1808–1812.

Lacepède, [Bernard Germain de]. *Tableau des Divisions, Sous-divisions, Ordres et Genres des Mammifères*. Paris: Plassan, Year IX of the Republic.

Lach, Donald F. *Asia in the Eyes of Europe: Sixteenth through Eighteenth Centuries*. Exhibition catalogue. Chicago: University of Chicago Library, 1991.

Lamarck, Jean-Baptiste. *Zoological Philosophy: An Exposition with Regard to the Natural History of Animals*. Translated by Hugh Eliot. New York and London: Hafner, 1963.

Leblanc, Ronald D. "A Russian Tarzan, or 'Aping' Jocko?" *Slavic Review* 46, no. 1 (1987): 70–86.

Lee, Maurice S. *Slavery, Philosophy, and American Literature, 1830–1860*. Cambridge: Cambridge University Press, 2005.

Leguat, François. *De Gevaarlyke en Zeldzame Reysen van den heere François Leguat*. Utrecht: W. Broedelet, 1708.

[Leguat, François]. *The Voyage of François Leguat of Bresse*, vol. 2. London: Hakluyt Society, 1891.

———. *Voyage et Avantures de Francois Leguat, et de ses Compagnons en deux Isles desertes des Indes Orientales*, vol. 2. Amsterdam: Jean Louis de Lorme, 1708.

Lehrer, Steven. "Introduction." In Frank Buck and Steven Lehrer, *Bring 'Em Back Alive: The Best of Frank Buck*. Lubbock: Texas Tech University Press, 2000.

Lévi-Strauss, Claude. *Totemism*. Translated by Rodney Needham. Boston: Beacon Press, 1963.

Linden, Eugene. "Can Animals Think?" *Time Magazine,* August 29, 1999. http://www
    .time.com/time/magazine/article/0,9171,30198-1,00.html. Accessed December 21, 2010.

———. *The Octopus and the Orangutan: More True Tales of Animal Intrigue, Intelligence
    and Ingenuity.* New York: Dutton, 2002.

———. *The Parrot's Lament: And Other True Tales of Animal Intrigue, Intelligence and
    Ingenuity.* New York: Plume, 2000.

Lindfors, Bernth. "Ethnological Show Business: Footlighting the Dark Continent." In
    Freakery: Cultural Spectacles of the Extraordinary Body, ed. Rosemarie Garland
    Thomson, 207–218. New York: New York University Press, 1996.

———. "P. T. Barnum and Africa." *Studies in Popular Culture* 7 (1984): 18–27.

Link. Directed by Richard Franklin. Cannon Screen Entertainment, 1986. VHS, 108 min.

Linnaeus, Carolus. *Systema Naturae per Regna Tria Naturae, Secundum Classes, Ordines,
    Genera, Species cum Characteribus, Differentiis, Synonymis, Locis,* vol. 1. Stockholm:
    Laurentius Salvius, 10th ed., 1758.

———. *Systema Naturae Sistens Regna Tria Naturae, in Classes et Ordines, Genera et
    Species Redacta.* Stockholm: Gottfr. Kiesewetter, 6th ed. 1748.

———. *Systema Naturae sive Regna Tria Naturae Systematice Proposita per Classes,
    Ordines, Genera, & Species.* Leiden: Theodorum Haak, 1735.

Logan, J. R. "Five Days in Naning, with a Walk to the Foot of Gunong Datu in Rembau."
    Journal of the Indian Archipelago and Eastern Asia 3 (1849): 24–41, 278–287, 402–
    412, 489–493.

[Long, Edward]. *The History of Jamaica: or, General Survey of the Antient and Modern
    State of that Island,* vol. 2. London: Lowndes, 1774.

Lovejoy, Arthur O. "Monboddo and Rousseau." *Modern Philology* 30, no. 3 (February
    1933): 275–296.

Lumholtz, Carl. *Through Central Borneo: An Account of Two Years' Travel in the Land of
    the Head-hunters between the Years 1913 and 1917.* Singapore: Oxford University Press,
    1991. Originally published New York: Charles Scribner's Sons, 1920.

Lysons, Daniel. *Collectanea: or, A Collection of Advertisements and Paragraphs from
    Newspapers, Relating to Various Subjects. Publick Exhibitions and Places of Amusement,
    1661–1840,* vol. 1. Unpublished scrapbook, British Library, shelfmark C.103.k.11.

Mackenzie, John M. *The Empire of Nature: Hunting, Conservation and British Imperialism.*
    Manchester: Manchester University Press, 1988.

MacKinnon, John. "The Future of Orang-utans." *New Scientist,* June 23, 1977, 697–699.

———. *In Search of the Red Ape.* New York: Holt, Rinehart and Winston, 1974.

———. "The Orang-utan in Sabah Today: A Study of a Wild Population in the Ulu Segama
    Reserve." *Oryx* 11 (1971): 141–191.

———. "Orang-utans in Sumatra." *Oryx* 12 (1973): 234–242.

———. *Tanjung Puting National Park: Management Plan for Development.* Bogor: World
    Wildlife Fund Indonesia Programme, WWF Project no. 1523, 1983.

Malone N., A. Fuentes, and F. J. White. "Variation in the Social Systems of Extant

Hominoids: Comparative Insight into the Social Behavior of Early Hominins." *International Journal of Primatology* 33 (2012): 1251–1277.

Maple, Terry L. *Orang-utan Behavior.* New York: Van Nostrand Reinhold, 1980.

Marazano, Richard, and Chris Lamquet. *Eco-Warriors: Orangutan.* 2 vols. Bielefeld: Splitter, 2010.

Markham, Rosemary J. "Breeding Orangutans at Perth Zoo: Twenty Years of Appropriate Husbandry." *Zoo Biology* 9 (1990): 171–182.

Marsden, William. *The History of Sumatra, Containing an Account of the Government; Laws, Customs, and Manners of the Native Inhabitants, with a Description of the Natural Productions.* London: Longman, Hurst, Rees, Orme, and Brown, 1811.

Matton, Sylvain, ed. *Tintinnabulum Naturae: Rêveries d'un Individu Semi Homme Semi Bête engendré d'une Négresse et d'une Orang-Outang, suivi de Pensées métaphisiques lancées dans le Tourbillon et de quelques Poésies et Pièces fugitives, par un Solitaire de Champagne.* Paris and Milan: SÉHA and ARCHÈ, 2002.

Mazel, M. "Van een Aap in 1777." *Die Haghe* 9 (1909): 361–380.

McClintock, Ann. *Imperial Leather: Race, Gender and Sexuality in the Colonial Contest.* London: Routledge, 1995.

McGrady, Patrick M. *The Youth Doctors.* New York: Coward-McCann, 1968.

McKay, Margaret. "Peacock, Monboddo, and the Swedish Connection." *Notes and Queries* (December 1990): 422–424.

Meijer, Miriam Claude. "The Century of the Orangutan." *XVIII: New Perspectives on the Eighteenth Century* 1 (2004): 62–78.

———. *Race and Aesthetics in the Anthropology of Petrus Camper (1722–1789).* Amsterdam: Rodopi, 1999.

Merke, F. *Geschichte und Ikonographie des endemischen Kropfes und Kretinismus.* Bern: Han Huber, 1971.

Mettrie, Julien Offray de la. *Man a Machine; and, Man a Plant.* Indianapolis: Hackett, 1994.

Miles, H. Lyn. "How Can I Tell a Lie? Apes, Language, and the Problem of Deception." In *Deception: Perspectives on Human and Nonhuman Deceit,* ed. Robert W. Mitchell and Nicholas S. Thompson, 246–266. Albany: State University of New York Press, 1986.

———. "Language and the Orangutan: The Old 'Person' of the Forest." In *The Great Ape Project: Equality Beyond Humanity,* ed. Paola Cavalieri and Peter Singer, 42–57. New York: St. Martin's Griffin, 1993.

Miles, M. "Goitre, Cretinism and Iodine in South Asia: Historical Perspectives on a Continuing Scourge." *Medical History* 42 (1998): 47–67.

Miller, Charles. "An Account of the Island of Sumatra &c." *Philosophical Transactions of the Royal Society of London* 68 (1778): 160–179.

Miller, Charles C. *Black Borneo.* New York: Modern Age Books, 1942.

Miller, Lillian B., Sydney Hart, and Toby A. Appel. *The Selected Papers of Charles Willson Peale and His Family.* Vol. 2. New Haven, Conn: Yale University Press, 1988.

Milton, Oliver. "The Orang-utan and Rhinoceros in North Sumatra." *Oryx* 7 (1964): 177–184.

Mr. Smith. Created by Stan Daniels and Edwin Weinberger. Paramount Television, NBC Productions, 1983. Television series.

Mitchell, Robert W. "Humans, Nonhumans and Personhood." In *The Great Ape Project: Equality beyond Humanity,* ed. Paola Cavalieri and Peter Singer, 237–247. New York: St. Martin's Griffin, 1993.

——. "The Natural History of Poe's Orangutan." *Dark Romanticism: History, Theory, Interpretation* 31, nos. 1–2 (1998): 32–34.

Mitman, Gregg. *Reel Nature: America's Romance with Wildlife on Film.* Cambridge, Mass.: Harvard University Press, 1999.

Mizelle, Brett. "'Man Cannot Behold It Without Contemplating Himself': Monkeys, Apes and Human Identity in the Early American Republic." *Pennsylvania History* 66 (1999): 144–173.

Monkey Island, or, Harlequin and the Loadstone Rock. Unpublished script, British Library. Lord Chamberlain's Plays, vol. III, 1824.

Montagu, M. F. Ashley. *Edward Tyson, M.D., F.R.S. 1650–1708 and the Rise of Human and Comparative Anatomy in England: A Study in the History of Science.* Philadelphia: American Philosophical Society, 1943.

Montgomery, Sy. *Walking with the Great Apes: Jane Goodall, Dian Fossey, Biruté Galdikas.* 2nd ed. White River Junction, Vt.: Chelsea Green, 2009.

Moore, John Robert. "Poe, Scott, and 'The Murders in the Rue Morgue.'" *American Literature* 8, no. 1 (1936): 52–58.

Moran, Francis III. "Of Pongos and Men: 'Orangs-Outang' in Rousseau's 'Discourse on Inequality.'" *Review of Politics* 57, no. 4 (Autumn 1995): 641–664.

Morris, Desmond. *The Naked Ape: A Zoologist's Study of the Human Animal.* London: Jonathan Cape, 1967.

Morris, Ramona, and Desmond Morris. *Men and Apes.* London: Hutchinson, 1966.

Moulins, Maurice de. *Morok l'Orang-outang.* Paris: Jules Tallendier, 1930.

Nelson, Sherry V. *The Extinction of Sivapithecus: Faunal and Environmental Changes Surrounding the Disappearance of a Miocene Hominoid in the Siwaliks of Pakistan.* Boston: Brill, 2003.

Nénette. Directed by Nicolas Philibert. Les Films d'Ici/Forum des Images, 2010. DVD, 70 min.

Niekerk, Carl. "Man and Orangutan in Eighteenth-Century Thinking: Retracing the Early History of Dutch and German Anthropology." *Monatshefte* 96, no. 4 (2004): 477–502.

Ninh, Bao. *The Sorrow of War.* London: Minerva 1994. Originally published in Vietnamese in 1991.

[Noble, Charles Frederick]. *A Voyage to the East Indies in 1747 and 1748.* London: T. Becket and P. A. Dehondt, 1762.

Noell, Anna Mae. *The History of Noell's Ark Gorilla Show*. Orlando: Noell's Ark Publisher, 1979.

Noltie, H. J. *Raffles' Ark Redrawn: Natural History Drawings from the Collection of Sir Thomas Stamford Raffles*. London: British Library and Royal Botanic Garden Edinburgh, 2009.

Obeyesekere, Gananath. *Cannibal Talk: The Man-eating Myth and Human Sacrifice in the South Seas*. Berkeley: University of California Press, 2005.

"Observations on Equivocal Generation." *Boston Medical and Surgical Journal* 31, no. 3 (1844): 49–56.

Ohnuki-Tierney, Emiko. *The Monkey as Mirror: Symbolic Transformations in Japanese History and Ritual*. Princeton, N.J.: Princeton University Press, 1987.

"The Orang Outang." *New-England Magazine* 1, no. 6 (December 1831): 497–500.

Osborn, Henry Fairfield. "The Influence of Bodily Locomotion in Separating Man from the Monkeys and Apes." *The Scientific Monthly* 26, no. 5 (May 1928): 385–399.

Osborne, Michael A. "Zoos in the Family: The Geoffroy Saint-Hilaire Clan and the Three Zoos of Paris." In *New Worlds, New Animals: From Menagerie to Zoological Park in the Nineteenth Century*, ed. R. J. Hoage and W. A. Deiss, 33–42. Baltimore: Johns Hopkins University Press, 1996.

Oskamp, D. L. *Naauwkeurige Beschryving van den grooten en kleinen Orang-outang, gelyk ook van den Gibbon, de twee Aapsoorten, die het meest naar den Mensch gelyken*. Amsterdam: J. C. Sepp, 1803. Manuscript completed in 1794.

Passions. Created by James E. Reilly. Outpost Farm Productions/NBC Studios/NBC Universal Television, NBC Productions, 1999–2008. Television series.

Patterson, Nick, Daniel J. Richter, Sante Gnerre, Eric S. Lander, and David Reich. "Genetic Evidence for Complex Speciation of Humans and Chimpanzees." *Nature* 441, no. 7097 (2006): 1103–1108.

Peacock, Thomas Love. *Melincourt, or Sir Oran Haut-Ton*. London: Macmillan, 1896. First edition, 1817.

Peary, Gerald. "Missing Links: The Jungle Origins of King Kong." In *The Girl in the Hairy Paw: King Kong as Myth, Movie, and Monster*, ed. Ronald Gottesman and Harry Geduld, 37–42. New York: Avon, 1976. http://geraldpeary.com/essays/jkl/kingkong-1.html. Accessed December 26, 2012.

Peluso, Nancy Lee, and Emily Harwell. "Territory, Custom, and the Cultural Politics of Ethnic War in West Kalimantan, Indonesia." In *Violent Environments*, ed. Nancy Lee Peluso and Michael Watts, 83–116. Ithaca, N.Y.: Cornell University Press, 2001.

Penebangan Liar di Taman Nasional Tanjung Puting: Update Laporan "Final Cut" = *Illegal Logging in Tanjung Puting National Park: An Update on "The Final Cut" Report*. Bogor: Telapak, 2000.

Perkins, L. "Conservation and Management of Orang-utans *Pongo pygmaeus* ssp." *International Zoo Yearbook* 36 (1998): 109–112.

Perkins, Lorraine A., and Terry L. Maple. "North American Orangutan Species Survival Plan: Current Status and Progress—in the 1980s." *Zoo Biology* 9 (1990): 135–139.

Perry, L. Curtis. *Apes and Angels: The Irishman in Victorian Caricature.* Washington, D.C.: Smithsonian Institution Press, 2nd ed., 1997.

Perry, W. J. *The Megalithic Culture of Indonesia.* Manchester: Manchester University Press, 1918.

Peterson, Dale, and Jane Goodall. *Visions of Caliban: On Chimpanzees and People.* Boston: Houghton Mifflin, 1993.

Piepers, M. "Door welke Maatregelen kan tot eene Rationeele Bescherming der Inheemsche Dieren- en Plantenwereld in Nederlandsch Indië worden gekomen?" *Tijdschrift voor Nederlandsch Indië* new series 25, no. 1 (1896): 38–72.

Pieters, Florence F. J. M. "Introduction." In *Wonderen der Natuur: In de Menagerie van Blauw Jan te Amsterdam, zoals gezien door Jan Velten rond 1700.* Amsterdam: ETI Digital, 1998, 37–47.

———. "De Menagerie van Stadhouder Willem V op het Kleine Loo te Voorburg." In *Een Vorstelijk Dierentuin: de Menagerie van Willem V,* ed. B. C. Sliggers and A. A. Wertheim, 39–59. Zutphen: Walburg Instituut, 1994.

———. "Notes on the Menagerie and Zoological Cabinet of Stadholder William V of Holland, directed by Arnout Vosmaer." *Journal of the Society for the Bibliography of Natural History* 9, no. 4 (1980): 539–563.

Pieters, Florence, and Kees Rookmaaker. "Arnout Vosmaer, Topcollectioneur van Naturalia en zijn *Regnum animale.*" In *Een Vorstelijk Dierentuin: de Menagerie van Willem V,* ed. B. C. Sliggers and A. A. Wertheim, 11–38. Zutphen: Walburg Instituut, 1994.

Pinker, Steven. *The Blank Slate: The Modern Denial of Human Nature.* New York: Viking, 2002.

Pitt, G. D. *Phillip Quarl or the Mariner and His Monkey.* Unpublished script, British Library. Lord Chamberlain's Plays, vol. CXLI, 1847.

Planet of the Apes. Directed by Franklin J. Schaffner. APJAC Productions/Twentieth Century Fox, 1968. DVD, 112 min.

Plato. "Protagoras." In J. Wright, *The Phædrus, Lysis, and Protagoras of Plato,* 132–220. London: John W. Parker, 1848.

The Playhouse. Directed by Buster Keaton and Edward F. Cline. Joseph M. Schenck Productions. First National Pictures, 1921. DVD 22 min.

Poe, Edgar Allan. "Hop-Frog" [also published as "Hop-Frog; Or, the Eight Chained Ourangoutangs"]. In *The Works of the Late Edgar Allan Poe,* Vol. 2, ed. Rufus Wilmot Griswold, 455–464. New York: Redfield, 1857.

———. "The Murders in the Rue Morgue." *Graham's Lady's and Gentleman's Magazine* 18 (1841): 166–179.

Pogorel'skii, Antonii. *The Double, or My Evenings in Little Russia.* Translated by R. Sobel. Ann Arbor, Mich.: Ardis, 1988.

———. "Journey in a Stagecoach." In *The Double, or My Evenings in Little Russia*. Translated by R. Sobel. Ann Arbor, Mich.: Ardis, 1988, 106–125.

[Polo, Marco]. *The Travels of Marco Polo: A Modern Translation by Teresa Waugh*. London: Sidgwick and Jackson, 1984.

Pougens, Charles [de]. *Jocko: Anecdote détachée des Lettres inédites sur l'Instinct des Animaux*. Paris: P. Persan, 1824.

Pratchett, Terry. *The Light Fantastic*. New York: Harper Torch, 1986.

Querido, A. "Epidemiology of Cretinism." In *Endemic Cretinism*, ed. Basil S. Hetzel and Peter O. D. Pharoah, 9–18. Goroka: Institute of Human Biology, 1971.

Radermacher, J. C. M. "Beschrijving van het Eiland Borneo, voor zoover hetzelfde tot nu toe bekend is." *Verhandelingen van het Bataviaasch Genootschap, der Konsten en Wetenschappen* 2 (1784): 107–148.

———. "Proeve nopens verschillende Gedaante en Coleur der Menschen." *Verhandelingen van het Bataviaasch Genootschap, der Konsten en Wetenschappen* 2 (1784): 213–228.

Radford, Thomas. "The Value of Embryonic and Fœtal life, Legally, Socially, and Obstetrically Considered." *British Record of Obstetric Medicine & Surgery* 1 (1848): 6–11, 53–56, 77–89.

Raffles, Sophia Hull. *Memoir of the Life and Public Services of Sir Thomas Stamford Raffles*. Singapore: Oxford University Press, 1991. Originally published London: James Duncan, 1835.

Raloff, Janet. "Caste-off Orangs." *Science News* 147, no. 12 (March 25, 1995): 184–185, 189.

Raphael, Ray. "America's Disastrous Invasion of Canada." *American History* 44, no. 6 (February 2010): 37.

Read, Bernard E. *Chinese Material Medica: Animal Drugs*. Beiping: Peking Natural History Bulletin, 1931.

Reed, Cindi. "Hear No Evil. See No Evil. Film No Evil." *Vegas Seven,* August 11, 2010. http://weeklyseven.com/news/2010/august/12/hear-no-evil-see-no-evil-film-no-evil. Accessed December 1, 2011.

Reed, David L., Jessica E. Light, Julie M. Allen, and Jeremy J. Kirchman. "Pair of Lice Lost or Parasites Regained: The Evolutionary History of Anthropoid Primate Lice." *BioMed Central Biology* 5, no. 7 (2007). http://www.biomedcentral.com/1741-7007/5/7. Accessed December 22, 2012.

Reeve, Henry. "Some Account of Cretinism." *Philosophical Transactions of the Royal Society of London* 98 (1808): 111–119.

Reichenbach, Herman. "A Tale of Two Zoos: The Hamburg Zoological Gardens and Carl Hagenbeck's Tierpark." In *New Worlds, New Animals: From Menagerie to Zoological Park in the Nineteenth Century,* ed. R. J. Hoage and W. A. Deiss, 51–62. Baltimore: Johns Hopkins University Press, 1996.

Reid, Mayne. *The Castaways: A Story of Adventure in the Wilds of Borneo*. New York: Sheldon, 1870.

Rétif de la Bretonne, [Nicolas-Edmé]. *La Découverte australe par un Homme-volant ou le Dédale français; Nouvelle très-philosophique* [The Southern Discovery by a Flying Man or the French Daedalus; a Very Philosophical Novel]. Leipsick [*sic*, i.e., Paris]: n.p., 1781.

Reynolds, V. "On the Identity of the Ape Described by Tulp 1641." *Folia Primatologica* 5 (1967): 80–87.

Richards, Robert J. *The Tragic Sense of Life: Ernst Haeckel and the Struggle over Evolutionary Thought*. Chicago: University of Chicago Press, 2008.

Richardson, Edgar P., Brooke Hindle, and Lillian B. Miller. *Charles Willson Peale and His World*. New York: H. N. Abrahams, 1983.

Rijksen, H. D., and E. Meijaard, *Our Vanishing Relative: The Status of Wild Orang-utans at the Close of the Twentieth Century*. Dordrecht: Kluwer, 1999.

Rise of Planet of the Apes. Directed by Rupert Wyatt. Twentieth-Century Fox/Dune Entertainment, 2011. DVD, 105 min.

Ritvo, Harriet. *The Animal Estate: The English and Other Creatures in the Victorian Age*. Cambridge, Mass.: Harvard University Press, 1987.

Robida, Albert. *The Adventures of Saturnin Farandoul*. Translated by Brian M. Stableford. Encino, Calif.: Black Coat Press, 2009. Originally published as *Le roi des singes*. In *Voyages très extraordinaires de Saturnin Farandoul dans les 5 ou 6 Parties du Monde, et dans tous les Pays connus et même inconnus de Monsieur Jules Verne*. Paris: Librairie illustrée [1879].

Robinson, Michael H. "Foreword." In *New Worlds, New Animals: From Menagerie to Zoological Park in the Nineteenth Century*, ed. R. J. Hoage and W. A. Deiss, vii–xi. Baltimore: Johns Hopkins University Press, 1996.

Rodwell, G. Herbert. *The Ourang Outang and His Double, or, The Runaway Monkey*. London: J. Dicks, 1842.

Roger, Jacques. *Buffon: A Life in Natural History*. Ithaca, N.Y.: Cornell University Press, 1997.

Röhrer-Ertl, Olav. "Zur Erforschungsgeschichte und Namengebung beim Orang-Utan Pongo satyrus (Linnaeus, 1758); synon. Pongo pygmaeus (Hoppius, 1763)." *Spixiana* 6, no. 3 (1983): 301–332.

Roland, Marcel. *La Conquête d'Anthar*. Paris: Pierre Lafitte, 1913.

———. *Le Déluge futur*. Paris: Fayant, 1910.

———. *Le Presqu'homme*. Paris: Albert Mericant, 1907. http://www.mediatheque-fm6 .ma/index2.php?option=com_docman&task=doc_view&gid=174&Itemid=78. Accessed October 21, 2009.

Römer, L. S. A. M. von. "Dr. Jacobus Bontius." *Bijblad op het Geneeskundig Tijdschrift van Nederlandsch-Indie,* bijblad 1 (March 1932): 1–33.

Rookmaaker, L. C. "A Living Orang Utan in Uppsala in 1785." *Zoologische Garten* 59, no. 4 (1989): 275–276.

Rooy, Piet de. "In Search of Perfection: The Creation of a Missing link." In *Ape, Man,*

Ape-Man: Changing Views since 1600, ed. Raymond Corbey and Bert Theunissen, 195–207. Leiden: Department of Prehistory, Leiden University, 1995.

Rosenthal, Mark, Carol Tauber, and Edward Uhlir. *The Ark in the Park: The Story of Lincoln Park Zoo.* Urbana: University of Illinois Press, 2003.

Rothfels, Nigel. *Representing Animals.* Bloomington: Indiana University Press, 2002.

———. *Savages and Beasts: The Birth of the Modern Zoo.* Baltimore: Johns Hopkins University Press, 2002.

Rousseau, G. S. *Enlightenment Crossings: Pre- and Post-modern Discourses, Anthropological.* Manchester: Manchester University Press, 1991.

———. "Madame Chimpanzee." In *Enlightenment Crossings: Pre- and Post-modern Discourses, Anthropological.* Manchester: Manchester University Press, 1991, 198–209.

Rousseau, Jean-Jacques. *Discourse on the Origin of Inequality.* Oxford: Oxford University Press, 1994.

Russon, Anne E. "Evolutionary Reconstructions of Great Ape Intelligence." In *The Evolution of Thought: Evolutionary Origins of Great Ape Intelligence,* ed. Anne E. Russon and David R. Begun, 1–14. Cambridge: Cambridge University Press, 2004.

———. "The Nature and Evolution of Intelligence in Orangutans (*Pongo pygmaeus*)." *Primates* 39, no. 4 (October 1998): 485–503.

Russon, Anne E., and Kim A. Bard. "Exploring the Minds of the Great Apes: Issues and Controversies." In *Reaching into Thought: The Minds of the Great Apes,* ed. Anne E. Russon, Kim A. Bard, and Sue Taylor Parker, 1–20. Cambridge: Cambridge University Press, 1996.

Ryder, Richard D. *Animal Revolution: Changing Attitudes towards Speciesism.* Oxford: Blackwell, 1989.

St. John, Spenser. *Life in the Forests of the Far East: or Travels in Northern Borneo.* 2 vols. London: Smith, Elder & Co., 2nd ed., 1862.

Salt, Henry S. *Animals' Rights: Considered in Relation to Social Progress.* London: Macmillan, 1894.

Savage, Thomas S. "Notice of the External Characteristics and Habits of *Troglodytes gorilla,* a New Species of Orang from the Gaboon River," and Jeffreys Wyman, "Osteology of the Same." *Boston Journal of Natural History* 5, no. 4 (1847): 417–441.

Schaik, Carel P. van, and Cheryl D. Knott. "Geographic Variation in Tool Use on *Neesia* Fruits in Orangutans." *American Journal of Physical Anthropology* 114 (2001): 331–342.

Schaik, Carel P. van, and Jan A.R.A.M. van Hooff. "Toward an Understanding of the Orangutan's Social System." In *Great Ape Societies,* ed. William C. McGrew, Linda F. Marchant, and Toshisada Nishida, 3–15. Cambridge: Cambridge University Press, 1996.

Schaller, George B. "The Orang-utan in Sarawak." *Zoologica* 46 (1961): 73–82.

Schaller, George B., Nguyen Xuan Dang, Le Dinh Thuy, and Vo Thanh Son. "Javan Rhinoceros in Vietnam." *Oryx* 24 (1990): 77–80.

Schlegel, Herm, and Sal. Müller, "Bijdragen tot de Natuurlijke Historie van den Orang-oetan (Simia satyrus)." In C. J. Emminck, *Verhandelingen over de Natuurlijke Geschiedenis der Nederlandsche Overzeesche Bezittingen*. Leiden: J. Luchtmans and C.C. van der Hoek, 1839–1844.

Scholl, Aurélien. *Fleurs d'Adultère*. Paris: E. Dentu, 1880.

Schultz, Adolph H. "Characters Common to Higher Primates and Characters Specific for Man." *The Quarterly Review of Biology* 11, no. 3 (September 1936): 259–283.

Schuster, Gerd, Willie Smits, and Jay Ullal. *Thinkers of the Jungle*. [Königswinter, Germany]: H. F. Ullmann, 2008.

Schwartz, Jeffrey H. *The Red Ape: Orang-utans and Human Origins*. Boston: Houghton Mifflin, 1987.

Scott, Charles Payson Gurley. "Malayan Words in English, Part II." *Journal of the American Oriental Society* 18 (1897): 49–124.

Scott, Walter. *Count Robert of Paris*. Edinburgh: Robert Cadell, 1829.

Sellers, Charles Coleman. *Charles Willson Peale*. Vol. 2, *Later Life, 1790–1827*. Philadelphia: American Philosophical Society, 1947.

Sens, Angelie. "Dutch Debates on Overseas Man and His World, 1770–1820." In *Colonial Empires Compared: Britain and the Netherlands, 1750–1850*, ed. Bob Moore and Henk F. K. van Nierop, 77–93. Aldershot: Ashgate, 2003.

Sheridan, Frank. "The Young Marooner, or An American Robinson Crusoe." *Brave and Bold Weekly* 314 (December 26, 1908). http://special.lib.umn.edu/clrc/hess/pdf/Brave_and_Bold_Weekly_314.pdf. Accessed May 25, 2011.

Sherratt, Andrew. "Darwin among the Archaeologists: The John Evans Nexus and the Borneo Caves." *Antiquity* 76, no. 291 (2002): 151–157.

[Sibly, Ebenezer]. *An Universal System of History, including a Natural History of Man; the Orang-outang; and the Whole Tribe of Simia*. London: [Ebenezer Sibly], [1795].

Silva, G. S. de. "The East-Coast Experiment." In *Conservation in Tropical South East Asia*. Bangkok: IUCN 1965, 229–302.

Sliggers, B. C., and A. A. Wertheim, eds., *Een Vorstelijk Dierentuin: de Menagerie van Willem V* [Zutphen]: Walburg Pers, 1994 [bilingual text, Du and Fr].

Smellie, William. *The Philosophy of Natural History*, vol. 1. Edinburgh: Heirs of Charles Elliot, 1790.

Smith, Dale. *What the Orangutan Told Alice*. Nevada City, Calif.: Deer Creek Publishing, 2001.

Smith, Paul. *Clint Eastwood: A Cultural Production*. London: UCL Press, 1993.

Smits, W. T. M. Heriyanto, and W. S. Ramono, "A New Method for Rehabilitation of Orangutans in Indonesia: A First Overview." In *The Neglected Ape*, ed. Ronald D. Nadler, Birute F. M. Galdikas, Lori K. Sheeran, and Norm Rosen, 69–78. New York: Plenum Press, 1995.

Snigurowicz, Diana. "Sex, Simians, and Spectacle in Nineteenth-Century France; or, How to Tell a 'Man' from a Monkey." *Canadian Journal of Natural History* 34, no. 1 (1999): 52–81.

Somerset, C. A. *Bibboo, or, The Shipwreck*. Unpublished script, British Library. Lord Chamberlain's Plays, vol. CI, 1842.

Sorenson, John. *Ape*. London: Reaktion, 2009.

Spalding, Linda. *A Dark Place in the Jungle*. Chapel Hill, N.C.: Algonquin Books of Chapel Hill, 1999.

Spencer, Frank. "From Pithekos to Pithecanthropus: An Abbreviated Review of Changing Scientific Views on the Relationship of the Anthropoid Apes to Homo." In *Ape, Man, Ape-Man: Changing Views since 1600,* ed. Raymond Corbey and Bert Theunissen, 13–27. Leiden: Department of Prehistory, Leiden University, 1995.

Stanford, Craig B. *Planet without Apes*. Cambridge Mass.: Harvard University Press, 2011.

Steinmann, A. "De Dieren op de Basreliefs van de Boroboedoer." *Tijdschrift voor Indische Taal-, Land- en Volkenkunde* 74 (1934): 101–122.

Stiles, Ch. Wardell, and Mabelle B. Orleman. "The Nomenclature for Man, the Chimpanzee, the Orang-Utan, and the Barbary Ape." *Hygienic Laboratory Bulletin* 145 (1927): 1–66.

Stump, Jordan. "Translator's Preface." In Jules Verne, *The Mysterious Island*. New York: Modern Library, 2004, xxvii–xxxiii.

Sugardjito, Jito, and Asep S. Adhikerana. "Measuring Performance of Orangutan Protection and Monitoring Unit: Implications for Species Conservation." In *Indonesian Primates,* ed. Sharon Gursky-Doyen and Jatna Supriatna, 9–22. New York: Springer, 2010.

A Summer to Remember. Directed by Robert Michael Lewis. Inter Planetary Pictures, 1985. VHS, 93 min.

Sureng, Gaun Anak. "Six Orang Stories." *Sarawak Museum Journal* 9, nos. 15–16 (1960): 452–457.

Sutlive, Vinson H., and Joanne Sutlive. *The Encyclopaedia of Iban Studies: Iban History, Society, and Culture*. Kuching: Tun Jugah Foundation, 2001.

Suvin, Darko. *Victorian Science Fiction in the UK: The Discourses of Knowledge and Power*. Boston: G. K. Hall, 1983.

Swadling, Pamela, with Roy Wagner and Billai Laba. *Plumes from Paradise: Trade Cycles in Outer Southeast Asia and Their Impact on New Guinea and Nearby Islands until 1920*. Coorparoo, Australia: Papua New Guinea National Museum in association with Robert Brown & Associates, 1996.

Swift, Jonathan. *Gulliver's Travels*. Rev. ed. New York: Penguin Classics, 2003. First edition 1726.

Tarzan the Ape Man. Directed by John Derek. Metro-Goldwyn-Mayer/Svengali, 1981. VHS, 107 min.

Taylor, Millie. *British Pantomime Performance*. Bristol: Intellect, 2007.

Taylor, Nick. "Heart to Heart." *New York Magazine,* July 13, 1987, 44–48.

Taylor, Rowan. "A Step at a Time: New Zealand's Progress toward Hominid Rights." *Animal Law Review* 7 (2000–2001): 35–43.

Teasdale, Harvey. *The Life and Adventures of Harvey Teasdale, the Converted Clown and Man Monkey.* 18th ed. Sheffield: General Printing and Publishing Co., 1879.

Thompson, C. Pelham. *The Dumb Savoyard and His Monkey.* In *Duncombe's Edition of The British Theatre,* vol. 6. London: J. Duncombe, 1825.

———. *Jack Robinson and His Monkey.* London: Thomas Hailes Lacy, 1829.

Thornbury, Walter, and Edward Walford. *Old and New London: Westminster and the Western Suburbs.* London: Cassell, Petter, & Galpin, 1891.

Thornhill, Randy, and Craig Palmer. *A Natural History of Rape: Biological Bases of Sexual Coercion.* Cambridge, Mass.: MIT Press, 2000.

Tiedemann, Frederick. "On the Brain of the Negro, Compared with That of the European and the Orang-Outang." *Philosophical Transactions of the Royal Society of London* 126 (1836): 497–527.

Torgovnik, Marianna. *Gone Primitive: Savage Intellects, Modern Lives.* Chicago: University of Chicago Press, 1990.

Travis, Cheryl Brown. *Evolution, Gender and Rape.* Cambridge, Mass.: Massachusetts Institute of Technology, 2003.

Tudor, Andrew. *Monsters and Mad Scientists: A Cultural History of the Horror Movie.* Oxford: Basil Blackwell, 1989.

Tulp, Nicolaes. *Geneeskundige Waarnemingen.* Leiden: Jurriaan Wishoff, 1740.

———. *Observationes Medicae.* Leiden: Georg Wisshoff, 6th ed., 1739.

Tyson, Edward. *Orang-Outang sive Homo Sylvestris or, the Anatomy of a Pygmie Compared with that of a Monkey, an Ape, and a Man.* London: Thomas Bennet, 1699.

———. "A Philological Essay Concerning the Pygmies of the Ancients." In Edward Tyson, *Orang-Outang sive Homo Sylvestris or, the Anatomy of a Pygmie Compared with that of a Monkey, an Ape, and a Man.* London: Thomas Bennet, 1699.

Veltre, Thomas. "Menageries, Metaphors and Meanings." In *New Worlds, New Animals: From Menagerie to Zoological Park in the Nineteenth Century,* ed. R. J. Hoage and W. A. Deiss, 19–29. Baltimore: Johns Hopkins University Press, 1996.

Verissimo, Luis Fernando. *Borges and the Eternal Orangutans.* Translated by Margaret Jull Costa. London: Random House, 2005. First published in Spanish 2000.

Verne, Jules. *The Mysterious Island.* Translated by Jordan Stump. New York: Modern Library, 2004. Originally published as *L'Île mystérieuse.* Paris: Pierre-Jules Hetzel, 1874.

Vevers, Gwynne, comp. *London's Zoo: An Anthology to Celebrate 150 Years of the Zoological Society of London.* London: Bodley Head, 1976.

Vint, Cheryl. "Simians, Subjectivity and Sociality: *2001: A Space Odyssey* and Two Versions of *Planet of the Apes.*" *Science Fiction Film and Television* 2, no. 2 (2009): 225–250.

Voronoff, Serge. *Rejuvenation by Grafting.* London: George Allen & Unwin, 1925.

Vosmaer, A. *Beschryving van de zo zeldzaame als zonderlinge Aap-soort genaamd Orang-outang, van het Eiland Borneo.* Amsterdam: Pieter Meijer, 1778.

Wadley, Reed L., ed. *Histories of the Borneo Environment: Economic, Political and Social Dimensions of Change and Continuity.* Leiden: KITLV Press, 2005.

Wallace, Alfred Russel. *The Malay Archipelago: The Land of the Orang-utan, and the Bird of Paradise; a Narrative of Travel, with Sketches of Man and Nature.* New York: Harper, 1869. Also London: Macmillan, 1869.

———. *My Life: A Record of Events and Opinions.* New York: Dodd, Mead, 1905.

Ward, David C. *Charles Willson Peale: Art and Selfhood in the Early Republic.* Berkeley: University of California Press, 2004.

Warren, Kristin S. "Speciation and Intrasubspecific Variation of Bornean Orangutans, *Pongo pygmaeus pygmaeus*." *Molecular Biology and Evolution* 18, no. 4 (2001): 472–480.

Wasserman, Renata R. Mautner. "Re-Inventing the New World: Cooper and Alencar." *Comparative Literature* 36, no. 2 (1984): 130–145.

Watson, Elizabeth E., Simon Eastael, and David Penny. "*Homo* Genus: A Review of the Classification of Humans and the Great Apes." In *Humanity from African Naissance to Coming Millennia*, ed. Philip V. Tobias et al., 311–323. Florence: Firenze University Press, 2001.

Weindling, Paul J. *Health, Race and German Politics between National Unification and Nazism 1870–1945.* Cambridge: Cambridge University Press, 1989.

Weiner, J. S. *The Piltdown Forgery.* Oxford: Oxford University Press, 2nd ed., 2003.

White, Charles. *An Account of the Regular Gradation in Man, and in Different Animals and Vegetables; and from the Former to the Latter.* Bristol: Thoemmes, 2001 [1795].

Whiten, Andrew, and Richard W. Byrne, eds. *Machiavellian Intelligence II: Extensions and Evaluations.* Cambridge: Cambridge University Press, 1997.

Wich, Serge A., et al. "Distribution and Conservation Status of the Orang-utan (*Pongo* spp.) on Borneo and Sumatra: How Many Remain?" *Oryx* 42 (2008): 329–339.

Wich, S. A., R. W. Shumaker, L. Perkins, and H. De Vries. "Captive and Wild Orangutan (*Pongo* sp.) Survivorship: A Comparison and the Influence of Management." *American Journal of Primatology* 71 (2009): 680–686.

Wills, Christopher. *The Runaway Brain: The Evolution of Human Uniqueness.* London: HarperCollins, 1993.

Wilson, Edward O. *Sociobiology: The New Synthesis.* Cambridge, Mass.: Belknap Press of Harvard University Press, 1975.

Winter, Marian Hannah. *The Theatre of Marvels.* Translated by Charles Meldon. New York: Benjamin Bloom, 1962.

Wise, Steven M. *Drawing the Line: Science and the Case for Animal Rights.* Cambridge, Mass.: Perseus Books, 2002.

Wolters, O. W. *Early Indonesian Commerce: A Study of the Origins of Śrīvijaya.* Ithaca, N.Y.: Cornell University Press, 1967.

Womack, Sarah Whitney. "The 'Most Correct Creature': Colonial Man, Colonial Beast and Colonial Culture." In *Regional Animalities: Cultures, Natures, Humans and Animals in Southeast Asia,* ed. Lucy Davis, 84–101. Singapore: Focas, 2007.

Wonderen der Natuur: In de Menagerie van Blauw Jan te Amsterdam, zoals gezien door Jan Velten rond 1700. Amsterdam: ETI Digital, 1998.

Wonders, Karen. *Habitat Dioramas: Illusions of Wilderness in Museums of Natural History.* Uppsala: Acta Universitatis Upsaliensis, 1993.

Woolf, Paul. "The Movies in the Rue Morgue: Adapting Edgar Allan Poe for the Screen." In *Nineteenth-Century American Fiction on Screen,* ed. R. Barton Palmer, 43–61. Cambridge: Cambridge University Press, 2007.

Wrangham, Richard, and Dale Peterson. *Demonic Males: Apes and the Origins of Human Violence.* New York: Houghton Mifflin, 1996.

Wright, R. V. S. "Imitative Learning of a Flaked Tool Technology—The Case of an Orangutan." *Mankind* 8, no. 4 (1972): 296–306.

Wurmb, F. Baron van. "Beschryving van de groote Borneoosche Orang Outang of de Oost-Indische Pongo." *Verhandelingen van het Bataviaasch Genootschap der Kunsten en Wetenschappen* 2 (1784): 245–261.

Wynne, Clive D. L. "Rosalià Abreu and the Apes of Havana." *International Journal of Primatology* 29 (2008): 289–302.

Yanni, Carla. *Nature's Museums: Victorian Science and the Architecture of Display.* London: Athlone, 1999.

Yeager, Carey P. "Orangutan Rehabilitation in Tanjung Puting National Park, Indonesia." *Conservation Biology* 11, no 3. (June 1997): 802–805.

Yerkes, Robert M. "Ideational Behavior of Monkeys and Apes." *Proceedings of the National Academy of Sciences of the United States of America* 2, no. 11 (Nov. 15, 1916): 639–642.

Yerkes, Robert M., and Ada W. Yerkes. *The Great Apes: A Study of Anthropoid Life.* New Haven, Conn.: Yale University Press, 1929.

Yule, Henry, and A. C. Burnell. *Hobson-Jobson: A Glossary of Colloquial Anglo-Indian Words and Phrases, and of Kindred Terms, Etymological, Historical, Geographical and Discursive.* 2nd ed. London: Murray, 1903.

Zimmerman, Eberh. Aug. Guilielm. *Specimen Zoologiae Geographicae, Quadrupedum Domicilia et Migrationes Sistens.* Leiden: Theodore Haak, 1777.

Zoetmulder, P. J., and S. O. Robson. *Old Javanese-English Dictionary, part 2: P–Y.* The Hague: Martinus Nijhoff, 1982.

Zuckerman, Lord. "The Rise of Zoos and Zoological Societies." In *Great Zoos of the World: Their Origin and Significance,* ed. Solly Zuckerman, 3–26. Boulder, Colo.: Westview, 1980.

☙ Index

Flaubert, Gustave: "Quidquid volueris"
(1837), 141–142
Florey, Robert: *Edgar Allan Poe's Murders
in the Rue Morgue* (film 1932), 172–173
food, orangutans as, 86, 210
Forster, Georg (German naturalist), 42
Fossey, Dian (American primatologist), 229
fossils, 209, 233, 234
Frankfurt Zoological Society, 229
Franklin, Richard: *Link* (film 1986),
174–175
frauds using orangutan body parts, 60–61,
97, 234
freakishness, 166–167
Frémiet, Emanuel (sculptor): *An Orangutan
Strangling a Native of Borneo,* 71, 95
Frey, Regina (Swiss naturalist), 229, 230
friendship, interspecies, 178, 188, 239–241;
children–orangutan intimacy, 179–181,
182
Fuegans, 46, 241, 276n. 35
Fu Manchu (orangutan), 195
Furness, William (American anthropologist), 236

Gabriel and Rochefort: *Jocko, ou le singe de
Brésil* (stage play 1825), 156–161; imita-
tions and productions, 156, 159, 160
Galdikas, Biruté (Canadian primatologist),
90, 229, 238
Galen (Roman physician), 233
Garson, Harry: *The Beast of Borneo* (film
1934), 170–171
gentling (non-punitive training method),
187–188, 197
Geri (orangutan), 202
Gesner, Konrad von (Swiss naturalist), 17
gibbons, 124, 173, 216, 218, 243, 247, 256n. 50
Glassbrenner, Adolf (German satirist), 50
Going Ape (film 1981), 179
Goldsmith, Oliver: *History of Animals*
(1774), 52–53, 101
Goodall, Jane (British primatologist), 182,
229
gorilla, 107, 131, 177, 199; behaviour, 26;
evolutionary relationships, 4, 51, 233–
235, 248, 254n. 44, 274n. 9; in fiction,
51, 118–120, 168, 171, 175; genus *Gorilla,*

3, 30, 49, 229, 248; sex with humans,
51, 171, 173, 274n. 8
Great Ape Protection and Cost Savings
Act, 274
great ape rights, 36, 179, 231, 244–247
Great Chain of Being, 86
Greenpeace, 224
Greenwood, James: *The Adventures of
Reuben Davidger* (1865), 148
Groves, Colin (Australian primatologist),
222
Gulluliou (fictional orangutan), 117–118,
142–143
Gundul (orangutan), 90
Gunung Palung reserve (Kalimantan),
216, 225

habitat, orangutan, 3, 85, 209
habitat diorama, 81–85, 170
Haeckel, Ernst (German biologist), 233
Hagenbeck, Carl Jr. (German zoo-keeper),
186–188, 191, 206, 228
Haggerty, Melvin (American psychologist),
236
hair, orangutan, 15, 16, 17, 22, 32, 36, 46,
79, 97, 121, 131, 198, 204, 211; absent
from chest, 12, 20; color, 4, 8, 10, 65,
102, 106, 114, 151, 252n. 16; texture, 11,
23; used in ritual, 86
hairy woman trope, 17
Hall, James: *Gone Wild* (1995), 150–154
Hamilton, Alexander (sea captain), 32, 56,
105
hands, 43, 50, 60, 106, 131, 166, 182, 205;
detached, 22, 96, 97–98; shaken, 177,
187, 192, 198; use of, 14, 15, 32, 34, 41,
45, 63, 71, 90, 93, 138, 199, 204, 236
Haraway, Donna, 229
Harrisson, Barbara (German art historian
and naturalist), 218–219; orangutan
rehabilitation, 228
Hauff, Willhelm: "Der Affe als Mensch:
Der junge Engländer" (1827), 147–148
helplessness, trope of, 230–231
Hensler, Karl Friedrich: *Der Orang Utang,
oder de Tigerfiest* (stage play 1791), 157,
158
Hesse, Elias (early traveler), 31

or *The Run Away Monkey* (stage play
1830s), 164–165
Ourangus outangus, 25, 103
Owen, Richard (British clergyman), 73,
232–233

paintings and drawings of orangutans, 11, 21,
27, 58–59, 63, 66, 68, 69, 72, 79, 81, 130, 165,
170–171, 210. *See also* advertisements
Pallavicini, Mr. (Batavia harbor master),
23
Pan, 3, 248; *P. troglodytes*, 12
Pastrana, Julia (sideshow exhibit), 166
patriotic dramas, 161–163
PAWS (Performing Animals Welfare
Society), 201
Peacock, Thomas Love, 37, 138–139; airs
environmental questions, 262n. 7; ambivalent use of Monboddo, 113; anticipates
multiculturalism, 56; *Melincourt* (1817),
37, 110–114, 123–124, 132, 145, 150
Peale, Charles Willson (American taxi-
dermist), 62, 78, 82; educational aims
of, 62–63, 66
Perth Zoo, 227
PETA (People for the Ethical Treatment
of Animals), 182, 201
pets, orangutans as, 4, 66, 96, 219, 207, 224–
225, 228, 246; in fiction, 132–133, 170
Philadelphia Zoo, 227
Philibert, Nicolas: *Nénette* (film 2010),
183–184
*Phillip Quarl, or the Mariner and His
Monkey* (stage play c. 1847), 163
Piepers, M. C. (Dutch colonial official), 213
Piltdown Man hoax, 234
Pithecanthropus alalus, 233
Pithecus satyrus, 25
Planet of the Apes (film 1968), 175–177
Playhouse, The (film 1921), 169
Pliny (Gaius Plinius Secundus), 1, 2, 3, 14
Poe, Edgar Allan: "Hop-Frog; Or, the
Eight Chained Ourangoutangs" (1849),
53; "Murders in the Rue Morgue" (1841),
145–147, 148, 150, 154–155, 165, 264n.
24; symbolic readings of, 147. *See also*
Florey, Robert: *Edgar Allan Poe's Murders
in the Rue Morgue*

Pogorel'ski, Antonii: "Journey in a Stagecoach," 135, 144
Polo, Marco (early traveler), 97
Pongo (fictional orangutan), 164
Pongo, 3, 25–26, 92, 248; *P. abelii*, 222,
226; *P. borneo*, 25; *P.p. abelii*, 222; *P.p.
morio*, 226; *P.p. wurmbii*, 222, 226, *P.
pygmaeus*, 7, 12, 19, 26, 151
pongo (term), 25–26, 34, 36, 48, 92, 117, 118,
134, 164–165, 207
population estimates (orangutan), 4, 211,
219
population growth (human), 125, 221,
249
Pougens, Charles de: *Jocko* (1824), 133–
134, 158
Pratchett, Terry: *Discworld* series (1983+),
123
prices: paid for orangutans, 34, 98, 100, 101,
102, 179, 197
pronoun use, 153–154, 255n. 29
Punch, 241, 242
Pyrard de Laval, François (early traveler),
35, 54, 56

race hierarchy theory, 46–49, 233–234;
orangutans co-opted to, 46–47, 49–
50, 51, 56, 63
Radcliffe (orangutan), 202
Radermacher, J. C. M. (Dutch naturalist),
91, 96
Raffles, Sir Stamford (British naturalist
and colonial administrator), 12, 72,
73, 98
Rajah (orangutan), 188–190
"ranga" (redhead), 8
Rango (film 1931), 170
Rannee (orangutan), 102
rape by orangutans: of humans, 31, 88, 90,
142–143, 158, 172, 173; of other orangutans, 153, 239
Regan, Tom (animal rights philosopher),
244
rehabilitation, 228–231
Reid, Mayne: *The Castaways* (1870), 141,
148
reproduction (orangutan), 209. *See also*
hybridization

Voronoff, Serge (monkey-gland experimenter), 51, 143–144, 171

Vosmaer, Aernout (Dutch menagerie director), 39–40, 45, 47, 56, 58, 96; explores humanness of orangutans, 40–42, 59, 223, 237, 255n. 29; savaged in "orangutan war," 42–43, 59

Wahid, Abdurrachman (President of Indonesia), 223

Wallace, Alfred Russel (Scottish naturalist), 93, 101, 134; fondness for orangutans, 105–106

Wanariset rehabilitation centre (Kalimantan), 230

Warner, Lyman (US showman), 7

Wasserman, Renata, 108

Watson, Elizabeth E. (New Zealand taxonomist), 235

White, Charles (English physician), 49

"Wild Man from Borneo": as expression, 8; as stage show, 7–8, 79

Willem V, Prince of Orange, 39, 58, 92

Wilson, E. O. (sociobiologist), 239

Woodward, Arthur Smith (Piltdown hoax accomplice), 234

Woolf, Paul, 173

work and orangutans, 4, 35, 53–55, 63, 88, 130, 157, 190–191

World Wildlife Fund (Worldwide Fund for Nature), 196, 221, 225

wretchedness trope, 52–54, 56, 211

Wurmb, Baron Friedrich von (German naturalist), 92

xing xing, 211

Yerkes, Robert, (American psychologist), 217, 236, 237, 251n. 2

Yerkes Primate Research Center, 181

Yudhoyono, Susilo Bambang (President of Indonesia), 225

Zaius, Dr. (fictional orangutan), 176, 177

Zimmerman, Eberhard (German biogeographer), 103

zoos, 55, 72, 76–79, 83, 183, 185–208, 225, 227; captive breeding programs, 226–228; origins, 58; scientific and educative aspirations of, 72, 185, 189, 196, 206–207; zoo "characters," 191–194, 204–205. *See also* menageries

About the Authors

Robert Cribb is professor of Asian history at the Australian National University. His work has focused on modern Indonesia and includes studies of national identity, mass violence, crime, justice systems, and conservation politics. He is author of *Gangsters and Revolutionaries* (1991), *Historical Dictionary of Indonesia* (1992), *Modern Indonesia* (with Colin Brown, 1995), and *Historical Atlas of Indonesia* (2000). He is editor, with Li Narangoa, of *Imperial Japan and National Identities in Asia, 1895-1945* (2003) and, with Michele Ford, of *Indonesia beyond the Water's Edge: Managing an Archipelagic State* (2009).

Helen Gilbert is professor of theatre at Royal Holloway, University of London, where she presently leads a large transnational project on contemporary indigenous performance in the Americas, the Pacific, South Africa, and Australia. Her books include *Performance and Cosmopolitics* (with Jacqueline Lo, 2007), *Sightlines: Race, Gender and Nation in Contemporary Australian Theatre* (1998), and *Postcolonial Drama: Theory, Practice, Politics* (1996).

Helen Tiffin is a leading scholar in postcolonial theory and literary studies. She has been professor of English at the University of Tasmania and the University of Queensland in Australia and was also professor of English at Queen's University in Canada, where she held a Canada Research Chair. She is the author of *The Empire Writes Back: Theory and Practice in Postcolonial Literatures* (1989; with Bill Ashcroft and Gareth Griffiths), regarded as a foundational text in postcolonial theory. Her research and teaching interests include the history of colonial and postcolonial settler societies, literatures in English, Caribbean studies, literary theory, and, more recently, the literary and cultural representation of animals.

Production Notes for Cribb, Gilbert,
and Tiffin | *Wild Man from Borneo*

Jacket design by Mardee Melton

Text Design by Integrated Composition Systems,
Spokane, Washington, with text and display type
in Garamond Premier Pro

Composition by Integrated Composition Systems

Printing and binding by Sheridan Books, Inc.

Printed on 60 lb. House White, 444 ppi.